5th International Conference on Environment Pollution and Prevention (ICEPP 2017)

IOP Conference Series: Earth and Environmental Science
Volume 120

Singapore
14 – 16 December 2017

ISBN: 978-1-5108-5950-0
ISSN: 1755-1307

Printed from e-media with permission by:

Curran Associates, Inc.
57 Morehouse Lane
Red Hook, NY 12571

Some format issues inherent in the e-media version may also appear in this print version.

Printed by Curran Associates, Inc. (2018)

For permission requests, please contact the Institute of Physics
at the address below.

Institute of Physics
Dirac House, Temple Back
Bristol BS1 6BE UK

Phone: 44 1 17 929 7481
Fax: 44 1 17 920 0979

techtracking@iop.org

Additional copies of this publication are available from:

Curran Associates, Inc.
57 Morehouse Lane
Red Hook, NY 12571 USA
Phone: 845-758-0400
Fax: 845-758-2633
Email: curran@proceedings.com
Web: www.proceedings.com

Table of contents

Volume 120

2017 5th International Conference on Environment Pollution and Prevention (ICEPP 2017)

14–16 December 2017, Singapore

Accepted papers received: 7 February 2018
Published online: 12 March 2018

Preface

Water Quality Assessment and Water Resource Management

Environmental and Chemical Engineering

IOP Conf. Series: Earth and Environmental Science **120** (2018) 011001 doi:10.1088/1755-1315/120/1/011001

PREFACE

It is our great pleasure to welcome you to 2017 5th International Conference on Environment Pollution and Prevention (ICEPP 2017) which will be held in Singapore, 14-16 December, 2017. ICEPP 2017 is dedicated to issues related to Green Energy Technology.

The major goal and feature of the conference is to bring academic scientists, engineers, and industry researchers together to exchange and share their experiences and research results, and discuss the practical challenges encountered and the solutions adopted. Professors from Korea, Singapore and USA are invited to deliver keynote speeches regarding latest information in their respective expertise areas. It will be a golden opportunity for students, researchers and engineers to interact with the experts and specialists to get their advice or consultation on technical matters, sales and marketing strategies.

This proceeding present a selection from papers submitted to the conference from universities, research institutes and industries. All papers were subjected to peer-review by conference committee members and international reviewers. The papers selected depended on their quality and their relevancy to the conference. The volume tends to present to the readers the recent advances in the field of Environment Pollution and Prevention and various related areas, such as Energy Management, Environmental Risk Assessments, Environmental Sustainability and Development, Pollution and Health Issues , Water Pollution and Treatment, Wastewater Management and Treatment , Environmental Safety Regulations , Air pollution and treatment , Environmental Protection, etc.

We would like to thank all the authors who have contributed to this volume, and also the organizing committee, reviewers, speakers, chairpersons, sponsors and all the conference participants for their support to ICEPP 2017.

Nanyang Executive Centre in Nanyang Technological University, Singapore
December 21, 2017

IOP Conf. Series: Earth and Environmental Science **120** (2018) 011002 doi:10.1088/1755-1315/120/1/011002

Organizing Committee

Conference Chair

Prof. James T. Anderson, West Virginia University, USA

Prof. Young Sun Mok, Jeju National University, Jeju, Korea

Program Chair

Assoc. Prof. M. A. Qader Maraqa, United Arab Emirates University, UAE

Assoc. Prof. Emmanuel Iyiegbuniwe, Department of Public Health, Western Kentucky University, USA

Organizing Chair

Ms Alice Lin, Hong Kong Chemical, Biological & Environmental Engineering Society, Hong Kong

International Technical Committee

Prof. AKBAR ESMAEILI, Islamic Azad University,Tehran, Iran

Prof. Bharat Bhushan Gulyani, BITS Pilani, Dubai Campus (UAE)

Prof. Catalina Iticescu,"Dunarea de Jos" University of Galati, Romania

Prof. Hiroshi Uechi, Osaka Gakuin University, Japan

Prof. Shankar Gargh, Environmental Disasters Research Institute, India

Prof. Farid A. Badria, Mansoura University, Egypt

Prof. Kyoungphile Nam, Seoul National University, Korea

Assoc. Prof. Eiji HARAMOTO,University of Yamanashi, Japan

Assoc. Prof Alaa El-Din Hamid Sayed,University of Tokyo, Japan

Assoc. Prof. Seema Jilani, University of Karachi, Pakistan

Assoc. Prof. Mehmet Burcin MUTLU,Anadolu University, Turkey

Assoc. Prof. Neeru Bansal, CEPT University, India

Assoc. Prof. Yubao Wang（Kevin Wong)，Xi'an Jiaotong University，China

Dr. Parnuch Hongsawat, King Mongkut's University of Technology North Bangkok, Thailand

Dr. Susira Bootdee, King Mongkut's University of Technology North Bangkok, Thailand

Dr. Rajeev Kumar Mishra,Delhi Technological University, India

IOP Conf. Series: Earth and Environmental Science **120** (2018) 011003 doi:10.1088/1755-1315/120/1/011003

Peer review statement

All papers published in this volume of *IOP Conference Series: Earth and Environmental Science* have been peer reviewed through processes administered by the proceedings Editors. Reviews were conducted by expert referees to the professional and scientific standards expected of a proceedings journal published by IOP Publishing.

IOP Conf. Series: Earth and Environmental Science **120** (2018) 012001 doi:10.1088/1755-1315/120/1/012001

Industrial Development and Challenges of Water Pollution in Coastal Areas: The Case of Surat, India

Neeru Bansal

Associate Professor, Faculty of Planning, CEPT University,
Ahmedabad-380009, India
Email: bansal.neeru@cept.ac.in

Abstract. Industrialisation plays an important role in the economic development of a country, however, pollution is the inevitable price paid for this development. Surat, a major industrial hub in western India, is located on the bank of the river Tapi and extends up to the Arabian Sea. The city is characterised by the presence of a number of creeks (known as 'khadis' in local language). This paper focusses on the industrial development in Surat and the challenges faced by the city due to water pollution. A constant deterioration in the quality of surface water resources has been observed due to discharge of treated or partially treated effluents from the industries. The problem of water pollution becomes critical due to increase in frequency of flooding, risks faced by the city due to climate change and the ineffective environmental governance. The paper provides insights into the challenges faced by the city and the learnings can lead to adoption of policy initiatives and other measures which can effectively address these challenges.

1. Introduction

Indian economy has grown at a high rate after liberalisation in 1991. Liberalisation has reduced the control exercised by the central government and has left a larger scope for state-level initiatives to attract industries. Gujarat state, in India (where Surat is located) has taken major initiatives to promote industries and is now one of the most industrialised state in the country. Within Gujarat, Surat is the second major urban area and has experienced high rate of industrial growth. It is located on the bank of the river Tapi and extends up to the Arabian Sea through gulf of Khambhat. Being a coastal city, it is characterised by the presence of a number of creeks (known as 'khadis' in local language). The city has an average elevation of 13 meters. The city is vulnerable to floods and has experienced two major river floods and a creek flood in recent times. The areas lying in flood plain zone of river Tapi in Surat have experienced high rates of growth in population and density [1]. During the rainy season, high tides often inundate settlements located along several creeks. Surat is a hub of textile industry and diamond industry. It has the maximum share of investments (22.9 percent) in large scale manufacturing units in the state and is at number two, with a share of 23.04 percent, when investments in micro, small and medium scale manufacturing units (MSMEs) are considered [2].

Industrial development though important for the economic development, often leads to environmental degradation due to problems of pollution and resource depletion. In the course of industrial growth, natural resources get exploited and are at the receiving end as they often get abused for private gains. Pollution too seems to be the inevitable price paid for this development [3]. The pollution problems manifest as problems of air quality, disposal of solid wastes and discharge of industrial effluents. Industrial effluents are the source of some of the most toxic pollutants and the most seriously polluted environments. They contribute significantly to the overall water pollution in an area and are considered to be more hazardous than sewage [4].

This paper focusses on the industrial development in Surat and the challenges faced by the city due to water pollution. The data for polluting industries, pollution control facilities and quality of water bodies has been collected from Gujarat Pollution Control Board (GPCB), which is the nodal agency at the state level for enforcing environmental legislation for the protection of the environment. It gives permission to all polluting industries and the pollution control facilities and directs them to comply with the applicable standards. It also regularly monitors quality of the water bodies as well as of the treated effluents from the pollution control facilities. To comment on the performance of the pollution control facilities, the data has been compared with the standards prescribed to these by the regulatory authority under environmental legislation. The data for quality of water bodies has been compared with the standards set by the Central Pollution Control Board (CPCB) based on their 'designated best use' [5]. According to this concept, out of the several uses a water body is put to, the use that demands the highest quality of water is termed as the 'designated best use.' The discussions have been carried out with the key personnel representing concerned government departments, operators of the common pollution control facilities, professionals and NGOs working in the environment field to get better insights into the Case.

The paper has been organised as follows: Section 1 introduces the problem and the key inquiry and also details out the methodology followed. Section 2 focusses on the details of industrial development in Surat and Section 3 discusses the performance of pollution control facilities set up to treat industrial effluents. The quality of water bodies receiving industrial discharges has been discussed in Section 4. The next section details out the challenges faced by the city due to water pollution. Section 6 covers the concluding remarks and the policy inferences following from the discussion in the paper. It also mentions some of the measures which can be adopted to address the challenges faced by the city.

2. Industrialisation in Surat

In Surat, a large number of MSMEs are located in the industrial estates, developed by Gujarat Industrial Development Corporation (GIDC), a government undertaking, which provides developed land and infrastructure to set up industrial units. There are also a few industrial estates developed by private companies which accommodates some of the industrial units. Many of the small industrial units are scattered all across the city. Hazira area, which is right on the coast and is also a port, is home to a number of large scale modern industries; prominent among these are Reliance, Essar Steel, Essar Power, Larsen & Toubro, Adani, National Thermal Power Corporation, KRIBHCO, Oil and Natural Gas Corporation etc. The details of the polluting industries in Surat, as registered with GPCB, upto July 2014, is presented in table 1.

Table 1. Registered polluting industries in Surat.

Category	Small	Medium	Large	Total
Red	1202	157	268	1627
Orange	169	16	20	205
Green	253	16	18	287
Total	1624	189	306	**2119**

Source: GPCB

Note: Industrial units are classified into large, medium and small category based on the investment as per the definition given under Micro Small Medium Enterprises (MSME) Act, 2006. CPCB classifies the industries into Red, Orange and green categories based on their pollution potential. Red means the most polluting, orange means the medium polluting and green means the less polluting industries. The same classification is followed by GPCB.

IOP Conf. Series: Earth and Environmental Science **120** (2018) 012001 doi:10.1088/1755-1315/120/1/012001

Important point to note from the above table is that the most polluting industries (red category) and belonging to small-scale, forms more than 50 percent of the registered polluting units. The small and medium scale industries account for 40 percent of industrial production, employ limited pollution control technologies and are responsible for an estimated 70 percent of the total industrial pollution load nationwide [6]. The location of industrial areas, common facilities for effluent treatment and the monitoring points for water quality are shown in figure 1. It can be seen that most of these are located along one creek or the other. These creeks finally flow to the gulf of Khambhat and then meet the Arabian Sea.

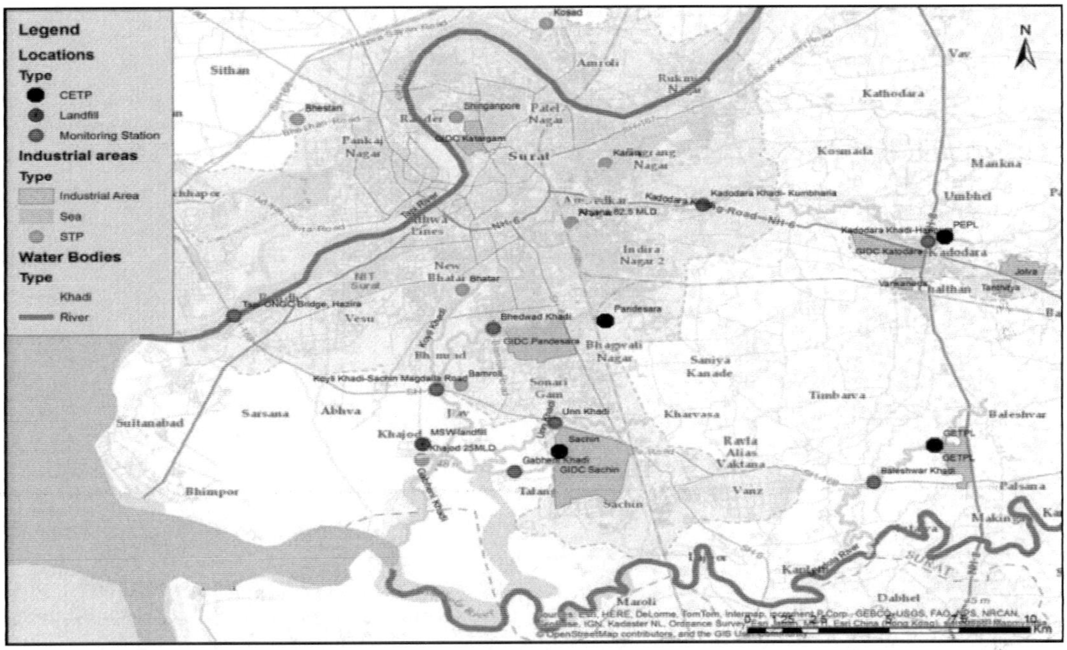

Figure 1. Location of the Creeks, Industrial Areas, CETPs and Monitoring Points.
Source: Author, details mapped using Google image

3. Environmental infrastructure & its performance

The large-scale industries in Hazira have their own effluent treatment plants and discharge treated effluent directly into the sea. The small-scale industries, which forms a major share of the polluting industries, do not have the financial, managerial and technical capacity to treat the industrial waste water [7]. In addition, there are space constraints faced by these units to set up the individual effluent treatment plants. Till early 2000, most of the industries in Surat used to discharge untreated effluents into the water bodies. After that, common facilities for treatment—common effluent treatment plants (CETPs) were set up, mainly as an intervention enforced by the Courts and pressure from the civil society [2]. The industries which are scattered in Surat city, discharge their treated/untreated effluents into either the sewerage network or the storm water drainage which finally meet any of the creeks. However, it is not possible to segregate the share of the industrial effluents from these discharges. The analysis in this paper, is therefore, limited to the performance of the CETPs in reducing the pollution load. There are five operational CETPs in Surat with a combined capacity of 310.5 MLD to treat industrial waste water [8]. These CETPs have been set up for the industries located in GIDC Pandesara, GIDC Sachin, Gujarat Eco Textile Park Ltd. (GETPL–

private industrial park), GIDC Kadodara and the industrial clusters in Vareli, Jolva, Chaltan, Tantithaiya villages. The details of functional CETPs are listed below in table 2. The industrial effluents after being treated in the CETPs are discharged into any of the creeks.

Table 2. Details of functional CETPs in Surat.

S No	Name of CETP	Location	Nature of member units	Discharge point	Mode of reception of effluent	Installed capacity (MLD)
1	**Pandesara Infrastructure Ltd**	GIDC Pandesara	Textile units & chemical units	Bhedwad creek	Network of pipeline	100
2	**Sachin Infra Environment Ltd(SIEL)**	GIDC Sachin	Textile units	Unn creek	Network of pipeline	50
3	**Globe Enviro Care Ltd (GECL)**	GIDC Sachin	Chemical units	Unn creek	Tankers	0.5
4	**Palsana Enviro Protection Ltd (PEPL)**	Umbhel, Surat	Textile units	Kadodara creek	Network of pipeline	100
5	**Gujarat Eco Textile Park Ltd (GETPL)**	Palsana, Surat	Textile units	Baleshwar creek	Network of pipeline	60

Source: The table has been compiled from the data collected from different sources.

All the CETPs mentioned in the above table have physico-chemical treatment as primary treatment and biological treatment as secondary treatment to reduce the pollution load. There are five parameters (pH[1], NH_3-N[2], BOD[3], TCOL[4] and DO[5]) on which quality of water bodies is classified based on their 'designated best use' by the CPCB. Out of these five, only three parameters—pH, NH_3-N and BOD—are being measured for the CETPs by the regulatory authority. NH_3-N is not being measured for some of the CETPs which are treating homogenous industrial effluents from the textile units. The effectiveness of the CETPs in treating pollution, as discussed in the following paragraphs is based on these parameters. Since the treatment is with the help of bacteria, the BOD becomes the most important parameter to comment on the effectiveness of the CETPs. As some of the CETPs are treating heterogeneous effluents, COD[6], another important parameter, has also been included in the analysis to study the effectiveness.

The CETPs have been prescribed 'inlet standards' for the effluents to be treated by the CETP and 'outlet standards' for the treated effluents from the CETP. CETP Pandesara, PEPL and GETPL have been given an inlet standard of 400 mg/l of BOD whereas CETP GECL, which is established for treating chemical

[1] A figure expressing the acidity or alkalinity of a solution on a logarithmic scale on which 7 is neutral, lower values are more acid, and higher values more alkaline.

[2] NH_3-N (Ammonical Nitrogen) is highly toxic to fish and other aquatic life and its value in the water bodies should be $\leq$ 1.2 mg/l.

[3] BOD (Bio-chemical Oxygen Demand) is the amount of dissolved oxygen needed by aerobic biological organisms to break down organic material present in a given water sample at certain temperature over a specific time period.

[4] TCOL is a measure of the total coliform bacteria present in the water body.

[5] DO is a measure of the amount of oxygen dissolved in the water body.

[6] COD (Chemical Oxygen Demand) indicates the amount of oxygen which is needed for the oxidation of all organic substances in the water.

effluents in GIDC Sachin, has been given an inlet standard of 1500 mg/l of BOD. CETP SIEL has not been prescribed any inlet standard. All the CETPs have been given the same outlet standard for the treated effluent— pH-6.5 to 8.5, NH_3-N-50 mg/l, BOD-30 mg/l and COD-250 mg/l; irrespective of the fact that the inlet standards for the effluents are different for the CETPs. The details of the treatment for all the CETPs has been summarised and presented in table 3.

Table 3. Summary of performance of the CETPs.

CETPs	pH in treated effluent	NH_3-N in treated effluent	Inlet standard for BOD	Actual inlet BOD	BOD in treated effluent	% Reduction in BOD	Inlet standard for COD	Actual inlet COD	COD in treated effluent
Prescribed standard for treated effluent	6.5–8.5	50			30				250
CETP-Pandesara	7.5	9	400	340	107	67	1000	1239	419
CETP-SIEL	7.3	Not monitored	-	319	107	64	Not Given	1152	403
CETP-GECL	7.4	48	1500	566	89	78	4000	2097	379
CETP-PEPL	7.8	Not monitored	400	251	97	60	1000	891	358
CETP-GETPL	7.8	Not monitored	400	221	70	67	1000	837	284

Source: The table has been compiled from the analysis done.

Note: All the standards are in (mg/l) except for pH.

It is observed that the pH and NH_3-N for the treated effluent from all the CETPs meets the prescribed standards, irrespective of whether the nature of inlet effluents is homogenous (effluents are from the same industrial sector) or heterogeneous (effluents are from varied industrial sectors like textiles, chemicals etc.). For the most critical parameters—BOD, none of the CETP is able to meet the prescribed standards of 30 mg/l for the treated effluents. The biological treatment (secondary treatment) employed by the CETPs should result in approx. 85 percent reduction in the organic load. There is not even one CETP which is able to consistently reduce organic pollution load by the expected 85 percent. In terms of exceedance factor of the treated effluent above the permissible standard, it is maximum for the CETP Pandesara and CETP SIEL at 3.6. The minimum exceedance factor is 2.3 for the CETP GETPL. The COD being another important parameter, the analysis has highlighted that the results for the COD reduction by the CETPs are comparable to the BOD reduction. There is not even one CETP which is able to meet the prescribed standard of 250 mg/l for COD in the treated effluent.

In Surat, as for the performance of the pollution control facilities is concerned, these are not able to treat the effluent and meet the permissible norms. The 'one-size-fits-all' approach has been adopted for treating the pollution. Though the effluents are different and the inlet norms given to the various CETPs are different, the same process has been adopted for treating the pollution and the standards given for the treated effluent are also the same. This is quite baffling as the reduction in pollution load is directly proportionate to the inlet values and the abatement technology used. Given the fact that the all the CETPs have employed similar abatement methods and have been prescribed similar standards for treated effluent, the pollution levels will be directly proportionated to the inlet values. This leads to conclude that the standards are prescribed randomly by the regulator without considering the abatement efficiencies of the

IOP Conf. Series: Earth and Environmental Science **120** (2018) 012001 doi:10.1088/1755-1315/120/1/012001

systems. Though the CETPs are regularly monitored by the regulator, still no action has been taken against them for non-compliance with the prescribed standards.
The treated effluents from all the CETPs are discharged into one creek or the other. The ineffective treatment of the effluents by the CETPs and the presence of pollution way above permissible levels has direct association with poor quality of the water bodies receiving industrial discharges. The quality of the receiving water bodies has been discussed in the following section.

4. Quality of waterbodies receiving industrial discharges

The quality of water bodies in Surat has been analysed in this section. It is realised that the state has not done any classification of water bodies (including all rivers) based on 'designated best use.' In general, it is an accepted norm that if the interventions for pollution control are effective, then the minimum water quality standards which a water body should meet (bathing standards) is (pH between 6.5–8.5; DO-5 mg/l or more; BOD-5 days 20^0 C 3 mg/l or less; NH_3-N-1.2 mg/l or less and TC-500 MPN/100 ml or less). These are indicators of pollution, of whether the water can sustain aquatic life. The presence of harmful, faecal related bacteria, viruses and protozoa in the water bodies are responsible for causing illnesses. The analysis of water quality is presented in table 4.
The water quality in the river Tapi is regularly monitored as is done for other major rivers in India. The quality in the river is observed to be good except for coliforms. However, in case of Surat, the industrial effluents are being discharged into numerous creeks and not the main river Tapi. It, therefore, becomes necessary to analyse the water quality of these creeks, but the irony of the situation is that the monitoring of creeks is not regular and has been discontinued lately, therefore, no recent data base is available to know their quality. Important parameter like dissolve oxygen which is a direct indicator of the health of a water body has not being measured at all for creeks by the regulatory authority. The value of BOD has been observed to be high for creeks, indicating high levels of pollution in these creeks. From creeks, the waste water flows to Arabian sea through gulf of Khambhat and nobody notices the pollution. The water pollution goes unnoticed as it gets camouflaged in various creeks along the banks of which industrial areas and CETPs are located. This is serious as Surat city is prone to floods and the presence of pollutants in water bodies above the acceptable levels can have serious implications on human health, livestock, agriculture, property etc.

5. Challenges faced due to pollution

A number of pollution related challenges are stemming from rapid industrial development and presence of high pollution levels in the water bodies receiving industrial discharges which are spread across the city. Following the conventional way, all the studies by the government agencies—Central Pollution Control Board [9], Central Water Commission [10] and Comptroller and Auditor General [11], have analysed the quality of water in the Tapi. The river has the same importance in Surat as the Sabarmati has in Ahmedabad or the Ganga in Varanasi and the Yamuna in Delhi. However, the analysis in this paper points out that in case of Surat, the industrial effluents are being discharged not into the Tapi but into the numerous creeks, which are a typical character of Surat's geography and the quality of water in these creeks is not meeting with the desirable values. From the creeks, the effluent flows to the Arabian Sea and nobody notices the pollution. This is really serious as Surat city is prone to floods and has experienced two major river floods and one creek flood in the last decade. The results of a study [1] show that 71,000 households in Surat are vulnerable to creek floods and when river floods occur, as many as 4,50,000 households are at threat. Ebenstein (2012) has estimated that the deterioration of water quality by a single grade point (on a six-grade scale) increases the digestive cancer death rate by 9.7 per cent [12]. The presence of slums near creeks make these people more vulnerable as they do not have alternative means of shelter, livelihood or finances to cope with the disasters. *"Those with the least resources have the least capacity to adapt and are the most vulnerable."*

In case of Hazira, where a number of very large-scale companies are located, the whole industrial area has developed by reclaiming section of the flood plain and consequently narrowing down the mouth of the Tapi, where it meets the sea. Blocking the network of natural drains by filling up, for the purpose of developing land for industries, may have a direct impact on the city and its people. Major part of Hazira falls under low elevation coastal area (less than 10 m elevation) exposing industries to considerable risk to water related hazards. Most of the current development activities have been planned without taking into consideration a potential threat from the sea level rise. The area faces a high vulnerability to inundation due to sea level rise. The risk becomes manifold as huge investments have been made in the area in infrastructure and industrial development.

Table 4. Quality of the water bodies.

Waterbody	Period	No of readings	pH	DO	BOD	NH_3-N	TCOL	Remarks
Standards as per designated best use			6.5–8.5	5	3	1.2	500	
Unn creek	July 08-May 10	124	7.8	-	9	5.6	14	Receive discharge from CETP SIEL & GECL
Gabheni creek	July 08-May 10	124	6.7	-	298	13.3	107	Located south to GIDC Sachin
Koyli creek	July 08-Sep 13	9	7.9	-	113	8.4	58	Bhedwad **creek**and Kadodara **creek**meet this **creek**
Baleshwar creek	No 11-Oct 13	45	8.1	-	14	2.0	18	Receive discharge from CETP GETPL
Bhedwad creek	July 09-July 10	10	7.1	-	230	6.3	93	Receive discharge from CETP Pandesara
Kadodara creek	Jun 08-May 13	8	8.0	-	96		58	Receive discharge from CETP PEPL
Tapi river (Swami Vivekanand bridge)	May 08-Jun 14	17	7.93	6.56	1	0.51	71623	Main river flowing through Surat
Tapi river (ONGC bridge-Hazira)	Jun 08-Jun 14	21	8.07	5.95	2	0.48	6264	Main river flowing through Surat

Source: GPCB

The shoreline in gulf of Khambhat has been experiencing changes. A study by Gupta (2014) for the period 1996-2004 has observed that a large area of the gulf shows a serious threat due to changing shoreline [13]. The entire coast has undergone accretion including coastal region around the Tapi and the Hazira port. The main reasons behind this accretion are attributed to the strong tidal currents and a large load of sediments brought through the major rivers. The sediments as well as pollutants under the influence of strong tidal currents undergo dispersion and settle within the Gulf of Khambhat resulting in siltation at a rapid rate. The anthropogenic shoreline change, including ports and harbour development, have added a major new factor into the complexity of natural processes determining coastal landform changes. There have been no studies so far for this area which have estimated the levels of pollutants present in the sediments deposited in the gulf and its impact on the eco-systems.

6. Conclusions & policy inference

Surat may have to prepare to live with the challenges it faces due to its location. The preparedness to deal with the challenges faced by the city has to be multi-dimensional in nature. This would mean understanding all future risks and responding to the changing risks. There are no land use rules in place in Surat to prevent urban development in current and future high-risk zones [14]. The coastal area may, in fact, become some of the most sought-after real estate. The urban planning process comes in actual operation much after the areas have already experienced growth of buildings, roads and commercial activities. This defeats the fundamental purpose of land use planning and ultimately results in legitimising the unplanned growth. Land use planning and the enforcement of development rules based on risk categorization is needed and the local body and development authorities must plan and implement keeping these risks in mind. The current approach of development happening first and then legalisation of development should be replaced with planning first, followed by development. Resettlement of all slums located on creeks is needed immediately and care needs to be taken to prevent further settlements coming up near the risky zones. Surat having a high share of industrial investments, the detailed data on risk zones will help the investors in taking informed decisions while locating their business.

Surat is unique in its character in terms of scale and composition of industries and in terms of the water bodies which receive the industrial effluents. In most of the cities, the waste water is discharged into a single water body e.g. in Ahmedabad, all the domestic sewage and industrial effluents are discharged into the Sabarmati, whereas in Surat, it is discharged into a number of creeks. The water pollution goes undetected as it gets distributed in the various creeks near the industrial areas. However, the approach by the regulator is the same throughout the state i.e. regulator monitors the main river and do not consider the location specific details. There is no comprehensive mechanism to address local-level environmental issues. There is a need to move away from the conventional way of monitoring of the water bodies by supplementing it with site-specific requirements. The case in specific for Surat highlights the fact that the conventional way of monitoring the main river in a city may not always represent the complete picture. The approach which needs to be adopted for Surat has to be different from that adopted for Ahmedabad, i.e., a differentiated approach, even within the same state to address local environmental concerns.

The feeble treatment by the common effluent treatment facilities and the ineffective enforcement by the regulator have resulted in the degradation of the quality of water bodies. The analysis in Section 3 has highlighted that even after many years of establishment, the CETPs have not been able to meet the prescribed standards. At the same time, for the MSMEs—due to the constraints faced by them—there are no other options but to rely on the CETPs [2]. Therefore, it is important to enforce improvement in their working so as to comply with the prescribed norms. Punitive measures should be taken against the defaulters to deal with non-compliance. The existing approaches to ensure environmental compliance needs to be supplemented by other approaches like heavy fines for values above the prescribed standards [15]; public disclosure of information about the pollution and the polluters [16]; supplementing environmental standards with taxes above permissible values [17] etc. The fines under the Water Act have

not been revised since 1974 and are merely capped at Rs. 10,000. This could be amended so that it acts as a deterrent. Stringent enforcement would be required to ensure environmental compliance and improve quality of water bodies.
The challenges faced by the city due to industrial development and the resultant water pollution can be addressed to some extent if the inferences from the discussion in the paper can be incorporated into the policies adopted for the city.

7. References

[1] Bhat G K, Karanth A, Dashora L and Rajasekar U 2013 Addressing flooding in the city of Surat beyond its boundaries *Environment and Urbanisation* pp 429–41
[2] Bansal N 2017 *Industry and environment: The case of Gujarat* (Ahmedabad: Dissertation CEPT University)
[3] Adams W M 2001 *Green development: Environment sustaibnability in third world countries* (London: Routledge)
[4] Parthasarathy R and Raja S 2013 Urban–rural water nexus: The case of Gujarat In Narain V, Goodrich C G, Chourey J and Prakash A (Eds.) *Water in a globalizing world: State, markets and civil society in South Asia (*Routledge: Taylor and Francis) pp 222–38)
[5] Central Pollution Control Board *Water quality criteria* Retrieved from http://www.cpcb.nic.in/Water_Quality_Criteria.php
[6] OECD 2006 *Environmental compliance and enforcement in India: Rapid assessment.* Retrieved from http://www.oecd.org/environment/environmentinemergingandtransitioneconomies/37838061.pdf
[7] Kathuria V and Turagaga R M 2014 Small industry in India: Are CETPs an appropriate response for controlling pollution In Keshab Das (Ed.) *Globalization and Standards: Issues and Challenges in Indian Business* (New Delhi: Springer) pp 245–62
[8] Gujarat Pollution Control Board 2014–15 *Annual Report* (Gandhinagar)
[9] Central Pollution Control Board 2012 *Polluted river stetches in India* (New Delhi)
[10] Central Water Commission 2014 *Status of trace and toxic metals in Indian rivers* (New Delhi)
[11] Comptroller and Auditor General 2011 *Performance audit of water pollution in India* (New Delhi)
[12] Ebenstein A 2012 The consequences of industrialization: Evidence from water pollution and digestive cancers in China *The Review of Economics and Statistics* **94**(1) pp 186–201
[13] Gupta M 2014 Monitoring Shoreline changes in Gujarat *Journal of Remote Sensing and Possitioning* pp 27–37
[14] Trust, S. C. (2007). Surat 2006 Floods: An Analysis of Causes. Consequences and Management. Surat: Surat Citizens' Council Trust.
[15] World Bank 2000 *Greening Industry: new roles for communities, markets and governments* (New York: Oxford University Press)
[16] Tietenberg T 1998 Disclosure strategies for pollution control *Environmental and Resource Economics* **11**(3–4) pp 587–602
[17] Xenarios S and Bithas K 2012 The use of environmental policy instruments for urban wastewater control: Evidences from an international survey *Environmental Policy and Governance* **22** 14–26

IOP Conf. Series: Earth and Environmental Science **120** (2018) 012002 doi:10.1088/1755-1315/120/1/012002

Impact of Leachate Discharge from Cipayung Landfill on Water Quality of Pesanggrahan River, Indonesia

Eki Noerfitriyani[1], Djoko M. Hartono, Setyo S. Moersidik, Irma Gusniani

1 Environmental Engineering Study Program, Civil of Engineering Department, Universitas Indonesia, Depok, West Java, 16424, Indonesia
E-mail: eki.noerfitriyani@ui.ac.id

Abstract. The landfill operation can cause environmental problems due to solid waste decomposition in the form of leachate. The evaluation of environmental impacts related with solid waste landfilling is needed to ensure that leachate discharge to water bodies does not exceed the standard limit to prevent contamination of the environment. This study aims to analyze the impact of leachate discharge from Cipayung Landfill on water quality of Pesanggrahan River. The data were analyzed based on leachate samples taken from influent and effluent treatment unit, and river water samples taken from upstream, stream at leachate discharge, and downstream. All samples were taken three times under rainy season condition from April to May 2017. The results show the average leachate quality temperature is 34,81 °C, TSS 72.33 mg/L, pH 7.83, BOD 3,959.63 mg/L, COD 6,860 mg/L, TN 373.33 mg/L, Hg 0.0016 mg/L. The BOD_5/COD ratio 0.58 indicated that leachate characteristics was biodegradable and resemble intermediate landfill due to the mixing of young leachate and old leachate. The effluent of leachate treatment plant exceeds the leachate standard limit for BOD, COD, and TN parameters. Statistical results from independent T-test showed significant differences ($p<0,05$) between upstream and downstream influenced with leachate discharge for DO parameter.

1. Introduction

The population growth in urban areas leads to the high volume of solid waste generated. Inevitably, this development has a significant impact on the increasing amount of solid waste disposed at Cipayung Landfill, Depok City. Depok City had the population over 2 million persons [1]. It generates solid waste 1,200 tonnes/day mainly derived from households, and solid waste disposed to Cipayung Landfill reached 750 tonnes/day [2]. Cipayung Landfill has been operated since 1984 using a controlled landfill system [2]. The increasing amount of waste generation can cause problems in its operation. Moreover, an improper waste management is potentially harmful to human health and can lead to environmental degradation. One potential environmental issue driven by the decomposition of waste in landfills is leachate production. Leachate can be derived from rain water seepage, runoff of waste decomposition, and moisture content of the waste itself [3]. The characteristics of leachate produced depend on several factors, including the degree of stabilization of the dumped waste, solid waste collection system, the type and composition of the discarded waste [4], the solubility of solid waste content [5], site and hydrological factors, solid waste compaction, as well as landfill design and operations [6]. Leachate is a liquid manifestation of solid waste and has been considered a serious polluter that affects natural resources such as water bodies and groundwater, as well as human health. Leachate is a contaminated liquid derived from the bottom of solid waste disposal facilities such as landfills containing dissolved organic compounds and inorganic compounds and suspended solids. The leachate composition depends on the nature of solid waste, chemical, and biochemical processes for the decomposition of waste materials, as well as the total moisture content of solid waste [7]. Cipayung Landfill has a leachate treatment plant for treating leachate from landfill using a stabilizing pond treatment system. Stabilization ponds are large basins in which wastewater is treated by natural processes involving microorganisms, and are considered as the most appropriate method for wastewater treatment in developing countries where the climate is favorable for its operation [8]. Stabilization pond system can be classified based on the types of biological activity, i.e. anaerobic, facultative, and maturation ponds [9]. Treated leachate is then discharged to the Pesanggrahan River. In Indonesia, the quality standards for liquid waste disposal are stated in Indonesia's Ministry of Environment and Forestry Law No. 59 of 2016 to ensure that the disposal of waste to recipient water bodies will not cause damage to the aquatic ecosystem and its potential economic resources, as well as public health affairs. Of the several contaminants carried in leachate, organic matter, nitrogen, and heavy metals are the primary substances of concern. In general, the concentration of the substances contained in the leachate depends on the biochemical decomposition of solid waste in landfills [10]. The main objectives of this study are to analyze leachate characterization, evaluate the performance of leachate treatment plant, and analyze impact of leachate discharge from Cipayung Landfill site on water quality of Pesanggrahan River, Indonesia. The results of leachate characterization of Cipayung Landfill treatment plant is expected to be useful for landfill management, especially to improve the existing leachate treatment system and prevent contamination of discharged leachate to recipient water bodies.

2. Method

2.1. Description of study area

This study was conducted in Cipayung Landfill, located in Depok, Indonesia. The climate in Depok City is influenced by monsoon season and is relatively similar, characterized by fairly small rain differences. The conditions of rainfall in all Depok areas are practically the same, with the average rainfall of 3,332 mm/year or 278 mm/month and the mean temperature of 26°C [11]. The leachate treatment system applied in Cipayung Landfill is a stabilization pond system. Figure 1 shows that the Cipayung Landfill stabilization pond consists of an anaerobic pond (AP), facultative pond (FP), and maturation pond (MP). The leachate treatment plant has a pretreatment unit equalization tank that is followed by the stabilization pond system.The treated leachate of the treatment plant is discharged to Pesanggrahan River as recipient water body. Pesanggrahan River is classified as Class III water body [12], based on Government Regulation No. 82 of 2001 is a group of water bodies with designated for fisheries and livestock activity.

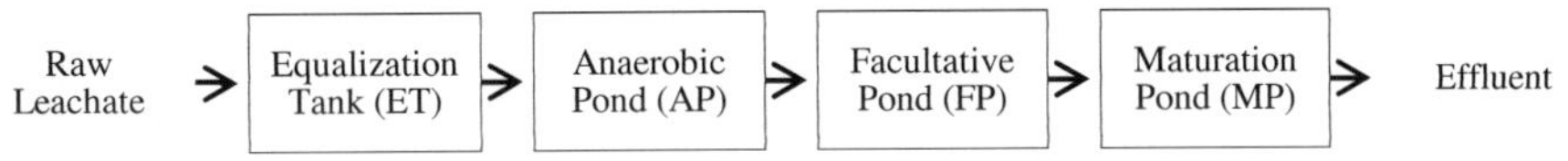

Figure 1. Schematic flow diagram of leachate treatment plant Cipayung Landfill.

2.2. Data collection and analysis

In this study, leachate samples were taken every two weeks under rainy season condition from April to Mei 2017. Table 1 describes the sampling locations and shows that samples were taken from leachate treatment plant and from Pesanggrahan River. Leachate samples were collected from inlet and outlet each treatment units to analyzed their characteristics and evaluated the performance of leachate treatment plant. River water samples were collected from upstream before leachate discharge, stream at leachate discharge, and downstream after leachate discharge to analyze the impact of leachate discharge from Cipayung Landfill on water quality of Pesanggrahan River. The sampling procedure follows SNI (Indonesian National Standard) number 6989.59:2008 which explains the wastewater sampling method, and SNI number 03.7016:2004 explains the river water sampling method. Impact of leachate discharge on river water quality were analyzed based on temperature (T), Total Suspended

Solids (TSS), (pH) value, Biological Oxygen Demand (BOD), Chemical Oxygen Demand (COD), Dissolved Oxygen (DO), Total Nitrogen (TN), and Mercury (Hg) parameters. Sampling analysis including *in-situ* and *ex-situ* measurement. The temperature is measured using a digital thermometer (SNI 6989.23:2005), TSS using gravimetric analysis (SNI 6989.3:2004), pH value using a digital pH meter (SNI 06-6989.11-2004), BOD_5 using Winkler with 5 days incubation time (SNI 6989.72.2009), COD using the close reflux titrimetric method (SNI 06-6989.2-2009), TN using spectrophotometer with Nessler (HACH 399), and Mercury using Atomic Absorption Spectrophotometry method (SNI 6989.78:2011). The methodology developed to study the impact of leachate discharge from Cipayung Landfill on water quality of Pesanggrahan River is divided into several steps, that is identification and characterization of leachate influent, performance evaluation based on removal efficiency, and identification of river water quality influenced with leachate discharge.

Table 1. Leachate and river water sampling locations.

Sampling point	Sampling location	Latitude; Longitude
S1	Anaerobic pond influent	6°25'11.60"S; 106°47'14.42"E
S2	Anaerobic pond effluent; Facultative pond influent	6°25'11.78"S; 106°47'13.70"E
S3	Facultative pond effluent; Maturation pond influent	6°25'12.09"S; 106°47'14.15"E
S4	Maturation pond effluent	6°25'11.94"S; 106°47'13.55"E
S5	Upstream of leachate confluence with Pesanggrahan River	6°25'32.20"S 106°47'20.00"E
S6	Leachate confluence with Pesanggrahan River	6°25'11.77"S 6°25'11.77"E
S7	Downstream of leachate confluence with Pesanggrahan River	6°25'1.30"S 106°47'10.69"E

2.3. Statistical analysis
Data were analyzed using IBM SPSS Statistics 23.0 software package. Data were presented in the form of mean and standard deviation. Impact of leachate discharge on river water quality was analyzed using Independent *t*-test at p-value ≤0,05 to compared the statistical differences between the means of upstream and downstream river water samples.

3. Result and Discussion

3.1. Leachate characterization
Cipayung Landfill has been operating since 1984 and is classified as a mature landfill. However, the measured average of raw leachate temperature of around 34.81°C, TSS of 72.33 mg/L, pH of 7.83, BOD of 3,959.63 mg/L, COD of 6,860 mg/L, TN of 373.33 mg/L, Hg of 0.0016 mg/L, and BOD/COD ratio of 0.58 indicate that the leachate characteristics resemble an intermediate landfill. This deviation is due to the mixing of young leachate and old leachate since the solid waste dumping system in Cipayung Landfill does not use cell system to produce leachate with the character between young and old leachate (intermediate). The comparison between the leachate characteristics in Cipayung Landfill and the typical leachate is shown in Table 2. Characteristics of Cipayung Landfill leachate tends to be alkali which is typical of leachate in Indonesia [13]. However, the pH value is still within the range of values 6–9 appropriate for biological life. While leachate temperature is in the range of 33,8°C–36,4°C, and still in accordance with the optimum temperature of microorganisms, which is equal to 25°C–35°C [14]. Mature leachates with BOD_5 concentrations 100-200 mg/L are less polluted than young leachates 2.000–30.000 mg/L [14]. The young landfill has BOD_5/COD ratio 0,4-0,6 indicated that biological activity in the acid phase of anaerobic degradation, while mature landfill with BOD_5/COD ratio 0,02-0,5 produced stabilized leachate indicated with low COD concentration and low biodegradability [14]. The BOD_5/COD ratio of Cipayung Landfill leachate is 0,58 so that leachate can be treated by biological processing [15].

Table 2. Comparison of Cipayung Landfill leachate characteristics and typical leachate.

Parameter	Cipayung Landfill (33 years)	Young landfill (<2 years)	Mature landfill (>10 years)
Temperature (°C)	34.81 ± 1.393	-	-
TSS (mg/L)	72.33 ± 2.517	200–2,000 [14]	100-400 [14]
pH	7.83 ± 0.058	4.5–7.5 [14]	6.6–7.5 [14]
BOD (mg/L)	3,959.63 ± 993.711	2,000–30,000 [14]	100–200 [14]
COD (mg/L)	6,860 ± 2,182.212	3,000–60,000 [14]	100–500 [14]
TN (mg/L)	373.33 ± 93350	500–1,500 [16]	50–200 [16]
Hg (mg/L)	0.0016 ± 0.00135	-	-
BOD_5/COD	0.58	0.4-0.6 [14]	0.02-0.5 [14]

Table 3. Average ± standard deviation and removal percentage of the parameters in the leachate treatment plant.

Pond type		TSS (mg/L)	pH	BOD (mg/L)	COD (mg/L)	TN (mg/L)	Hg (mg/L)
AP	Influent	72.33 ± 2.517	7.83 ± 0.058	3959.63 ± 993.711	6860 ± 2182.212	373.33 ± 93.350	0.0016 ± 0.00135
	Effluent	151 ± 6.00	8.0 ± 0.173	3894.77 ± 1064.242	9956.8 ± 530.288	793.33 ± 84.120	0.00076 ± 0.00033
	% Removal	**-110%**	**-2%**	**2%**	**-45%**	**-113%**	**53%**
FP	Influent	151 ± 6.00	8.0 ± 0.173	3894.77 ± 1064.242	9956.8 ± 530.288	793.33 ± 84.120	0.00076 ± 0.00033
	Effluent	32.33 ± 2.517	7.97 ± 0.058	3714.63 ± 258.802	6155.1 ± 1605.116	186.70 ± 70.00	0.0031 ± 0.00286
	% Removal	**79%**	**0%**	**5%**	**38%**	**76%**	**-307%**
MP	Influent	32.33 ± 2.517	7.97 ± 0.058	3714.63 ± 258.802	6155.1 ± 160.116	186.70 ± 70.00	0.0031 ± 0.00286
	Effluent	60.33 ± 2.517	8.07 ± 0.289	3399.13 ± 1418.785	9525.6 ± 2040.284	334.43 ± 94.318	0.00072 ± 0.00028
	% Removal	**-87%**	**-1%**	**8%**	**-55%**	**-79%**	**77%**
Total	**% Removal**	**17%**	**-3%**	**14%**	**-39%**	**10%**	**55%**
Leachate standard [17]		100	6 - 9	150	300	60	0,005

3.2. Performance evaluation of leachate treatment plant

The results obtained for each treatment unit and the overall leachate treatment plant are presented in Table 3. The quality of Cipayung Landfill leachate in each treatment unit shows that the removal efficiency does not comply with the design criteria, implying that the effluent quality of Cipayung Landfill leachate does not meet the quality standard based on [17] for BOD, COD, and TN parameters. The examination results of the anaerobic pond leachate samples state that the TSS concentration of 72.33 mg/L at the influent has been increased to 151 mg/L at the effluent since the mud at the base of the pond was too high. The TSS concentration of the facultative pond effluent was 32.33 mg/L, indicating that the removal of suspended solids at facultative pond processing unit reached 79%. Although the TSS concentration rose back to 60.33 mg/L at the maturation pond effluent, the overall removal efficiency of TSS concentration could reach 17%, and the quality of leachate discharged has fulfilled the requirement. The pH values of leachate in all treatment units are still within the standard quality range. The pH value of the influent leachate is 7.8–7.9 which tends to be alkali. Therefore, the leachate treatment does not require a neutralization process as a pre-treatment because pH 6 is the lowest limit for anaerobic processing [18]. The pH value of the anaerobic pond effluent has increased, indicating that the anaerobic processing has entered the final stage, raising the alkalinity that allows the methane microorganism to live. Volatile acids will be converted to methane and carbon dioxide (CO_2), while organic matter will decrease since its solubility will drop due to the boosted pH value. Supposedly, the pH value will dwindle back in the maturation pond because of the organic matter hydrolysis in the aerobic process, which resulted in the decrease of pH value and mineral dissolution. The rise of pH value to 8.07 in the effluent can be brought by the lack of oxygen concentration in the aerobic treatment within the maturation pond. The BOD concentration decreased toward the end of treatment with a removal efficiency of 14%. However, the leachate effluent has not met the leachate effluent standard. The high BOD concentration is caused by the organic residues contained in the anaerobic treatment that cannot be degraded in abundant quantity [19]. COD effluent concentration of 9,525.6 mg/L has not met the quality standard. The concentration of COD effluent is increased due to the absence of aeration and the presence of organic material residues. The concentration of TN decreased by 6% removal until the end of treatment. However, the effluent concentration of TN of 334.4 mg/L has not met the leachate standard. The concentration of ammonia nitrogen within the range of 25–30 mg/L can be toxic to anaerobic microorganisms because it inhibits their growth, affecting the removal of BOD and COD [20]. The high concentration of TN in the Cipayung Landfill leachate lessen the efficiency of BOD and COD removal.The concentration of Hg effluent was 0.0007 mg/L, and the removal efficiency was 56%. High concentration of heavy metals can be toxic to *Chlorella* species that are the main species in the stabilization ponds system. The pH value of >8 can cause the metal ions to precipitate, and the treatment process of stabilization ponds can run as usual [21].

3.3. River water quality

The water quality of Pesanggrahan River is shown in Table 4. Temperature is a parameter of aquatic physics that affects the chemical reaction and reaction rate, aquatic life, and the suitability of water use for a specific designation [22]. Increased temperatures can lead to increased viscosity, chemical reactions, evaporation, and volatilization, as well as decreased water solubility of gases, and increased metabolic rate of aquatic organisms [23]. It is known that the Pesanggrahan River water temperature is 26.92 °C at S5, 26.32 °C at S6, and 27.94 °C at S7. In general, the value of leachate temperature that enters the water body does not significantly affect the water temperature based on the *t*-test (p=0.56) and fulfill the standard of Class III water body. The condition of the Pesanggrahan River water temperature has been in accordance with the optimum temperature conditions for the growth of phytoplankton in the waters, i.e. between 20 °C to 30 °C [23]. The TSS consists of mud, fine sand, and microorganisms caused by soil erosion and is carried into the body of water [23]. While the concentration of TSS in leachate is dominated by dead bacteria. The entry of TSS into the waters can cause turbidity in the water, and the effect on the rate of photosynthesis phytoplankton decrease, and water productivity decreases. Such conditions may reduce the supply of dissolved oxygen in water bodies [23], but TSS concentrations in waters are naturally not toxic. The concentration of TSS in

Pesanggrahan River due to the leachate discharge has not significance influence on the fishery interest, where it is known that the TSS concentration of the Pesanggrahan River water is 78 mg/L at S5, increased at leachate discharge (S6) of 95.67 mg/L, and 69.67 mg/L at S7. In general, the concentration of TSS leach into the water body did not significantly affect the water temperature of Pesanggrahan River based on the *t*-test (p=0.77) and met the Class III water body standard of 400 mg/L. Waters will have a negative impact on fisheries if TSS concentrations are greater than 400 mg/L [24]. The pH value of Pesanggrahan River water is 7.1 at S5, 6.6 at S6, and 6.67 at S7. Pesanggrahan River pH values fluctuated from upstream, stream on the leachate discharge, and downstream after being influenced by leachate discharges, but still within the range of the Class III water body standard. The pH values fluctuated in water bodies may be affected by the amount of organic matter and inorganic materials in water bodies. High organic matter in water bodies can increase the pH value due to the decomposition process of organic matter releasing CO_2 and inorganic materials which generally contain high amounts of mineral acid [25]. Meanwhile, the pH value of leachate that enters the water body of 7.83 still meets the quality standard of leachate effluent pH, which is between 6 – 9. It shows that the leachate discharge does not significantly affect the pH water quality based on the *t*-test (p=0.452).

Table 4. Average ± standard deviation of Pesanggrahan River water quality and p-value showed differences between upstream (S5) and downstream (S7) using Independent t-test.

Parameters	Sampling Locations			Acceptable Limits [26]	*p*-value
	S5	S6	S7		
Temperature (°C)	26.92 ± 0.934	26.32 ± 0.926	27.94 ± 1.270	dev 3	0.56
TSS (mg/L)	78 ± 34.771	95.67 ± 25.968	69.67 ± 31.501	400	0.77
pH	7.10 ± 0.872	6.60 ± 0.100	6.67 ± 0.231	6 - 9	0.452
BOD (mg/L)	78.10 ± 21.913*	90.25 ± 72.761*	87.17 ± 59.351*	6	0.816
COD (mg/L)	111.6 ± 38.311*	129.5 ± 116.673*	133.5 ± 91.905*	50	0.723
DO (mg/L)	7.1 ± 0.498	5.93 ± 0.459	5.04 ± 0.821	3	0.02**
TN (mg/L)	1.43 ± 0.252	1.97 ± 0.058	1.90 ± 0.529	3	0.24
Hg (mg/L)	0.00098 ± 0.00018	0.00108 ± 0.00024	0.00070 ± 0.00039	0.002	0.378

* Exceed the standard limits for Class III water bodies quality
** Significant at $p \leq 0,05$

Organic matter is composed of carbon, oxygen, hydrogen, and some compounds containing nitrogen compounds [22]. According to Pratiwi [27], if the need for oxygen levels is not proportional to the addition of oxygen from the air as well as photosynthetic results of phytoplankton, it can lead to a decrease in dissolved oxygen concentration, and increase the concentration of BOD. BOD concentration in Pesanggrahan River flow exceeds Class III water quality standard 6 mg/L. Leachate discharge affects BOD concentration, that is by increasing the concentration of BOD at point S6, that is equal to 90.25 mg/L. This indicates that in the locations occur the process of decomposition of organic materials and oxidation of inorganic materials that require dissolved oxygen (deoxygenation). However, based on the *t*-test (p=0.816) showed that leachate discharge did not significantly affect BOD concentration in Pesanggrahan River. The COD concentration in the Pesanggrahan River flow increased after leachate discharge, i.e. 111.6 mg/L at S5, 129.5 mg/L at S6, and 133.5 mg/L at S7. The concentration of COD exceeds the standard of Class III water body of 50 mg/L. However, based on the *t*-test (p=0.723) showed that leachate discharge did not significantly influence COD concentration. COD concentrations are greater than BOD concentrations because the amounts of chemical compounds that can be chemically oxidized are greater than that of biological oxidation. DO serves as an indicator of water quality, which plays a role in the oxidation process and reduction of organic and inorganic materials [28]. The DO concentration depends on the mixing process, the movement of the water mass, the process of photosynthesis, and the effluent of waste entering the body of water [23]. The higher temperature and water level also affect the low concentration of DO due to low atmospheric pressure. DO concentrations decreased from upstream to downstream after being affected by leachate discharge from Cipayung Landfill. The DO concentration at point S5 of 7.1 mg/L, 5.93 mg/L at S6, and 5.04 mg/L at S7. Reduced DO concentration in downstream is influenced by increasing concentrations of organic and inorganic materials in river water. In the downstream after leachate discharge occurs the decomposition process of pollutants through the decomposition of organic materials, and oxidation of inorganic materials that require DO or called deoxygenation process. If the reaeration rate progress is lower than the rate of deoxygenation or self-validating ability, it can cause the DO concentration to decrease faster until it reaches anaerobic condition [23]. Leachate discharge affects DO concentrations based on the *t*-test (p=0.02). Pesanggrahan River TN concentration increased by leachate discharge, which was 1.43 mg/L at S5, 1.97 mg/L at S6, and 1.90 mg/L at S7. But still complied with the Class III water body standard based on [29], that is 3 mg/L. The increasing of TN concentration due to leachate discharge was not significant based on the *t*-test (p=0.24). The concentration of Hg at leachate discharge was under the standard of Class III water body and did not give a significant influenced the Pesanggrahan River water quality based on the *t*-test (p=0.378). The heavy metals contained in river water are lower than the heavy metals contained in sedimentary sediments where heavy metal deposits, and reach 99% of the total amount of heavy metals in the waters [30]. Overall, leachate discharge from Cipayung Landfill caused an increase in the concentration of pollutants in Pesanggrahan River, especially the concentration of COD and BOD. It is known that BOD and COD parameters exceed the standard of Class III water body, while the temperature, TSS, pH, DO, TN, and Hg meet below the Class III water body standard.

4. Conclusions

This study concern on environmental problems related with solid waste landfilling, that is the impact of Cipayung Landfill leachate on the Pesanggrahan River water quality. Based on this study, Cipayung Landfill classified as mature landfill, but the measured BOD/COD ratio 0.58 showed that this landfill had a high biodegradability through the anaerobic phase. The effluent of Cipayung Landfill leachate treatment plant exceeds the leachate standard limit based on [17] for BOD, COD, and TN parameters. Leachate characteristics had high concentration of organic chemicals and influenced the water quality of Pesanggrahan River indicated by the increasing of BOD and COD concentrations which resulted in significant decreasing of DO concentrations. To improve the quality of leachate effluent, it is necessary to make physical improvements by redesigning the treatment unit according to the design criteria to ensure that leachate discharge to water bodies does not exceed the standard limit to prevent contamination of the environment. Establishing a constructed wetland is recommended as well since the climatic condition of the study area implies that this method is suitable and has a high potential to improve the leachate effluent quality by reducingthe nutrients, suspended solids, algae, and BOD concentrations.

5. References

[1] Badan Pusat Statistik, 2014.
[2] "Buku Putih Sanitasi Kota Depok," Dinas Komunikasi dan Informatika Kota Depok, Depok, 2011.
[3] Renou, S., Givaudan, J., Poulain, S., Dirassouyan, F., Moulin, P, "Landfill Leachate Treatment: Review and Opportunity," *Journal of Hazardous Materials,* pp. 468-493, 2008.
[4] Fitzke, B., Blume, T., Wienands, H., Cambiella, A., "Hybrid Processes for the Treatment of Leachate from Landfills, Economic Sustainability and Environmental Protection in Mediterranean Countries through Clean Manufacturing Methods," *Springer,* pp. 107-126, 2013.
[5] Salem, Z., Hamouri, K., Djemaa, R., Allia, K., "Evaluation of Landfill Leachate Pollution and Treatment," *Desalination 220,* pp. 108-114, 2008.
[6] Baig, S., Coulomb, I., Courant, P., Liechti, P., "Treatment of Landfill Leachates: Lapeyrouse and Satrod Case Studies," *Ozone: Science & Engineering,* pp. 1-22, 1999.

[7] Fatta, D.; Papadopoulos, A.; Loizidou, M., "A Study on The Landfill Leachate and Its Impact on The Groundwater Quality of The Greater Area," *Environmental, Geochemical, Health,* pp. 175-190, 1999.
[8] Mara, D. D., Domestic Wastewater Treatment in Developing Countries, London: Earth scan, 2004.
[9] C.B. Gawasiri, Modern Design of Waste Stabilization Ponds in Warm Climates: Comparison with Traditional Design Methods, UK: University of Leeds, 2003.
[10] Jokela, J.P.Y., Rintala, J.A., "Anaerobic Solubilisation of Nitrogen from Municipal Solid Waste (MSW)," *Reviews in Environmental Science and Biotechnology 2,* pp. 67-77, 2003.
[11] Balai Data dan Informasi Sumber Daya Air Kota Depok, 2009.
[12] BPLHD, "Laporan Status Lingkungan Hidup Daerah Propinsi Daerah Khusus Ibu Kota Jakarta Tahun 2004-2005," BPLHD DKI Jakarta, 2005.
[13] Damanhuri, Enri, Padmi, Tri, "Diktat Kuliah Pengelolaan Sampah," in *Pengurugan (Landfilling) Sampah*, Bandung, Program Studi Teknik Lingkungan FTSL ITB, 2008, pp. 1-11.
[14] Tchobanoglous, G., Hilary Theisen, Samuel Vigil, Integrated Solid Waste Management, Singapore: McGraw Hill, 1993.
[15] Grady, C.P. Leslie. Lim, Henry C., Biological Wastewater Treatment Theory and Applications, USA: Marcel Dekker, 1980.
[16] J.F. Crawford, P.G. Smith, Landfill Technology, UK: British Publishing, 1985.
[17] Indonesia's Ministry of Environment and Forestry Law No. 59 of 2016.
[18] Pescod, Mara, 1988. [Online]. Available: http://stabilizationponds.sdsu.edu/
[19] Malina, Joseph F., Pohland, Frederick G., Design of Anaerobic Processes for The Treatment of Industrial and Municipal Wastes, USA: Technomic Publishing Company, Inc, 1992.
[20] Sergrist, 1997. [Online]. Available: http://stabilizationponds.sdsu.edu/.
[21] Murniwati, Camelia Indah, "Studi Evaluasi dan Perbaikan Instalasi Pengolahan Lindi TPA (Studi Kasus : TPA Suwung Kota Denpasar)," Tugas Akhir S1 Teknik Lingkungan Institut Teknologi Bandung, Bandung, 2012.
[22] E. Metcalf, Wastewater Engineering: Treatment and Reuse Fourth Edition, Singapore: McGraw-Hill, Inc, 2003.
[23] H. Effendi, Telaah Kualitas Air Bagi Pengelolaan Sumber Daya dan Lingkungan Perairan, Yogyakarta: Kanisius, 2003.
[24] D. Rahmawati, "Pengaruh kegiatan Industri terhadap Kualitas Air Sungai Diwak di bergas Kabupaten Semarang dan Upaya Pengendalian Pencemaran Air Sungai," Tesis Ilmu Lingkungan Program Pascasarjana Universitas Diponegoro, 2011
[25] H. P. Siradz, "Kualitas Air Sungai Code, Winongo, dan Gajahwong, Daerah Istimewa Yogyakarta," *Jurnal Ilmu Tanah dan Lingkungan Vol 8 (2),* pp. 121-125, 2008.
[26] Peraturan Pemerintah Republik Indonesia Nomor 82 Tahun 2001 tentang Pengelolaan Kualitas Air dan Pengendalian Pencemaran Air.
[27] Y. Pratiwi, "Penentuan Tingkat pencemaran Limbah Industri Tekstil Berdasarkan Nutrition Value Coefficient Bioindikator," *Jurnal Teknologi Vol 3 No.2,* pp. 129-137, 2010.
[28] Salmin, "Oksigen terlarut (DO) dan kebutuhan Oksigen Biologi (BOD) sebagai Salah Satu Indikator untuk Menentukan Kualitas Perairan," *Jurnal Oseana Vol 30 No. 3,* pp. 21-26, 2015.
[29] EPA, Parameters of Water Quality: Interpretation and Standards, 2001.
[30] Ademoroti, "Environmental Chemistry and Toxicology," *Fodulex press,* pp. 79-121, 1996.

Acknowledgment
The authors gratefully acknowledge Universitas Indonesia as the sponsor for this research under the project of PITTA 2017 funding scheme.

IOP Conf. Series: Earth and Environmental Science **120** (2018) 012003 doi:10.1088/1755-1315/120/1/012003

Characteristics of Leachate and Their Effect on Shallow Groundwater Quality (Case Study : TPA Cipayung, Depok)

Atika Widiastuti[1], Djoko M. Hartono, Setyo S. Moersidik and Irma Gusniani

1 Environmental Engineering Study Program, Civil Engineering Departement, University of Indonesia, Depok, 16424, Indonesia
E-mail: atikawidi11@gmail.com

Abstract. The problems arising from landfill activity is leaked leachate that is not absorbed well into leachate stabilization pond which furthermore contaminates shallow groundwater around landfill, include Cipayung landfill. The aims of this study is to determine the characteristics of leachate and their effect on shallow groundwater quality around landfill based on temperature, pH, Total Suspended Solids (TSS), Biological Oxygen Demand (BOD), Chemical Oxygen Demand (COD), Total Nitrogen (TN), Mercury (Hg), and fecal coliform. Data were analyzed based on leachate samples at influent point, effluent point, and 7 sampling points of residents's well with distance variation every 100 meters within 300 meters radius having leachate stabilization pond as benchmark. According to the standard of Indonesia's Ministry of Environment and Forestry law No. 59 of 2016, the results showed that leachate quality was still above the standard of BOD, COD, and Total Nitrogen parameters; 4178.0 mg/L, 70556.0 mg/L and 373.3 mg/L for influent point, and 3142.0 mg/L, 9055.2 mg/L, and 350 mg/L for the effluent point. Pollution Index of shallow groundwater is between lightly and moderately contaminated. This study showed that the further the distance between sampling point and leachate stabilization pond is, the lower the Polution Index is.

1. Introduction

One of the problems that arise from landfill activity is a possibility of leaked leachate that is not absorbed well into leachate stabilization pond, especially in areas with high rainfall. Leachate is a liquid that permeates through solid waste that contains dissolved and suspended elements or liquid that pass through the landfill which are mixed and suspended with substances or materials in the landfill. The characteristics of leachate is related to the level of environmental pollution that can be caused and influenced by various factors, such as the composition of waste, landfill age, seasons, hydrological conditions and activities in the landfill [1].

In practice, leachate in Cipayung landfill in Depok was not fully accommodated in the pipes and gathered in leachate stabilization pond, but leak into the soil and contaminate shallow groundwater in residential areas. Shallow groundwater is a groundwater located above the first waterproof layer, which in Depok City, the average depth of shallow groundwater is ranged from 10 meters to 30 meters [2]. Leaked leachate cause shallow groundwater used by residents to change color and is not safe for consumption, however this water is commonly used for daily needs such as drinking water consumptions, washing, bathing, toilet and other similar activities.

This study aims to determine the characteristics of Cipayung landfill leachate and their effect on shallow groundwater quality of residential around landfill based on quantitative measurement such as temperature, pH, Total Suspended Solids (TSS), Biological Oxygen Demand (BOD), Chemical Oxygen Demand (COD), Total Nitrogen, Mercury and Fecal coliform.

In this study, 7 water samples of shallow groundwater from residents' well and leachate samples at influent point and effluent point were taken once which procedure is according to SNI 6989.59: 2008 about Water and Wastewater. Determination of sampling location is using purposive sampling by taking 3 straight lines having leachate stabilization pond as benchmark, where line A is Kampung Bulak Barat area, line B is Kampung Benda Barat area, and line C is Pasir Putih area with distance variation of 100 meters, 200 meters and 300 meters within 300 meters radius.

The results of this study were descriptively comparatively analyzed; (1) Influent point and effluent point leachate data is compared to the standard of Indonesia's Ministry of Environment and Forestry law No. 59 of 2016 on quality standard for business and / or landfill activities to know the leachate characteristics, (2) Result data of shallow groundwater's samples are compared to Government Regulation No. 82 of 2001 on Water Quality Management and Water Pollution Control to determine the shallow groundwater quality of residential' well around the landfill, and (3) Shallow groundwater quality data will be calculated by Pollution Index (IP) according of Ministry of Environment Decree No. 115 of 2003.

2. Materials and methods

2.1. Description of study area

The location of this study is at influent point and effluent point of leachate stabilization pond in Cipayung landfill, and 7 sampling points at the groundwater of local residential area within 300 meters radius. The sampling points that were used in this study are considered to represent the condition of Cipayung landfill leachate, whereas the location that is selected in this study is based on groundwater flow obtained from interpolation of the groundwater surface point. Figure 1 shows the area of sampling location and Table 1 shows the sampling coordinates that were used in this study.

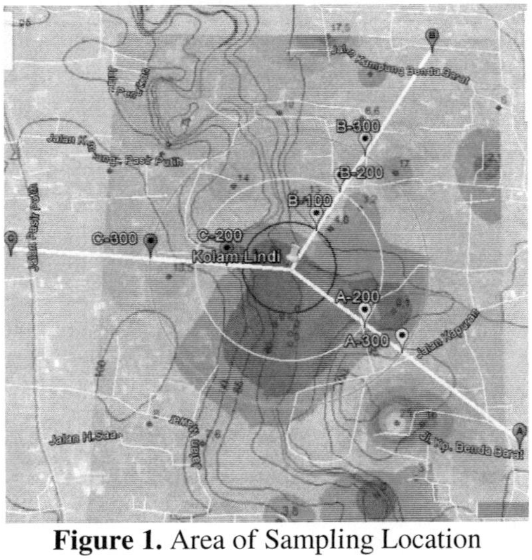

Figure 1. Area of Sampling Location

2.2. Data collection and analysis

Leachate sampling was conducted at the influent point and effluent point of stabilization pond and carried out on the same day as shallow groundwater sampling. The method used in groundwater sampling method is according to SNI 7989.58: 2008 about Water and Wastewater – Part 58: Groundwater Sampling Method. After that, pH and temperature were directly tested at the site, whereas for other parameters (TSS, BOD, COD, TN, Mercury and Fecal coliform), samples were preserved and stored for laboratory testing.

The method which used for each test is using pH meter for measuring pH value (SNI 6989.11 : 2004), thermometer for temperature measurement (SNI 6989.23 : 2005), TSS measurement according to SNI 6989.3: 2004, BOD measurement according to SNI 6989.72:2009, COD measurement according to SNI 6989.2:2009, TN measurement according to HACH 399, Mercury measurement according to SNI 6989.78:2011, and fecal coliform measurement according to Most Probable Number Method.

Table 1. Sampling Coordinates of Leachate and Shallow Groundwater

Sampling point	Sampling location	Latitude; Longitude
S1	Leachate stabilization pond influent	6°25'11.60"S; 106°47'14.42"E
S2	Leachate stabilization pond effluent	6°25'11.94"S; 106°47'13.55"E
A200	Line A area within 101-200 meters	6°25'14.99"S; 106°47'19.59"E
A300	Line A area within 201-300 meters	6°25'17.35"S; 106°47'22.02"E
B100	Line B area within 0-100 meters	6°25'08.97"S; 106°47'15.85"E
B200	Line B area within 101-200 meters	6°25'06.20"S; 106°47'17.45"E
B300	Line B area within 201-300 meters	6°25'05.40"S; 106°47'19.037"E
C200	Line C area within 101-200 meters	6°25'12.004"S; 106°47'9.564"E
C300	Line C area within 201-300 meters	6°25'11.480"S; 106°47'4.270"E

3. Results and disscussion

3.1. Characteristics of leachate

Leachate quality measurement from leachate stabilization is used as a basic consideration for pollutant sources of leachate into shallow groundwater. The influent point (S1) and effluent point (S2) are used as sampling points to determine leachate' quality changes before and after processing in stabilization pond, before being discharged into water body. Table 2 shows the result of parameter testing for leachate at influent and effluent point in stabilization pond.

Table 2. Result of Leachate Samples Testing

Parameter	Standard[3]	Influent Point (S1)	Effluent Point (S2)
Temperature (°C)	Deviation 3^0	34,8	33,8
TSS (mg/L)	100	70	63
pH	6 - 9	7,8	8,4
BOD (mg/L)	150	**4178,0**	**3142,0**
COD (mg/L)	300	**70556,0**	**9055,2**
TN (mg/L)	60	**373,3**	**350**
Hg (mg/L)	0.005	0,00161	0,00072
Fecal coliform	-	9500	18000

As the table shows, the values of BOD, COD and TN at S1 and S2 exceed the quality standard that is permitted. Cipayung landfill is a mature landfill because its age has been more than 10 years. This is evidenced by its low BOD_5/COD value ratio which is only 0,06, which normally exhibit BOD_5/COD value ratio from 0,05 to 0,2 [1]. It can happen because the composition of mature landfill generally contains humic and fulvic acid that can't be deciphered. The stabilization pond can reduce the levels of BOD, COD and fecal coliform present in leachate. As the table shows, the level of those parameters in effluent point is lower than influent point, but it still exceeds the standard quality limit, but for fecal coliform value, leachate stabilization pond does not decrease but increase it. Increasing the value of fecal coliform indicated that leachate water treatment has failed. Fecal coliform is included in the type of anaerobic bacteria that can breed without the presence of oxygen. Increasing number of fecal coliform at the effluent point is possible due to aerator in the leachate stabilization pond in Cipayung

landfill that is not used optimally. Supposedly, the aerator can create the stabilization pond rich in oxygen to prevent anaerobic bacteria growing, but it does not. So, the pond conditions become anoxic and let anaerobic bacteria, including fecal coliform to grow and breed. The BOD and COD value that still exceed the quality standard indicate that the amount of organic matter in leachate is significantly high, so it requires a lot of oxygen to be biodegrade by aerobic microorganisms and to oxidize organic material chemically (non-biodegradable) [4]. The decomposition of organic matter is a natural event, where if a water body is contaminated by organic matter then the bacteria will be able to dissolve the dissolved oxygen in the water during the biodegradable process, so that the natural water biota will die and the state of the water body becomes anaerobic, reyealing bad odor.
High TN value may be caused by irregular processing time on leachate stabilization pond, where the aerator in leachate stabilization pond is not operated on a daily basis. Lack of aeration causes aerobic bacteria die and the wastewater treatment process that is not optimal, also causes leachate water become odor and the color gets black [5].

3.2. Shallow groundwater quality

Testing of shallow groundwater use parameters of leachate quality standard, i.e. temperature, pH, TSS, BOD, COD, TN, Mercury, and added with fecal coliform. The equation of leachate and shallow groundwater samples parameters test was done so the results can be compared based on the same parameters. Table 3 shows the result of shallow groundwater quality testing.

Table 3. Result of Shallow Groundwater Samples Testing

Parameters	**pH**	**TSS**	**BOD**	**COD**	**TN**	**Mercury**	**Temperature (^{0}C)**	***Fecal Coliform***
Stardard	6-9	50	2	10	-	0,001	Deviation 3^0	100 MPN
A-200	7,2	8	**52,8**	**78,0**	5,6	0,000125	30,6	**2300**
A-300	7	1	**90,5**	**14,0**	2,8	0,00022	37,8	6,8
B-100	7,4	5	**63,2**	**84,7**	6,2	0,000031	28,8	4,5
B-200	6,9	7	**44,9**	**56,5**	5,7	undefined	28,3	6,1
B-300	7,2	1	**41,6**	**51,7**	2,5	undefined	31,7	0
C-200	6,6	6	**42,3**	**62,0**	5,2	0,000283	32,1	4,5
C-300	7,4	1	**22,7**	**30,0**	3,3	undefined	30,3	1,8

Figure 2 and Figure 3 shows the graphic of each other parameter. All standard that used for shallow groundwater quality comparison is standard of Government Regulation No. 82 of 2001 design of water class I. As the figure 1 shows, the normal temperature range is 24^0C – 30^0 C with 3^0C deviation, but as the graph shows, temperature exceeds the limit at some points. This high temperature can be caused by the natural influences of the depth of the water and seasons occurring at the time when the sampling is taken, as well as the human influence of waterproof surface construction and the release of warm water [6]. Overall, the pH value decreases with the distance from the benchmark to the furthest point at 300 m radius. Changes in pH values are still within the limit quality range, which is 6 -9. These different pH values correlate with the solubility of metal in the water, where the solubility of metal will be higher at low pH values or in acid conditions, Figure 2 shows the pH values variations.

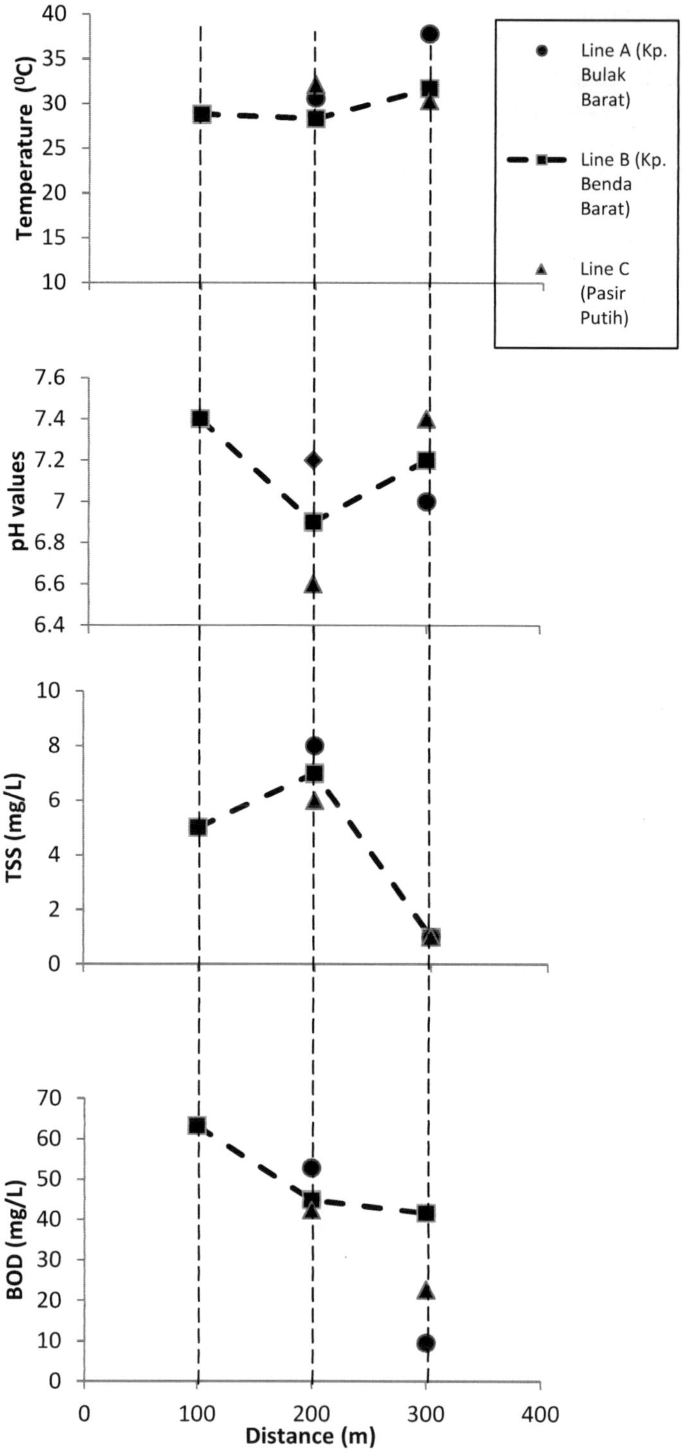

Figure 2. Correlation between Temperature, pH value, TSS and BOD parameters with distance

IOP Conf. Series: Earth and Environmental Science **120** (2018) 012003 doi:10.1088/1755-1315/120/1/012003

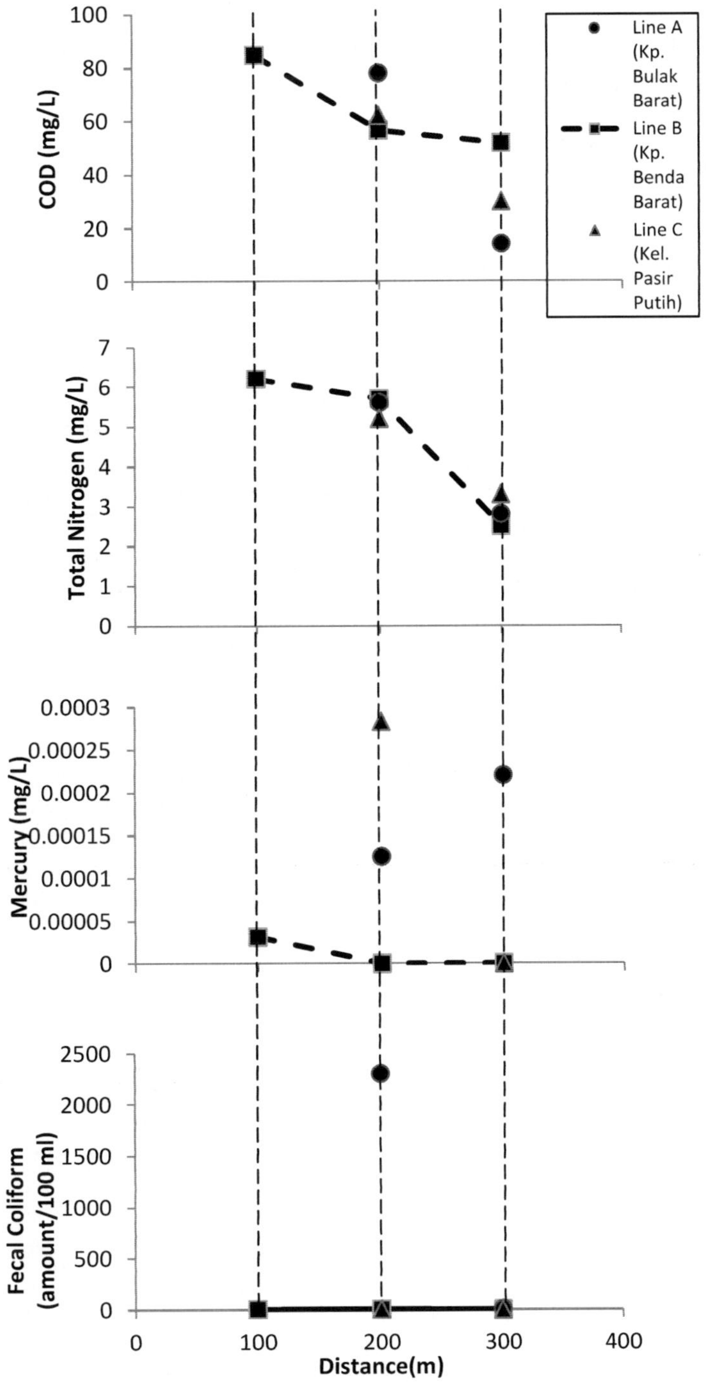

Figure 3. Correlation between COD, TN, Mercury, Fecal Coliform parameters with distance

Like the Figure 2 shows, TSS value fluctuates but is still below the permitted TSS standard of 50 mg/L, which means that the tested shallow groundwater quality is still good and not polluted by leaked leachate of Cipayung landfill. BOD values indicates the amount of oxygen needed by bacteria when stabilizing decomposable organic material under aerobic conditions, where the organic material serves as food for bacteria and energy obtained from oxidation process [7]. As the graph at Figure 3 shows, BOD values of all tests exceed the permitted quality standard of 2 mg/L, but it can be seen that even

all BOD values exceed the quality standard, the values are decreases as the distance gets further. COD values are always higher than BOD values, because COD values are derived from organic and inorganic materials of microbial degradation that accumulate in groundwater [8]. At the graph at Figure 3, COD value is decreasing along with the distance, just like BOD value. Although the values decrease, but COD concentration at all sampling points are exceed the permitted water quality standard. For the TN value, graphic at Figure 3 shows that all sampling point results are above the standard, but the value decreases as the distance gets further.
The content of mercury in shallow groundwater may be derived from leachate landfill whose garbage is mixed with hazardous and toxic garbage [9]. The graph at Figure 3 shows that mercury values of all sampling points are below permissible quality standards, indicating that shallow groundwater is not polluted by mercury. Apart from leaked leachate factor, mercury in shallow groundwater can also be caused by pH values in the water itself. At the acidic pH values, mercury can dissolve easily, and at the alkaline pH values, mercury may precipitate or slightly dissolve [10]. At some sampling point, such as B-200, B-300 and C-300, mercury values are undefined, this is probably due to alkaline pH values (above 7.0), so that the mercury precipitates and is undetectable. Fecal coliform is a bacterial indicator of pathogen contamination in water. This is because the number of colonies must be related with the presence of pathogenic bacteria. The standard quality of fecal coliform is 100 MPN/100 mL, which means that in 100 mL of sample water, there is maximum of 100 colony forming unit. The graph at Figure 3 shows that the fecal coliform index number exceeds the quality standard at point A-200, that is 2300 MPN/100 mL. It is probably because point A-200 is the source of water at waste management unit of Cipayung landfill, although there is no septic tank on this site, but the distance is closest to the leachate stabilization pond, also from landfill.

3.3. Pollution Index (PI)
The Pollution Index (PI) is used to determine the level of pollution relative to the permitted water quality parameters of Ministry of Environment Decree No. 115 of 2003. Water quality status is calculated by using the formula in accordance with the calculation of PI methods. PI calculation results of shallow groundwater quality status can be seen in Table 4 below. Overall, there are 2 points where the water quality status is lightly contaminated (range $1.0 < PI < 5.0$) and 5 points of moderately contaminated (range $5.0 < PI < 10$).

Table 4. Result of Shallow Groundwater Samples Testing

Point	PI	Status
A-200	6.140777	Moderately contaminaed
A-300	3.283034	Lightly contaminated
B-100	6.210206	Moderately contaminaed
B-200	5.67031	Moderately contaminaed
B-300	5.537416	Moderately contaminaed
C-200	5.625227	Moderately contaminaed
C-300	4.582264	Lighty contaminaed

4. Conclusions
The conclusions of this research are that the influent point and effluent point of leachate stabilization pond Cipayung landfill exceeded the quality standard on BOD, COD, and TN parameters, i.e. 4178.0 mg / L, 70556.0 mg / L and 373.3 mg / L for influent point and 3142.0 mg / L, 9055.2 mg / L, and 350

mg / L for the effluent point. While other parameters such as temperature, TSS, pH, and mercury are still below the quality standard. Fecal coliform increased from 9500 MPN / 100 ml at the point of influent to 18000 MPN / 100 ml at the effluent point. The further the distance between the sampling point and the leachate stabilization pond of Cipayung landfill is, the lower the Polution Index is, which are in between lightly and moderately contaminated with 2 points of lightly contaminated (range 1.0 < PI < 5.0, and 5 points of moderately contaminated (range 5.0 < PI < 10). Further processing of leachate Cipayung landfill is needed considering some sampling parameters still exceed the permitted quality standard. There is also a need for repairing on leachate pipe in Cipayung landfill to minimize the contamination caused by leaked leachate.

5. References

[1] Tchobanoglous, George et. al. 1993. "*Integrated Solid Waste Management Engineering Principles and Management Issues*". New York : Mc. Graw-Hill.

[2] Said, N. dan Widayat, W. 2014. "*Pemanenan Air Hujan dan Teknologi Pengolahan Air Hujan "Studi Kasus Kota Depok"*. Jakarta : Badan Pengkajian dan Penerapan Teknologi

[3] Republik Indonesia. 2016. *"Indonesia's Ministry of Environment and Forestry Law No. 59 of 2016 about standard of Bussiness and / or Landfill Activity"*. Jakarta : State Secretariat

[4] Yatim, E. dan Mukhlis. 2013. "*Pengaruh Lindi (Leachate) Sampah Terhadap Air Sumur Penduduk Sekitar Tempat Pembuangan Akhir (TPA) Air Dingin"*. Padang : Jurnal Kesehatan Masyarakat

[5] Ismail, M. 2011. *"Pemodelan Total Nitrogen Pada Sungai Pesanggrahan Akibat Input Lindi TPA Cipayung Kota Depok Berupa Beban Impuls"*. Depok : Universitas Indonesia.

[6] Erlinna, A. 2011. *"Pengaruh Keberadaan TPA Cipayung Depok Terhadap Kualitas Sumber Air Bersih di Wilayah Pemukiman Sekitarnya (Dengan Parameter Besi dan Mangan)"*. Depok : Universitas Indonesia

[7] Sawyer, C., McCartey, P., dan Parkin, G. 2003. "*Chemistry for Environmental Engineering and Science fifth edition"*. New York : McGraw-Hill

[8] Kusumawati, T. 2012. *"Kajian Degradasi Air Tanah Dangkal Akibat Air Lindi (Leachate) di Lingkungan Tempat Pembuangan Akhir Putri Cempo Surakarta"*. Surakarta : Sebelas Maret University.

[9] Sedigul, M. 2011. "*Pencemaran Air Tanah Dangkal Akibat Lindi TPA Sampah"*. Journal of Environmental Science

[10] Komala, P., Primasari, B., Rivai, F. 2008. "*Pengaruh Sistem Open Dumping di Lokasi Pembuangan Akhir (LPA) terhadap Kandungan Logam Berat pada Air Tanah Dangkal di Sekitarnya (Studi Kasus LPA Air Dingin, Padang)"*. Padang : Engineering Faculty of Andalas University.

Acknowledgments

The authors gratefully thanked Universitas Indonesia as the sponsor for this study research under the project of PITTA 2017 funding scheme.

IOP Conf. Series: Earth and Environmental Science **120** (2018) 012004 doi:10.1088/1755-1315/120/1/012004

Detection of Waterborne Protozoa, Viruses, and Bacteria in Groundwater and Other Water Samples in the Kathmandu Valley, Nepal

E Haramoto[1]

1 Interdisciplinary Center for River Basin Environment, Graduate Faculty of Interdisciplinary Research, University of Yamanashi, 4-3-11 Takeda, Kofu, Yamanashi 400-8511, Japan
E-mail: eharamoto@yamanashi.ac.jp

Abstract. In this study, the prevalence of various waterborne pathogens in water samples collected in the Kathmandu Valley, Nepal, and the applicability of *Escherichia coli* as an indicator of pathogen contamination in groundwater were assessed. Fifty-three water samples, including shallow groundwater and river water, were analyzed to examine the presence of protozoan (oo)cysts via fluorescence microscopy and that of viral and bacterial genomes via quantitative PCR. At least one of the seven types of pathogens tested (i.e., *Cryptosporidium*, *Giardia*, human adenoviruses, noroviruses of genogroups I and II, group A rotaviruses, and *Vibrio cholerae*) was detected in 68% (15/22) of the shallow dug well water samples; groundwater in the shallow dug wells was more contaminated compared with that in shallow tube wells (8/15, 53%). River water and sewage samples were contaminated with extremely high concentrations of multiple pathogens, whereas a tap water sample supplied by a water tanker tested positive for human adenoviruses and *V. cholerae*. The detection of host-specific *Bacteroidales* genetic markers revealed the effects of human and animal feces on groundwater contamination. The tested pathogens were sometimes detected even in *E. coli*-negative groundwater samples, indicative of the limitations of using *E. coli* as an indicator for waterborne pathogens in groundwater.

1. Introduction

High concentrations of waterborne pathogens, such as protozoa, viruses, and bacteria, are present in the feces of infected individuals; these pathogens are primarily transmitted via the fecal–oral route, including the consumption of contaminated food or water. An increasing number of studies have been reported on the prevalence of these pathogens in various water samples, mainly because of the rapid development of molecular biological methods; however, a limited number of studies have been conducted in developing countries.

Meanwhile, the Kathmandu Valley in Nepal has been plagued with waterborne diseases, which are considered as a serious public health concern; this issue is probably associated with the low coverage of drinking water and wastewater treatment systems. Because groundwater is the most important water source for drinking and domestic purposes in the valley [1], some recent studies have attempted to demonstrate the prevalence of waterborne pathogens, such as protozoa [2], [3], viruses [2], [4], and bacteria [5]–[7], in groundwater and other water samples. While these studies have contributed to reveal the prevalence of waterborne pathogens, most of them have examined one of the three major types of pathogens, limiting the assessment of the utility of conventional fecal indicator bacteria such as *Escherichia coli* for the management of microbiological safety of water. In addition, water sampling is sometimes conducted only once, which may not be sufficient to support the generalization of the obtained conclusions. In reality, the author of this study reportedly examined two protozoa and three viruses; however, the number of analyzed samples was small ($n = 10$) [2].

On the basis of this background, this study aimed to determine the prevalence of various waterborne pathogens in water samples collected in the Kathmandu Valley during different seasons. Recently, the electronegative membrane-vortex (EMV) method [8] was utilized for the simultaneous concentration of protozoa and viruses from single water samples. In particular, it was employed to detect seven pathogens (two protozoa, four viruses, and one bacterium). *E. coli* and total coliforms were examined to evaluate their applicability as the indicators of pathogen contamination of groundwater. In addition, sources of fecal contamination of water were estimated on the basis of the detection of host-specific *Bacteroidales* genetic markers.

2. Materials and methods

2.1. Collection of water samples

Figure 1 shows the locations of the sampling sites in the Kathmandu Valley. As described previously [4], water sampling was conducted in August 2010 (wet season) and May 2011 (dry season) with a total of 43 samples comprising 16 groundwater samples from nine shallow dug wells (SG06, 17, 18, 26, 31, and 34–37); 12 groundwater samples from six shallow tube wells (SG07, 10, 12, 23, 25, and 33); 13 river water samples from eight sites along the Bagmati River and its tributaries (RW01–08); one sample of tap water supplied by a water tanker (TW01); and one sewage sample collected from a sewage pipe (SW01, not shown in Figure 1).

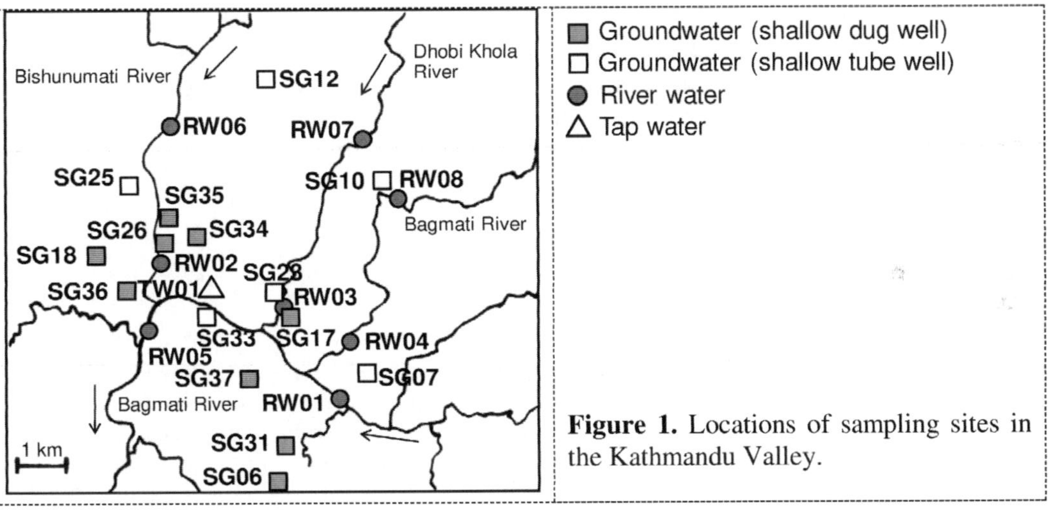

Figure 1. Locations of sampling sites in the Kathmandu Valley.

2.2. Detection of E. coli and total coliforms

The Colilert method (IDEXX Laboratories, Westbrook, ME, USA) was employed to quantify the concentrations of *E. coli* and total coliforms in the water samples according to the manufacturer's protocol, as described previously [2], [9]. Samples of groundwater from shallow dug wells have been tested in a previous study [9].

2.3. Detection of waterborne protozoa, viruses, and bacteria

2.3.1. Concentration. For all water samples excluding the sewage sample, the EMV method using a mixed cellulose ester membrane filter (pore size of 0.45 μm and diameter of 47 mm; Merck Millipore, Billerica, MA, USA) [8] was employed to simultaneously concentrate three types of waterborne pathogens (i.e., protozoa, viruses, and bacteria) in single water samples (50–3000 ml, depending on the sample type), as described previously [2], [4]. After the centrifugation of a 12-ml eluate at 2,000 ×

g for 10 min, a supernatant was obtained as a virus concentrate, whereas a pellet suspension was used to detect protozoa and bacteria. The sewage sample was directly subjected to centrifugation without any concentration step [4].

2.3.2. Detection of viruses. Viral DNA was extracted from virus concentrates using a QIAamp DNA Mini Kit (Qiagen, Hilden, Germany), whereas viral RNA was extracted using a QIAamp Viral RNA Mini Kit (Qiagen), followed by reverse transcription using a High Capacity cDNA Reverse Transcription Kit (Thermo Fisher Scientific, Waltham, MA, USA), as described previously [2, 4]. A Thermal Cycler Dice Real Time System TP800 (Takara Bio, Otsu, Japan) was used to perform quantitative PCR (qPCR) for human adenoviruses (HuAdVs) [10], noroviruses of genogroups I (NoVs-GI) and II (NoVs-GII) [11], and group A rotaviruses (RVAs) [12]. HuAdVs, NoVs-GI, and NoVs-GII were completely quantified using standard curves generated from 10-fold serial dilutions of standard samples, whereas semi-quantitative detection was conducted for RVAs based on threshold cycle values. All water samples, standard samples, and negative control were tested in duplicate.

2.3.3. Detection of protozoa. Eleven milliliters of the 12-ml pellet suspension was subjected to immunomagnetic separation (IMS) using a Dynabeads GC-Combo (Thermo Fisher Scientific). One half of the resulting IMS-purified sample (110 μl) was passed through a hydrophilic polytetrafluoroethylene membrane (pore size of 1.0 μm and diameter of 25 mm; Advantec, Tokyo, Japan), followed by direct fluorescent antibody staining using an EasyStain (BTF, North Ryde, Australia) according to the manufacturer's protocol. Fluorescence microscopy (BX60; Olympus, Tokyo, Japan) was employed for counting the number of *Cryptosporidium* oocysts and *Giardia* cysts.

2.3.4. Detection of bacteria. Bacterial DNA was extracted from the remaining portion of the pellet suspension using the QIAamp DNA Mini Kit, followed by qPCR for the *ctx*A toxic gene of *Vibrio cholerae* [13]. In addition, three host-specific *Bacteroidales* qPCR assays were performed to identify the sources of fecal contamination of the tested water samples: *gyr*B for human-specific assay [14], Pig2Bac for pig-specific assay [15], and BacR for ruminant-specific assay [16]. All qPCR runs were performed using a Thermal Cycler Dice Real Time System TP800.

3. Results and discussion

3.1. Detection of E. coli and total coliforms

Table 1 summarizes the results obtained from the detection of *E. coli* and total coliforms in the collected water samples. They were detected in almost all samples, except for a few shallow tube well water samples. The concentrations of *E. coli* and total coliforms in the groundwater samples from shallow dug wells were significantly greater than those from shallow tube wells (*t*-test, $P < 0.05$). This result is probably related to the more vulnerable structure of shallow dug wells than tube wells. According to the guidelines stipulated by the World Health Organization, *E. coli* should be absent in drinking water [17]; however, the results clearly indicated that shallow groundwater and water supplied by water tankers are not suitable for drinking purposes.

3.2. Detection of waterborne protozoa, viruses, and bacteria

Table 2 summarizes the results obtained from the detection of seven waterborne pathogens. In addition to the collected water samples, ten samples (six shallow dug well, three shallow tube well, and one river water samples) collected from the same locations in August–September 2009 were included in this table; the data for *Cryptosporidium* and *Giardia* have been reported previously [2], whereas those for viruses and *V. cholerae* has been analyzed herein.

At least one of the seven types of the tested pathogens was detected in 68% (15/22) of the shallow dug well water samples. Compared with the shallow dug wells, the shallow tube wells were less contaminated with pathogens (8/15, 53%); this result is in agreement with those obtained from the detection of fecal indicator bacteria (Table 1).

Table 1. Detection of *E. coli* and total coliforms in water samples.

Water sample	No. of tested samples	*E. coli*		Total coliforms	
		No. of positive samples (% positive)	Concentration[a] (log MPN[b]/100 ml)	No. of positive samples (% positive)	Concentration[a] (log MPN[b]/100 ml)
Shallow dug well water[c]	16	16 (100)	0.61–4.05	16 (100)	2.86–5.42
Shallow tube well water	12	7 (58)	0.00–1.43	11 (92)	0.93–3.96
River water	13	13 (100)	5.20–9.38	13 (100)	5.82–>9.38
Sewage	1	1 (100)	9.30	1 (100)	9.30
Tap water	1	1 (100)	2.04	1 (100)	3.31
Total	43	38 (88)		42 (98)	

[a] Range of concentrations among positive samples.
[b] MPN, most probable number.
[c] Reported previously [9].

Figure 2 shows the concentrations of the selected protozoa and viruses in the tested samples, clearly demonstrating that the river water samples contain a considerably higher number of pathogens compared with groundwater samples. The concentrations of pathogens at upstream sites (RW05–08) were sometimes lesser than those at densely polluted downstream sites (RW01–04) or even below the limit of detection.

Some shallow dug wells such as SG06 and SG37 were found to be highly contaminated with multiple pathogens, indicative of a serious risk of pathogen infections via contaminated groundwater. HuAdVs and *V. cholerae* were the only two pathogens found in the tap water supplied by a water tanker, which was also tested positive for *E. coli* and total coliforms. Tanker water is one of the major water sources in the Kathmandu Valley [1]; however, high levels of microbial contamination in water tanker samples were observed (unpublished data), indicative of the unsuitability of tanker water for drinking purposes.

Table 2. Detection of waterborne protozoa, viruses, and bacteria in water samples.

Pathogen	No. of positive samples (% positive)					
	Shallow dug well water (n = 22)	Shallow tube well water (n = 15)	River water (n = 14)	Sewage (n = 1)	Tap water (n = 1)	Total (n = 53)
Cryptosporidium	7 (32)[a]	0 (0)[a]	14 (100)[a]	1 (100)	0 (0)	22 (42)
Giardia	8 (36)[a]	1 (7)[a]	14 (100)[a]	1 (100)	0 (0)	24 (45)
HuAdVs	7 (32)	5 (33)	13 (93)	1 (100)	1 (100)	27 (51)
NoVs-GI	5 (23)	1 (7)	13 (93)	1 (100)	0 (0)	20 (38)
NoVs-GII	7 (32)	1 (7)	13 (93)	1 (100)	0 (0)	22 (42)
RVAs	1 (5)	0 (0)	10 (71)	1 (100)	0 (0)	12 (23)
V. cholerae	2 (9)	3 (20)	2 (14)	0 (0)	1 (100)	8 (15)
Total	15 (68)	8 (53)	14 (100)	1 (100)	1 (100)	39 (74)

[a] Including the data reported previously [2].

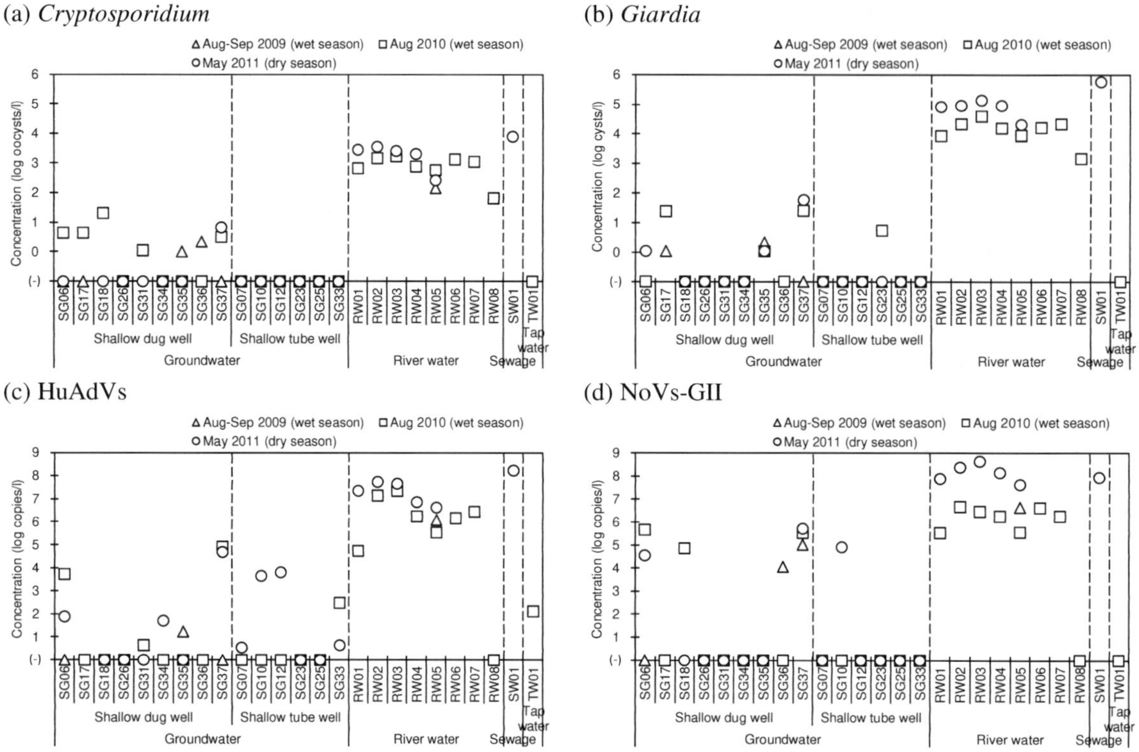

Figure 2. Concentrations of *Cryptosporidium*, *Giardia*, HuAdVs, and NoVs-GII in water samples.

3.3. Detection of host-specific Bacteroidales genetic markers

At least one of the three *Bacteroidales* genetic markers tested were detected in 18 (34%) of 53 samples, including the samples collected in August–September 2009; 14 (26%), 8 (15%), and 10 (19%) samples tested positive by human-, pig-, and ruminant-specific qPCR assays, respectively (Table 3). River water is clearly contaminated heavily with feces from humans and animals considering the low coverage of wastewater treatment plants and poor management of livestock wastewater.

Two shallow dug wells tested positive for the markers: the human-specific marker was detected in SG37 in August 2010, whereas both pig- and ruminant-specific markers were detected in SG35 in August–September 2009 and August 2010, respectively, indicative of the continuous animal fecal contamination of this well. The shallow tube well sample at SG10 in May 2011 tested positive for the ruminant-specific marker. These results emphasize the elimination of fecal waste from not only humans but also animals to improve the microbial quality of groundwater in the Kathmandu Valley. In addition, the EMV method is found to be effective for concentrating bacteria.

Table 3. Detection of host-specific *Bacteroidales* genetic markers in water samples.

Water sample	No. of tested samples	No. of positive samples (% positive)			
		*gyr*B (human)	Pig2Bac (pig)	BacR (ruminant)	Total
Shallow dug well water	22	1 (6)	2 (9)	2 (9)	3 (14)
Shallow tube well water	15	0 (0)	0 (0)	1 (7)	1 (7)
River water	14	12 (86)	6 (43)	6 (43)	12 (86)
Sewage	1	1 (100)	0 (0)	0 (0)	1 (100)
Tap water	1	0 (0)	0 (0)	1 (100)	1 (100)
Total	53	14 (26)	8 (15)	10 (19)	18 (34)

3.4. Relation between pathogens and E. coli

To assess the applicability of *E. coli* as an indicator of pathogen contamination of groundwater, shallow dug and tube well water samples were divided into four groups based on the *E. coli* concentrations. In general, the ratios of samples that tested positive for at least one of the seven types of pathogens tested and mean numbers of the types of pathogens detected in a single sample increased with increasing *E. coli* concentration, although the plots at a concentration of 1–10 most probable number (MPN)/100 ml did not follow this trend (Figure 3). In case a water sample contained *E. coli* at a concentration of >1000 MPN/100 mL, 73% of such samples tested positive for at least one type of pathogens, and on average, 2.9 types of pathogens were detected in a sample.

However, importantly, these pathogens were sometimes detected even in *E. coli*-negative samples with a positive ratio of 38%, indicating that *E. coli* is not suitable to confirm the absence of waterborne pathogens in the groundwater samples tested in this study. A similar result was obtained for total coliforms (data not shown).

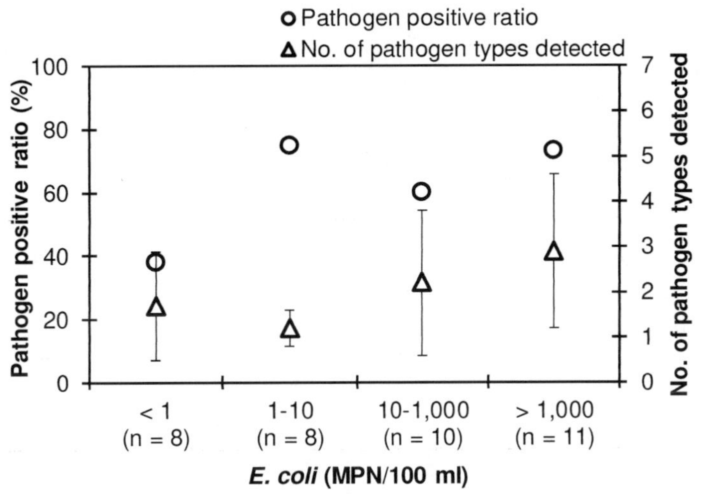

Figure 3. Relation between pathogens and *E. coli* in groundwater.

4. Conclusions

In summary, various types of waterborne pathogens, such as *Cryptosporidium*, *Giardia*, and HuAdVs, were successfully detected in the water samples in the Kathmandu Valley, Nepal. A higher prevalence of waterborne pathogens in the shallow dug wells compared with the shallow tube wells indicated a potentially higher risk of pathogen infections by drinking water from a shallow dug well. The limited applicability of *E. coli* as an indicator of the pathogen contamination of groundwater strongly emphasized the need for further studies related to the establishment of more suitable indicators. Furthermore, as not only human but also animal fecal contamination was identified in the tested water samples, future studies need to focus more on the prevalence and genetic analysis of zoonotic pathogens to evaluate the risk of infections from animals to humans.

5. References

[1] Shrestha S, Aihara Y, Bhattarai AP, Bista N, Rajbhandari S, Kondo N, Kazama F, Nishida K and Shindo J 2017 Dynamics of domestic water consumption in the urban area of the Kathmandu Valley: situation analysis pre and post 2015 Gorkha Earthquake *Water* **9** 222

[2] Haramoto E, Yamada K and Nishida K 2011 Prevalence of protozoa, viruses, coliphages and indicator bacteria in groundwater and river water in the Kathmandu Valley, Nepal *Trans. Roy. Soc. Trop. Med. Hyg.* **105** 711–6

[3] Shrestha S, Haramoto E, Sherchand JB, Hada S, Rajbhandari S and Shindo J 2016 Prevalence of protozoa and indicator bacteria in wastewater irrigation sources in Kathmandu Valley, Nepal:

cases from Kirtipur, Bhaktapur, and Madhyapur Thimi municipalities *J. Water Environ. Technol.* **14** 149–57

[4] Haramoto E and Kitajima M 2017 Quantification and genotyping of Aichi virus 1 in water samples in the Kathmandu Valley, Nepal *Food Environ. Virol.* **9** 350–3

[5] Ghaju Shrestha R, Tanaka Y, Malla B, Bhandari D, Tandukar S, Inoue D, Sei K, Sherchand JB and Haramoto E 2017 Next-generation sequencing identification of pathogenic bacterial genes and their relationship with fecal indicator bacteria in different water sources in the Kathmandu Valley, Nepal *Sci. Total Environ.* **601-2** 278–84

[6] Inoue D, Hinoura T, Suzuki N, Pang J, Malla R, Shrestha S, Chapagain SK, Matsuzawa H, Nakamura T, Tanaka Y, Ike M, Nishida K and Sei K 2014 High throughput DNA microarray detection of pathogenic bacteria in shallow well groundwater in the Kathmandu Valley, Nepal *Curr. Microbiol.* **70** 43–50

[7] Tanaka Y, Nishida K, Nakamura T, Chapagain SK, Inoue D, Sei K, Mori K, Sakamoto Y and Kazama F 2012 Characterization of microbial communities distributed in the groundwater pumped from deep tube wells in the Kathmandu Valley of Nepal *J. Water Health* **10** 170–80

[8] Haramoto E, Katayama H, Asami M and Akiba M 2012 Development of a novel method for simultaneous concentration of viruses and protozoa from a single water sample *J. Virol. Methods* **182** 62–9

[9] Shrestha S, Nakamura T, Malla R and Nishida K 2014 Seasonal variation in the microbial quality of shallow groundwater in the Kathmandu Valley, Nepal *Water Sci. Technol. Water Supply* **14** 390–7

[10] Heim A, Ebnet C, Harste G and Pring-Akerblom P 2003 Rapid and quantitative detection of human adenovirus DNA by real-time PCR *J. Med. Virol.* **70** 228–39

[11] Kageyama T, Kojima S, Shinohara M, Uchida K, Fukushi S, Hoshino FB, Takeda N and Katayama K 2003 Broadly reactive and highly sensitive assay for Norwalk-like viruses based on real-time quantitative reverse transcription-PCR *J. Clin. Microbiol.* **41** 1548–57

[12] Jothikumar N, Kang G and Hill VR 2009 Broadly reactive TaqMan assay for real-time RT-PCR detection of rotavirus in clinical and environmental samples *J. Virol. Methods* **155** 126–31

[13] Blackstone GM, Nordstrom JL, Bowen MD, Meyer RF, Imbro P and DePaola A 2007 Use of a real time PCR assay for detection of the *ctx*A gene of *Vibrio cholerae* in an environmental survey of Mobile Bay *J. Microbiol. Methods* **68** 254–9

[14] Lee CS and Lee J 2010 Evaluation of new *gyr*B-based real-time PCR system for the detection of *B. fragilis* as an indicator of human-specific fecal contamination *J. Microbiol. Methods* **82** 311–8

[15] Reischer GH, Kasper DC, Steinborn R, Mach RL and Farnleitner AH 2006 Quantitative PCR method for sensitive detection of ruminant fecal pollution in freshwater and evaluation of this method in alpine karstic regions *Appl. Environ. Microbiol.* **72** 5610–4

[16] Mieszkin S, Furet JP, Corthier G and Gourmelon M 2009 Estimation of pig fecal contamination in a river catchment by real-time PCR using two pig-specific *Bacteroidales* 16S rRNA genetic markers *Appl. Environ. Microbiol.* **75** 3045–54

[17] World Health Organization 2011 *Guidelines for drinking-water quality, fourth edition* (Gebeva: World Health Organization)

Acknowledgments

This study was supported by the Grant-in-Aid for the Global COE Program "Evolution of Research and Education on Integrated River Basin Management in Asian Region" from the Ministry of Education, Culture, Sports, Science and Technology of Japan. The author thanks Prof. Kei Nishida and Dr. Takashi Nakamura (University of Yamanashi, Japan) for their help in water sampling.

IOP Conf. Series: Earth and Environmental Science **120** (2018) 012005 doi:10.1088/1755-1315/120/1/012005

Identification of Important Parameter from Leachate Solid Waste Landfill on Water Quality, Case Study of Pesanggrahan River

R Yanidar[1*], D M Hartono[2], S S Moersidik[2]

1 Environmental Engineering Department, Trisakti University, as a Doctoral Student, Department of Civil Engineering, Faculty of Engineering, University of Indonesia, Depok 16242, Indonesia.
2 Environmental Engineering Departement, Faculty of Engineering, University of Indonesia, Depok 16242, Indonesia
E-mail: ramadhani@trisakti.ac.id

Abstract. Cipayung Landfill takes waste generation from Depok City approximately ± 750 tons/day of solid waste. The south and west boundaries of the landfill is Pesanggarahan River which 200m faraway. The objectives of this study are to indicate an important parameter which greatly affects the water quality of Pesanggrahan River and purpose the dynamic model for improving our understanding of the dynamic behavior that captures the interactions and feedbacks important parameter in river in order to identify and assess the effects of the treated leachate from final solid waste disposal activity as it responds to changes over time in the river. The high concentrations of BOD and COD are not the only cause significantly affect the quality of the pesanggrahan water, it also because the river has been contaminated in the upstream area. It need the water quality model to support the effectiveness calculation of activities for preventing a selected the pollutant sources the model should be developed for simulating and predicting the trend of water quality performance in Pesanggrahan River which can potentially be used by policy makers in strategic management to sustain river water quality as raw drinking water.

1. Introduction

Population increase and urbanization are challenging municipal authorities to manage solid waste. Landfill, one of several components of the waste management chain that needs more attention to reduce its environmental impact. Eventhough landfill is less expensive than other forms of waste treatment but it has make environmental problems. [1]. The leachate that produced by solid waste landfilling can caused surface water pollution because of the organic concentration.

Pesanggrahan River is a strategic river which 7.7 km length for Banten Province, DKI Jakarta, and also West Java Province in Indonesia. It is located within Kabupaten Bogor, Depok City, and Tangerang City in west Java Province, flow to South Jakarta, West Jakarta, and to the North Jakarta in DKI Jakarta Province, and flows to the estuary.

Depok City, West Java Province, Indonesia extends from longitude 106°43’00”-106°55’30” E and latitude 6°19’00” 6°28’00”S has an solid waste final disposal landfill (6°25'19.08"S 106°47'16.48"E) with an area of 11.6 Ha, which located in the sub-district of Cipayung. Cipayung Landfill takes waste generation from Depok City approximately ± 750 tons/day of solid waste. It has been operationalized since 1992. The south and west boundaries of the landfill is Pesanggarahan River which 200m faraway.

Based on those condition above, it need a identification the parameter which has a negative impact on the water quality of river pesanggrahan.
The objectives of this study are: (1) identification an important parameter which greatly affects the water quality of Pesanggrahan River (2) purpose the dynamic model for improving our understanding of the dynamic behavior that captures the interactions and feedbacks important parameter in river in order to identify and assess the effects of the treated leachate from final solid waste disposal activity as it responds to changes over time in the river.

2. Method

2.1 Study area

Pesanggrahan river in Depok City, West Java Province, Indonesia extends from longitude 106°43'00" - 106°55'30" E and latitude 6°19'00" - 6°28'00". To identify the water quality parameter, water samples were taken at 1 point before the landfill location and 3 points after the landfill location which is still within the administration area of Depok city. They are sampling location (SL) H which has 848 m before landfill area, SL B before the leachate outlet and SL_X and SL_W which located after landfill location, and the distance around 0.53 and 5.792 km from landfill location. (see figure 1).
Selected sampling stations located in Depok City, Indonesia. The water samplings were conducted in April and May 2017 by analyzing in situ for the parameter of DO, pH and temperature, and 23 parameters that are in accordance with raw water standards, by ex situ analysis to determine whether the Pesanggrahan River can still be used as a raw drinking water. The parameters analyzed are all physical and chemical parameters in raw water standard are color odor, taste turbidity and conductivity and TDS, TSS organic permangant, alkalinity, hardness, COD, BOD, Ca, Mg, CO2, Fe, Mn, Nitrat, Nitrit, Sulfat, BOD and COD.

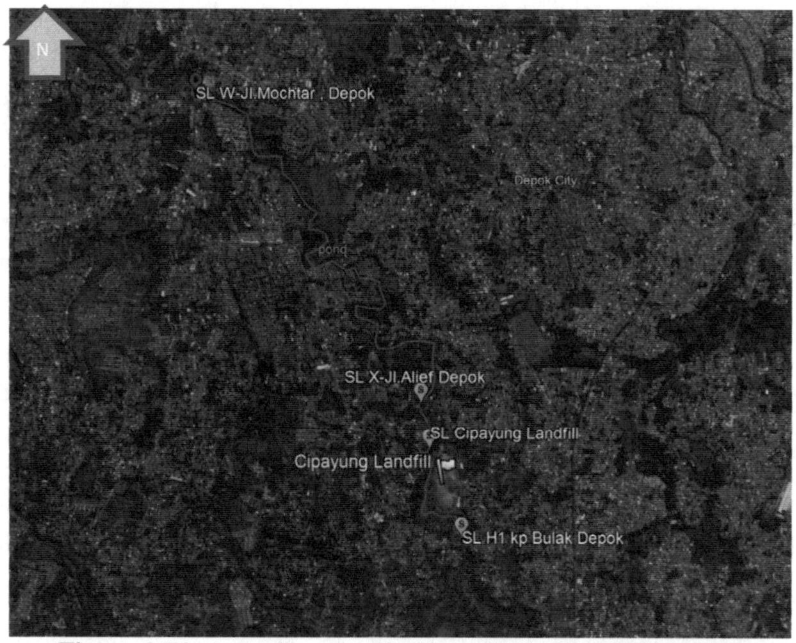

Figure 1 Water quality sampling location in Depok City Area

2.2 Statistical data analysis for proposing mechanistic water-quality model.

The data processing and statistical analysis for the parameters which exceed the class II standard were conducted using SPSS (ver. 23.0). Multivariate Analysis of Variance (Manova) analyzed the significance water quality parameters differences between four (4) sampling locations. Before running the Manova, the multivariate data must be normally distributed assumptions and independently sampled recommendation. The Kolmogorov-Smirnov and Levene's Test was applied to check the normality and verify homogeneity of variance across samples which explain the suitability and validity of the parameters being addressed through the Manova. The null hypothesis (Ho) of the Manova is the equality of variation mean for each parameter.

The results of Manova are presented as a basic to propose mechanistic water-quality model. To recognize the effects of leachate discharge into the river, the principle of mass balance was used because it is basic principle of water quality models [2]. It suggests using different loads from natural and anthropogenic pollutant that are contributed by the area between the upstream and downstream of the landfill location to build the water quality model.

3. Results and discussion

The results of the discharge and velocity measurements in sampling locations can be seen in Table 1 and the illustration of the transfer of water carrying leachate is illustrated in the chart in figure 2. Furthermore, the discharge measurements at the time showed a value of 3,59-3,95m^3/sec (Table 1). This is not a measurement that can show the amount of real discharge, because the discharge will be affected by the rain fall. Moreover, it need time series data debit to further analyze in detail.

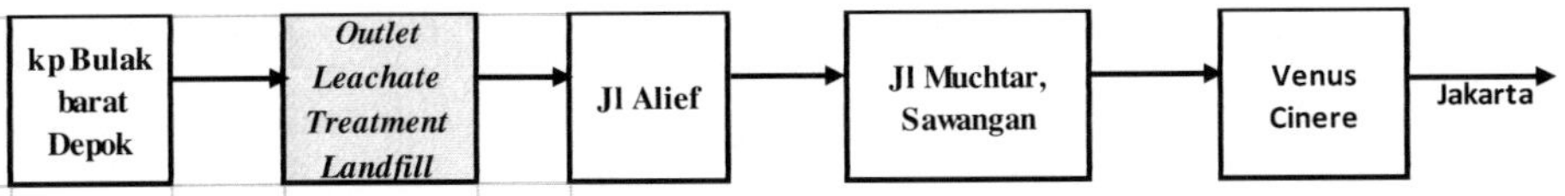

SL Number		SL_H		SL_B		SL_X		SL_W		SL_W_JKT
Distance from Landfill (m)			848		532		5,792		20,124	
	v (m/seconds)	0.67		0.68		0.68		1.08		1.09
Cummulative Time	T= (Seconds)			1,264.96		782.19		8,517.49		18,462.26
	(minutes)			21.08		13.04		141.96		307.70
	(hours)			0.35		0.22		2.37		5.13

Figure 2. Distance and velocity sampling location pesanggrahan river from Cipayung landfill in Depok City

Table 1. The results of the river's discharge and velocity measurement

Sampling Location (SL)	GPS	River's Width (cm)	Distance between SL (m)	Velocity (m/s)	Debit (m3/s)
Kampung Bulak barat Depok	6°25'32.10"S 106°47'20.1"E	8.2	850	0.67	3.59
Outlet Leachate Treatment	6°25'19.08"S 106°47'16.48"E		-	0.68	3.95
Jl Alief	6°25'01.20"S 106°47'10.8"E	11	540	0.68	3.6
Jl Muchtar, Sawangan	6°23'50.1"S; 106°46'18.4"E	8.2	5,260	1.08	

The water quality in table 2 shows almost all parameter meet the standards class II for raw drinking water standard, unless total suspended solid (TSS), and the organic parameters; BOD, COD and organic permanganate exceed the standards, therefore that 4 parameters can be regarded as an important parameter. The excessive BOD and COD loads indicate the level of water pollution which could damage the quality of river water. It causes low DO (dissolved oxygen) concentration and unsuitable life conditions in the river. Water quality assessments generally use the Biochemical Oxygen Demand (BOD) which is one of the most widely used criteria because it delivers information about the fraction of the organic load which is ready biodegraded in water. Especially, it indicates as the biodegradable fraction of an effluent as the ratio between BOD_5 and COD (chemical oxygen demand) in the water treatment plant [3].

Table 2. The results of the river's water quality

No	PARAMETER	Unit	STANDARD Raw water Class II	Mean SL_H Kp.Bulak	Mean SL_B Landfill Loc.	Mean SL_X Jl.Alief	Mean SL_W Jl.Sawangan
Phisics Parameter							
1	Color	Pt-Co	50	5.0	5.0	4.7	5.0
2	Odor	-	-				
3	Taste	-	-				
4	Turbidity	NTU	25	15.5	20.0	15.4	14.6
5	Conductivity	µS/cm	-	107.0	195.4	168.4	137.2
Chemistry Parameter							
6	pH	-	6,5 - 9,0	8.2	7.7	7.7	7.7
7	TDS	mg/L	1000	51.5	97.7	83.7	66.3
8	Organic Permanganat	mg/L $KMnO_4$	10	**91.7**	**58.4**	**86.5**	**103.2**
9	CO_2	CO_2	-	0.0	0.0	0.0	0.0
10	Alkalinity	mg/L $CaCO_3$	-	85.3	137.6	103.5	129.1
11	Hardness	mg/L $CaCO_3$	500	136.0	117.3	174.7	130.7
12	Ca	mg/L Ca^{2+}	-	38.4	32.0	36.3	29.9
13	Mg	mg/L Mg^{2+}	-	9.5	9.1	20.4	13.6
14	Cl	mg/L Cl^-	600	10.5	36.0	54.1	14.0
15	Fe	mg/L Fe	1.00	0.6	0.4	0.4	0.5
16	Mn	mg/L Mn	0.50	0.0	0.1	0.0	3.0
17	Nitrat	mg/L NO_3	10.0	0.1	0.1	0.1	0.2
18	Nitrit	mg/L NO_2	1.00	0.1	0.1	0.1	0.2
19	Sulfat	mg/L $SO4^{2-}$	400	29.1	31.8	28.8	30.6
20	TSS	mg/L	50	**78.0**	**72.3**	**69.7**	**90.3**
21	BOD	mg/L	3	**78.1**	**65.3**	**87.2**	**54.9**
22	COD	mg/L	25	**111.6**	**93.7**	**133.5**	**170.0**
Insitu Parameter							
23	pH	-	6,5 - 9,0	7.6	6.7	6.7	6.7
24	DO	mg/L	4	6.9	6.2	4.8	5.4
25	Temperatur	^0C	-	27.0	25.3	27.4	26.6

It needs to detect a main or interaction effect significantly for each concentration of parameters among sampling locations. The probabilities from the kolmogorov-smirnov test (table 3) below are greater than 0.05 (the typical alpha level) except the DO (see table 3). They indicate that the data are multivariate normally distributed. The results of Levene's Test also show homogeneity of variance across samples (table 4). Because of their cause, the Manova is suitable to check differences mean significantly for each parameter among sampling locations.

Table 3. The Kolmogorov-Sminorv Test for normality

		Organic permanganat (mg/L)	TSS (mg/L)	BOD (mg/L)	COD (mg/L)	DO (mg/L)
N		12	12	12	12	12
Normal Parameters[a,b]	Mean	84.955	77.583	85.074	127.201	6.915
	Std. Deviation	43.077	37.366	43.784	68.462	.438
Most Extreme Differences	Absolute	.138	.132	.155	.164	.353
	Positive	.130	.132	.105	.140	.353
	Negative	-.138	-.097	-.155	-.164	-.170
Test Statistic		.138	.132	.155	.164	.353
Asymp. Sig. (2-tailed)		.200[c,d]	.200[c,d]	.200[c,d]	.200[c,d]	.000[c]

Table 4. Lavene's Test of homogeneity of variance across samples

	F	*df1*	*df2*	*Sig.*
BOD (mg/L)	*3.358*	*3*	*8*	*.076*
COD (mg/L)	4.538	3	8	.039
Organic permanganat (mg/L)	.741	3	8	.557
TSS (mg/L)	2.520	3	8	.132
DO (mg/L)	3.223	3	8	.082

Table 5. The output of Manova Tes

Effect		Value	F	Hypothesis df	Error df	Sig.	Partial Eta Squared	Noncent. Parameter	Observed Power[d]
Intercept	Pillai's Trace	1.00	2881.63[b]	5.0	4.0	0.00	1.000	14408.13	1.000
	Wilks' Lambda	.00	2881.63[b]	5.0	4.0	0.00	1.000	14408.13	1.000
	Hotelling's Trace	3602.03	2881.63[b]	5.0	4.0	0.00	1.000	14408.13	1.000
	Roy's Largest Root	3602.03	2881.63[b]	5.0	4.0	0.00	1.000	14408.13	1.000
SL	Pillai's Trace	1.57	1.32	15.0	18.0	0.28	.524	19.83	.528
	Wilks' Lambda	.03	2.05	15.0	11.4	0.11	.699	26.55	.554
	Hotelling's Trace	16.69	2.97	15.0	8.0	0.06	.848	44.52	.690
	Roy's Largest Root	15.70	18.84[c]	5.0	6.0	0.00	.940	94.19	.999

a. Design: Intercept + SL
b. Exact statistic
c. The statistic is an upper bound on F that yields a lower bound on the significance level.
d. Computed using alpha = .05

Table 5 indicate that $p > 0.0005$, therefore, It accepts Hnull, that the means are equal for each response variable. It means there are not interaction effect of TSS and organic pollutant concentration (Organis Permanganate, BOD and COD) and sampling locations or differences of the concentration of parameters significantly among sampling location. This results describe that high concentrations of BOD and COD from leachate treatment plant from landfill are not the only cause significantly affect the quality of the pesanggrahan water. The BOD and COD concentration in LS before the landfill location showed a high value of 78.1 mg/L BOD and 111.6 mg/L COD. BOD mixture between BOD leachate and BOD river based on the mas balance principle is 78,5 mg/litre. Since the leachate discharge (0,35 litre/seconds) is not significant compared to the large river flow (3,7m/seconds). Pollution occurs not only because of the leachate, but the river has been contaminated with garbage in the upstream area. Besides when water sampling was conducted, there were many garbage on the river that also play a role in pollution water of Pesanggrahan River.

Reaeration is the major source of oxygen supply for BOD assimilation in streams, and assuming a balance of other sources (of oxygen supply) and decline (for oxygen demand) [4] [5]. This manova result, which there are not significant differences between the upstream and downstream points, leads to conjecture an indication that there is no self-purification process or the incidence of pollutant loads along the Pesanggrahan River, and another contaminant besides treated leachate landfill.
To know the effect of BOD leachate load on Pesanggrahan River is need model for developing understanding of the dynamic behaviour BOD dan DO parameter in River. The Biological Oxygen Demand (BOD) parameter is generally used to indicate the level of waste water pollution. BOD is quantified using dissolved oxygen (DO) (in mg/L) as the common, but it is not a constituent. Instead, it is a surrogate considered to quantify the potential of oxygen consumption by bacteria to break down organic carbon in the water. [6]. Therefore the Biochemical Oxygen Demand (BOD) typifies more precisely water and the biodegradable organic quality of the hydro system. Determination of BOD concentration is significant to trace the pollution flow from the upper to lower reach in the river [7]
Effluent dischargers from leachate landfill is an important issue in river water quality management. It is considering the role played in the decision-making process and in the implementation of any proposed waste load allocation program in Depok and DKI Jakarta region therefore Environmental management need model to predict the fluctuations river water quality.
The population growth and their activities makes the threat solid waste and leachate that related with polluted water more and more serious. A proper water resources management is important because water is a transportation medium for many pathogens that can make health risk seriously. Consequently, it needs a tool to simulate the temporal and spatial progress of pollutant concentrations in the surface water. The water quality model supports to calculate the effectiveness of activities to prevent a selected the pollutant sources, not only in defining the necessities for meeting the water quality standards [8]
There are various water quality models which use reaeration as the major source of oxygen supply for BOD assimilation in streams, and assuming a balance of oxygen supply and decline (for oxygen demand) enthusiastically, which the rate of DO deficit expressed as dD/dt. Water quality data can be projected by modeling which were simulated to understanding of the dynamic behavior the BOD and DO concentration as an impact of solid waste landfill to pesanggrahan River as the equation (1) and (2) following :

$$\frac{dBOD}{dt} = A - k_1.BOD(t) \ldots.(1) \qquad [4]$$

$BOD_{out}(t)$ = the rate of BOD decrease (mg/litre /days)
k_1 = deoxygenation coefficient (/days)
BOD (t) = BOD concentration at time t (mg/litre)

$$\frac{dDO}{dt} = k_r\left[DO_{sat} - DO(t)\right] - k_1.BOD(t) \ldots.(2) \qquad [4]$$

k_a = reoxygenation rate (/days)
DO_{sat} = DO saturated (mg/L)

Every model has its boundaries and advantages for detailed conditions. The data availability for the accurate model application as well as research goal are basic to choose the appropriate modelling approach [9]. It will proposed conceptual river water quality system dynamics model System dynamics modelling (SDM) is a computer-based method which is well balanced and sensible in the modern theory of non-linear dynamics and feedback control concept. [10]. This model starts with the development of a dynamic hypothesis which will build qualitative and quantitative model, generally mentioned as a Causal Loop Diagram (CLDs) (see figure 3), nonlinear systems and analyst the internal relation of systems [11]. The discrete concept of time is based upon the discrepancy between time-points and fixed time intervals, while the continuous concept deals with changes over time, based on infinitesimal mathematics [12].

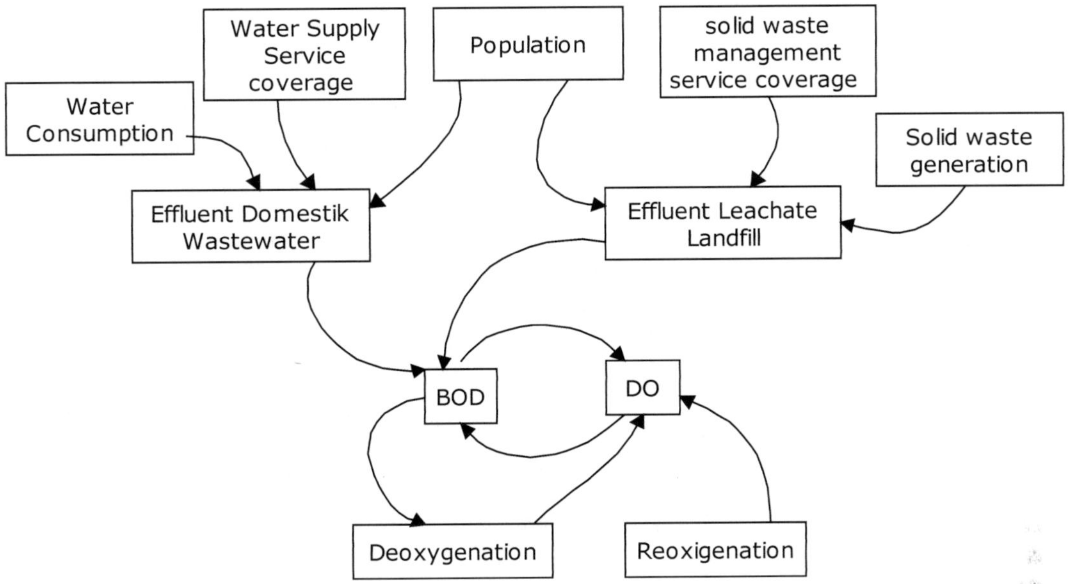

Figure 3. Causal Loop Diagram BOD and DO model.

The assumption of a river is made up by sequence of several reaches, according to scheme originally proposed by Thommann (1963), which the differential equations can be transformed into finite expressions [5]. To simplify the problem, it could be approached by developing a one-dimensional BOD-DO model with cross-sectional averaged and time-dependent assumption. The dynamic and quality terms are assumed to be constant over the entire cross section. It is only the longitudinal side is importance [5]. It is assumed sufficient because the river's lateral and vertical scale is much smaller than longitudinal ones [8]. The BOD and COD modeling can be disregarding some other aspects that can be pertinent to the problem of the approximating the river behavior. This assumption may be inconsistent with the natural behavior of the river and does not consider that pollutant concentration can vary within the length of each reach. Nevertheless, the approach of modelling can adopt for a first-glance estimation through considering the time effortlessly as a sequence of several steady state situation. It can simulate to describe any possibility how the pollutant behaves after being discharged in the stream

4. Conclusions

The high concentrations of TSS and Organic parameters occur not only because of the leachate, it also because the river has been contaminated by garbage in the upstream area. It needs the water quality model to support the effectiveness calculation of activities for preventing a selected the pollutant sources. The future study will build the BOD-DO system dynamics model. It should be developed for simulating and predicting the trend of water quality performance in Pesanggrahan River which can potentially be used by policy makers in strategic management to sustain river water quality as raw drinking water.

5. References

[1] S. P. T. P. B. M. J. S. M. J. &. K. V. T. S. 3. Gbanie, "Modelling landfill location using geographic information systems (GIS) and multi-criteria decision analysis (MCDA): case study Bo, Southern Sierra Leone. Applied Geography," *Applied Geography,* vol. 36, pp. 3-12., 2013.
[2] S. C. Chapra, Surface water-quality modeling, Singapore: Waveland press, 2008.

[3] Jouanneau, S., et al, "Methods for assessing biochemical oxygen demand (BOD): A review," *Water research 49,* pp. 62-82, 2014.
[4] M. a. J. J. W. Deaton, Dynamic modeling of environmental systems, Springer Science & Business Media, 2000.
[5] M. a. G. T. Benedini, Water quality modelling for rivers and streams, Springer Science & Business Media, 2013.
[6] C. H. L. W. S. L. S. W. &. L. C. F. Chen, "Technical challenges with BOD/DO modeling of rivers in Taiwan.," *Journal of Hydro-environment Research, 6(1),* pp. 3-8, 2012
[7] G. N. T. &. F. S. Ajeagah, "Monitoring of organic load in a tropical urban river basin (Cameroon) by means of BOD and oxydability measurements," *Ecohydrology & Hydrobiology, 10(1),* pp. 71-80, 2010.
[8] C. H. Y. H. L. &. M. X. I. E. XUE, "Development of integrated catchment and water quality model for urban rivers," *Journal of Hydrodynamics, Ser. B, 27(4),* pp. 593-603, 2015.
[9] I. &. W. P. Keupers, "Development and testing of a fast conceptual river water quality model.," *Water Research 113,,* pp. 62-71, 2017.
[10] J. Sterman, Business dynamics, system thinking and modeling for a complex world., Boston: McGraw-Hill, 2000.
[11] Z. &. W. S. Chen, "Application of system dynamics to water security research.," *Water Resources Management 28 (2),* pp. 287-300, 2014.
[12] O. S. R. A. &. P. M. G. Sahin, "Water security through scarcity pricing and reverse osmosis: a system dynamics approach," *Journal of cleaner production, 88,* pp. 160-171, 2015.

Acknowledgment
The authors gratefully thanked to University of Indonesia as the sponsor for this research under the project of PITTA 2017 funding scheme.

IOP Conf. Series: Earth and Environmental Science **120** (2018) 012006 doi:10.1088/1755-1315/120/1/012006

Integrity Model Application: A Quality Support System for Decision-makers on Water Quality Assessment and Improvement

D Mirauda[1,5], M Ostoich[2], F Di Maria[3], S Benacchio[4] and I Saccardo[4]

1 School of Engineering, Basilicata University, Viale dell'Ateneo Lucano 10, Potenza, 85100, Italy
2 Dipartimento Provinciale di Venezia, ARPAV, Via Lissa 6, Mestre-Venezia, 30171, Italy
3 Environmental Engineering Services srl - academic spin off, Basilicata University, Via Varco d'Izzo 28 - 85100 Potenza Italy
4 Servizio Osservatorio Acque Interne, ARPAV, Via Ospedale Civile 24, Padova, 35121, Italy
E-mail: domenica.mirauda@unibas.it

Abstract. In this paper, a mathematical model has been applied to a river in North-East Italy to describe vulnerability scenarios due to environmental pollution phenomena. Such model, based on the influence diagrams theory, allowed identifying the extremely critical factors, such as wastewater discharges, drainage of diffuse pollution from agriculture and climate changes, which might affect the water quality of the river. The obtained results underlined how the water quality conditions have improved thanks to the continuous controls on the territory, following the application of Water Framework Directive 2000/60/EC. Nevertheless, some fluvial stretches did not reach the "good ecological status" by 2015, because of the increasing population in urban areas recorded in the last years and the high presence of tourists during the summer months, not balanced by a treatment plants upgrade.

1. Introduction

In recent years, increasing emergency situations have occurred regarding the availability of water resources in many countries, due to various reasons: high water demand by human uses and by the production sector, such as industry and agriculture; irreversible deterioration in surface water quality; saline intrusion of coastal aquifers; contamination of groundwater by nitrates; and presence of extreme meteorological events caused by climate change. Until several decades ago, water was considered to be an unlimited resource but nowadays it appears that the amount of water available is scarce and that the quality status of rivers, lakes and wet areas is definitely at risk and, in some cases, even irreversible.

These alarming conditions, which are mainly due to the use of surface water bodies as receptors of urban and industrial wastewaters, have forced the European Union to take action concerning water management by applying specific measures to help reduce growing pressure on waters from the population, industry and agriculture, such as: Directive 91/271/EEC concerning urban wastewater treatment, Directive 91/676/EEC on water protection against pollution caused by nitrates of agricultural source, Directive 98/83/EEC focused on safeguarding human health from adverse effects of any contamination of water, and Directives 76/464/EEC, 80/68/EEC, 82/176/EEC, 84/156/EEC,

IOP Conf. Series: Earth and Environmental Science **120** (2018) 012006 doi:10.1088/1755-1315/120/1/012006

85/513/EEC, 86/280/EEC, 91/414/EEC and 98/8/EC about the pollution caused by certain dangerous substances discharged into the aquatic environment.

However, increasing problems related to water quantity and quality led to the development of an integrated approach for water management systems, including all water-related impacts. These efforts resulted in the amendment of Water Framework Directive (WFD) 2000/60/EC, which is among the most progressive water-related regulations worldwide. In fact, it proposes an innovative attitude to water protection and management able to combine the identification of quality objectives, the monitoring of water bodies, the definition of intervention measures and the control of pressure sources from the surrounding territory. With regard to the last aspect, the WFD indirectly suggests using, at river level, a framework that works as a decision support model, able to design and plan interventions aimed at mitigating the pressure sources, with a view to reaching environmental quality objectives.

Therefore, in the present paper a Decision Support System named "Integrity model" was applied to a sample river, the Bacchiglione in North-East Italy, in order to reduce its vulnerability to the phenomenon of pollution. Such approach, besides representing an important support tool for local decision-makers when choosing effective mitigation actions, can also become useful to communicate the quality status of the river to the population in a fast and simple way [1], [2].

The model, previously developed in the same river in the period 2001-2003 [3], has now been recalibrated on a period of nearly 10 years (2007-2015). The obtained results, compared with the classification of water quality status carried out according to the criteria of both Italian Decree n. 152/1999 and Decree n. 152/2006 (adopting the WFD), demonstrated the reliability and accuracy of the model over a long period of time. It was also possible to verify its sensitivity to the changes occurred in the river during the last years, thanks to the application of the WFD. In particular, the continuous controls of the territory, which have reduced the presence of illegal discharge points, and a series of upgrades to the treatment plants have modified the vulnerability scenarios of the river and improved its status.

2. Integrity model

The Integrity Model is based on the theory of influence diagrams [4]-[6], which is a simple graphical representation of a group of arrows and nodes.

For each node, i, on the graph it is necessary to establish the level of functional integrity, which depends on the level of physical integrity, y_i, as well as on the level of functionality, $\overrightarrow{x_j}$:

$$\overrightarrow{x_j} = \overrightarrow{\varphi_i}\left(y_i, \overrightarrow{x_j}, j \in P(i)\right) \tag{1}$$

y_i can be defined as a variable describing the physical condition of the node integrity, which has values between 0 and 1, where 1 indicates complete physical integrity and 0 its absence. The level of physical integrity is related to the stress vector, $\overrightarrow{\xi_i^k}$, through the vulnerability function, $f_i^k(\cdot)$, which is defined for each natural hazard k and for each node i of the territory considered, and it is expressed as $y_i = f_i^k \cdot \left(\overrightarrow{\xi_i^k}\right)$. The main role of this function is to provide a quantitative evaluation of the vulnerability level, which can be affected by a given stress caused by a natural hazard, k. In the present paper, the stress can be characterised, for example, by the flow of wastewater from treatment plants or by agricultural fertilizers reaching the water body after being washed off land.

In the simplest case, where the stress $\overrightarrow{\xi_i^k}$ may be a scalar quantity and described with a specific index, I, the level of physical integrity is simply a non–increasing monotonic function, included in the co–domain [0, 1]:

$$y_i = a\left(\frac{\exp(-\alpha I_i^2)}{1+\exp(-\alpha I_i^2)}\right) \tag{2}$$

where the parameters a and α define the shape of the vulnerability function. Besides, the level of physical integrity, y_i, depends directly on its intrinsic functionality vector, $\overrightarrow{x_i^0}$, which describes the condition of the related node functional integrity: $\overrightarrow{x_i^0} = \overrightarrow{\rho_i}(y_i) \cdot y_i$. y_i does not consider the possible functional interconnection with other nodes. As for y_i, the intrinsic functional integrity can assume

values between 0 and 1, where 1 indicates complete functional integrity and 0 the case in which the entity is non-functional.
The variable, $\vec{x_j}$, includes information related to the functionality of the other nodes that precede the node, *i*, and belong to *P(i)*. It is clear that also $\vec{x_j}$ can assume values between 0 and 1, where 1 indicates complete functional integrity and 0 that the entity has broken down.
The function, $\vec{\varphi_i}$, which links the above-described variables, is generally quite difficult to define, as it requires detailed knowledge of the relationship between the different nodes. Generally speaking, it represents a problem in terms of optimisation [4], [5]. When analysing territorial problems affecting the loss of functionality (e.g., pollution risk), it is reasonable to consider the optimisation of equation (1). This can be done by looking into the conditions of minimum system functionality, and what represents the maximum risk for the territory.
Therefore, maximum vulnerability of the territorial system is obtained when each identified entity reaches a condition of minimum functionality. Assuming that the level of intrinsic functionality and of physical integrity of the node, *i*, and the level of functionality of each node belonging to *P(i)* can be expressed by scalar quantities, equation (1) can be presented as follows:

$$x_i = min\left(y_i, w_{ij}(x_j), j : e_j \in P(i)\right) \quad (3)$$

where the function $w_{ij}(x_j)$ takes into account the dependence level between the functionality of node *i* and node *j*. The identification of the mathematical structure able to represent the various dependences between the nodes is actually the most crucial step in the proposed model. In this paper, it can be assumed as a monotonic non–decreasing function, with domain and co–domain belonging to the range 0-1, according to [7]:

$$w_{ij} = \left(1 - 0.1\alpha_{ij}\right)\frac{\left(1-\exp\left(-\alpha_{ij}x_j^2\right)\right)}{\left(1-\exp\left(-\alpha_{ij}\right)\right)} + 0.1\alpha_{ij} \quad (4)$$

where α_{ij} is a parameter characterising the link (*i, j*) and it has been calculated in relation to the existent correlations between the entities *i* and *j*.
In this paper, in equation (2) the coefficients *a* and α were computed hypothesising three integrity classes, low (0-0.4), mean (0.4-0.8) and high (0.8-1), for each index [3]. In equation (4), instead, the parameter α_{ij} was evaluated assigning values of 2, 4, or 8 to it, according to the correlation between the index values of the nodes (weak, medium-strong and strong, respectively).

3. Discussion of results

The River Bacchiglione is located in North-East Italy (figure 1) and its waters flow into the Adriatic Sea. It borders to the southwest with the tributary reservoir of the Agno-Gua, to the West with that of the Adige and to the north-east with that of the Brenta.
The river has a length of 118 km and crosses two major cities, Padova and Vicenza, deeply marking their economy since ancient times. It is characterised by seasonal variation of the flows: generally high levels in winter and low levels in summer.
In order to apply the model, the river was represented into nodes and arrows. The nodes correspond to single quality gauge stations along the river Bacchiglione, managed by the Regional Environmental Prevention and Protection Agency of Veneto (ARPAV). The arrows are homogeneous stretches of the river, which link the nodes. The quality status of the single node (gauge station) was assumed as representative of the status of the whole arrow (stretch) associated to it.

3.1. Evaluation of stresses

The indices used here had previously been discussed in [3] and are the Non-Treated Loads index and the Total Pollutant Reduction index.
The Non-Treated Loads index measures those loads not provided by treatment plants and that reach the water body without any reduction. It is defined as:

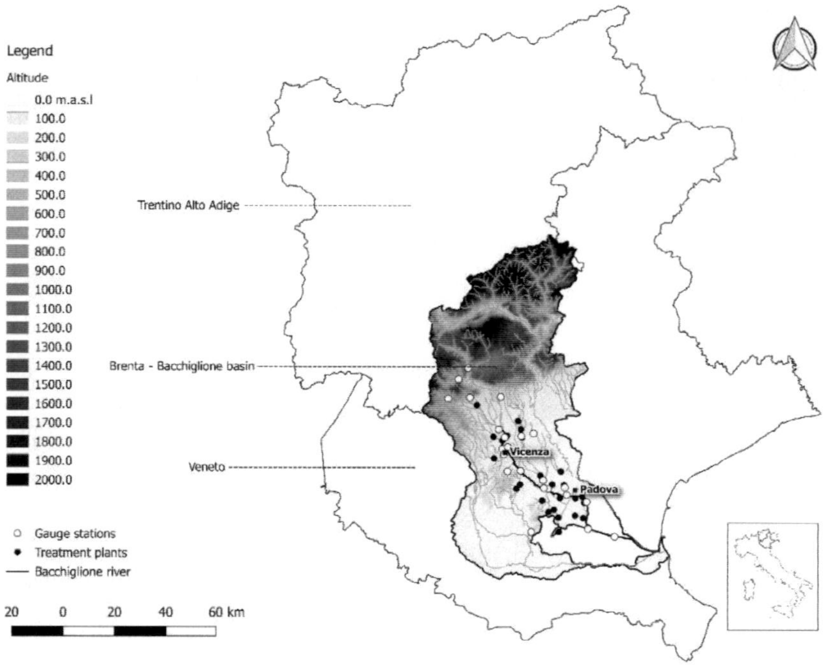

Figure 1. Brenta-Bacchiglione basin with gauge stations and treatment plants.

$$I_{NTL,COD} = 1 - \frac{\sum_{i=1}^{n} Q_{tr,p,i} c_{COD,i}}{Q_{R,j} \cdot c_{COD,j}} \tag{5}$$

where $Q_{tr,p,i}$ represents the flow rate discharged by the treatment plant, $c_{COD,j}$ is the *COD* concentration of each single discharge point, and Q_{Rj} and $c_{COD,j}$ are, respectively, the flow rate and the *COD* concentration measured at the downstream node of the fluvial reach to be analysed.

The Total Pollutant Reduction index allows evaluating the overall river response to the pollutant load and it is defined as:

$$I_{RED,COD} = \frac{Q_{R,i} \cdot c_{COD,i}}{P \cdot L_{SP,COD}} \tag{6}$$

where Q_{Rj} and $c_{COD,j}$ represent, respectively, the flow-rate and the *COD* concentration measured at the node located downstream of the considered river reach, while $L_{SP,COD}$ is the *COD* load provided by each single equivalent-inhabitant, and *P* is the population of the sub-basin flowing into the node. $I_{RED,COD}$ estimates the extent to which the hydrological environment is able to protect itself from anthropic pollution loads insisting on the specific sub-basin.

Only the plants with a potentiality of more than 2000 Equivalent Inhabitants (EI) have been here considered, releasing more pollution loads and water flow. The project wastewater flow data were used and assumed constant for the whole period of study 2007-2015.

For the calculation of the $I_{RED,COD}$, the specific *COD* load produced by single equivalent inhabitant (inh.) was assumed equal to 100 g/inh a day and the population of the sub-basin flowing into the node was estimated according to the census by the National Institute for Statistics (ISTAT) from 2007 to 2015. The flow discharge in each gauge station was calculated using expeditive mathematical models based on the entropy theory [8,]-[12].

3.2. Evaluation of the functionality integrity level

The application of the integrity model to the River Bacchiglione allowed the identification of different vulnerability scenarios. Figure 2 shows the vulnerability scenario in 2009 before the WFD 2000/60/EC application.

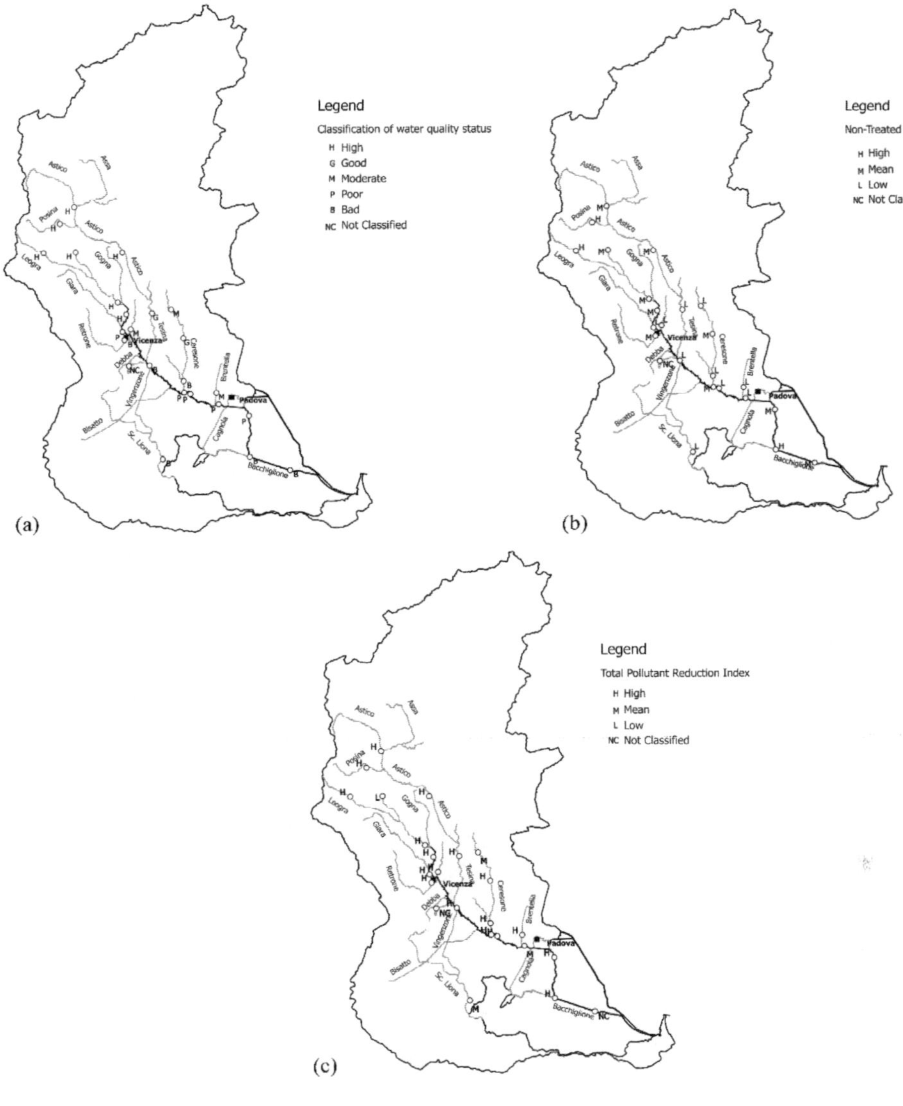

Figure 2. (a) Water quality classification according to Italian Decree n. 152/2006 (adopting the WFD); (b) level of functional integrity for the Non-Treated Loads index; (c) level of functional integrity for the Total Pollutant Reduction index (year 2009)

As represented by the figure, the upper part of the river is characterised by a "*High*" or "*Moderate*" quality status, also confirmed by the high values of the functional integrity level for both indices. Therefore, the good water quality recorded by ARPAV monitoring activities underlines the good auto-depuration capacity of the river and the absence of pressure sources from the surrounding territory. In the remaining part of the river and, in particular, in the fluvial stretch that goes from the gates of Vicenza to its mouth in the Adriatic Sea, the water environmental status changes and becomes "*Bad*" or *"Poor"*. The low functional integrity value of the $I_{NTL,COD}$ underlines the absent or inadequate treatment plants and the presence of unidentified discharge points along the river, while the high functional integrity linked to the second index, $I_{RED,COD}$, demonstrates the presence of civil pollution

IOP Conf. Series: Earth and Environmental Science **120** (2018) 012006 doi:10.1088/1755-1315/120/1/012006

loads only. This vulnerability scenario is explained by the numerous small-scale wastewater treatment plants in the area, as well as by a strong diffuse urbanisation and industrialisation. Therefore, being the investigated stretches in a very critical environmental situation, the implementation of appropriate structural interventions is needed, such as wastewater collection, well-projected depuration systems and efficient disinfection equipment.
Figure 3 shows the vulnerability scenario in 2013, where the upper part of the river looks unchanged, while in the stretch near Vicenza the water quality has improved from a "*Moderate*" to "*Good*" status.

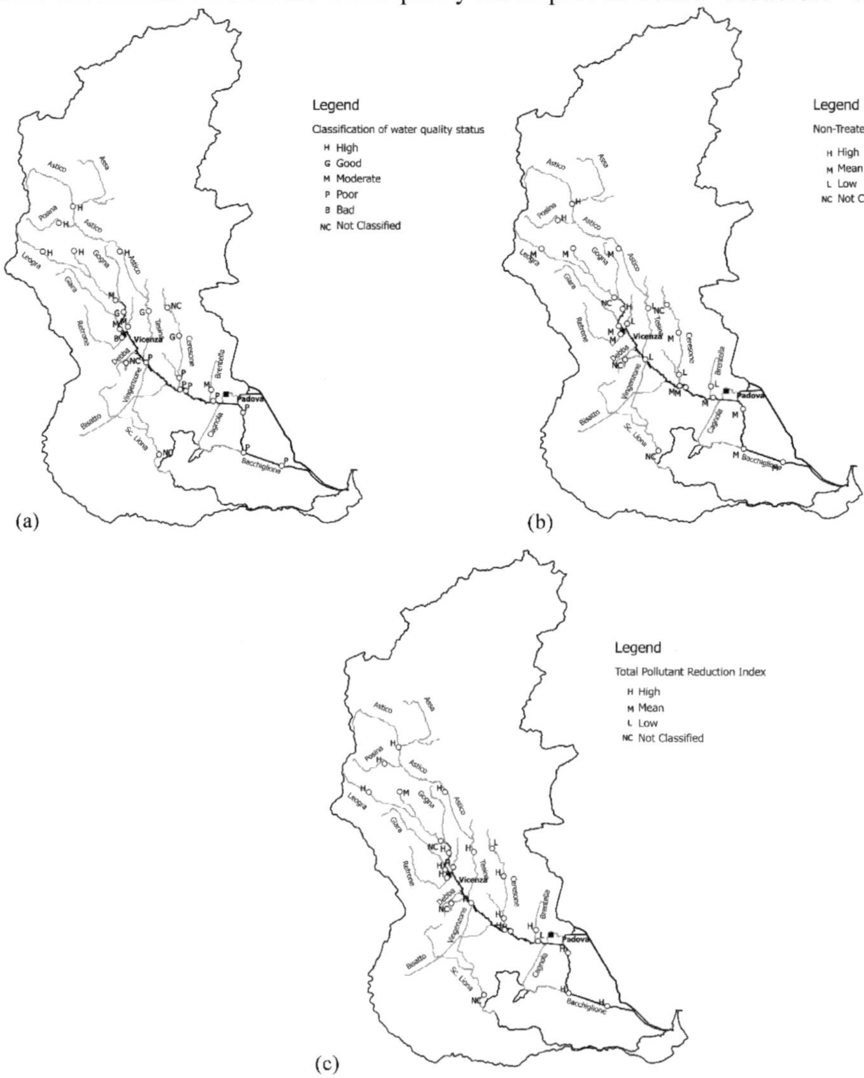

Figure 3. (a) Water quality classification according to Italian Decree n. 152/2006 (adopting the WFD); (b) level of functional integrity for the Non-Treated Loads index; (c) level of functional integrity for the Total Pollutant Reduction index (year 2013).

This is confirmed by high functional integrity for both indices. The WFD application to the river around 2010-2011, which led to a wide monitoring actions regarding the detection of non-controlled discharge points and to a series of upgrades of the treatment plants, has reduced the pollution phenomenon.
In the stretch that goes from the gates of Padova to its mouth, the situation is still very critical, as shown by the low functional integrity value of $I_{NTL,COD}$. In particular, despite continuous controls aimed at detecting and reducing the discharge points and different interventions on the treatment

plants, the river in its final stretch cannot mitigate pollution. This is also due to the increasing population in the last years and the high presence of tourists in the summer months.
Finally, figure 4 reports the vulnerability scenario of 2015, when all the rivers should have reached the "*Good*" quality status in compliance with the WFD.

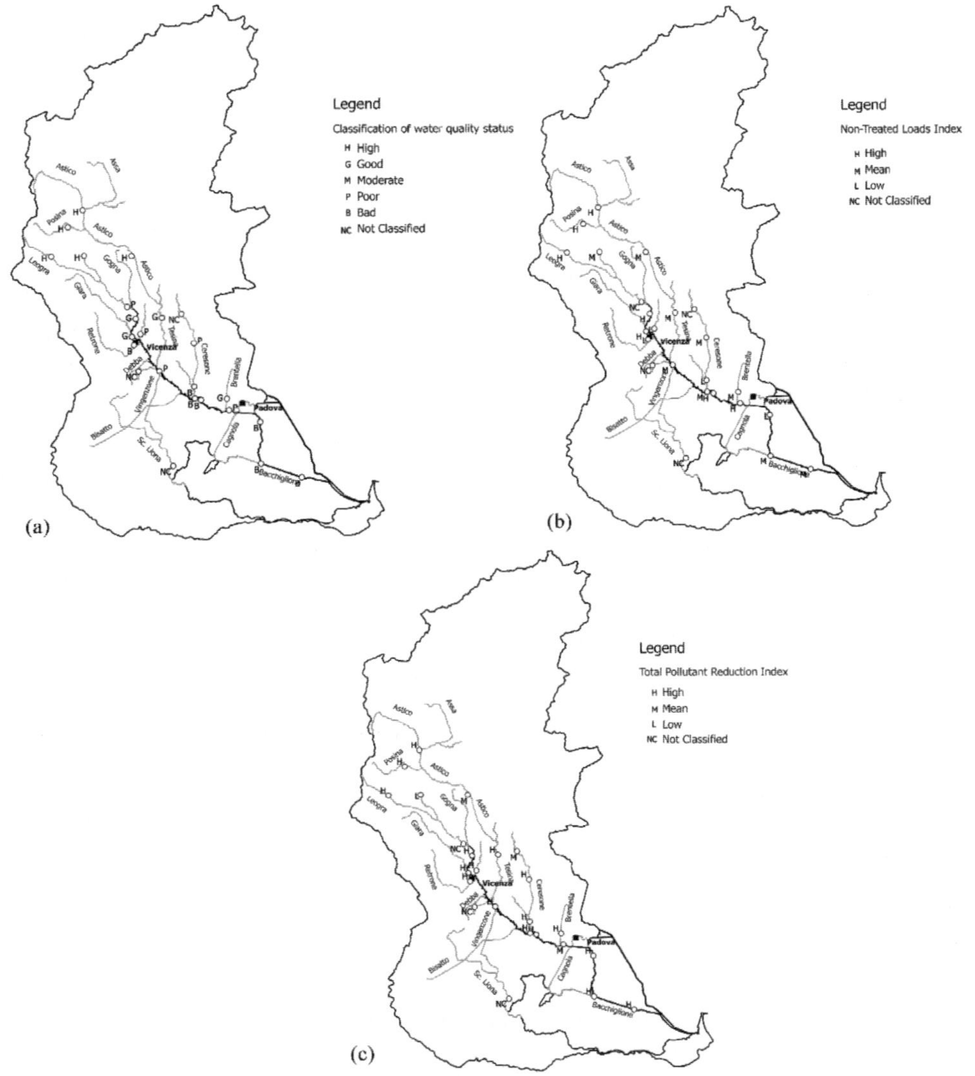

Figure 4. (a) Water quality classification according to Italian Decree n. 152/2006 (adopting the WFD); (b) level of functional integrity for the Non-Treated Loads index; (c) level of functional integrity for the Total Pollutant Reduction index (year 2015).

Actually, the status of the river has generally improved in nearly all the stretches and the values of both indices are high. Therefore, almost all the treatment plants have been correctly set, and the detailed monitoring activity has removed most of the illegal discharges and pollution loads from other sources. However, this must not distract from maintaining a policy of preventing and reducing the causes of pollution, in order to avoid the worsening of the water quality. The presence of stretches in a "*Bad*" or "*Poor*" status, such as the entrance to the city of Padova, should instead lead to a deeper control of the territory, aimed at reaching the "*Good*" water quality status.

4. Conclusions

The proposed model for the study of the River Bacchiglione vulnerability to the pollution phenomenon allows detecting the criticalities and the possible interventions needed to improve its water quality. Water body classification, performed according to the criteria of Italian Decree n. 152/2006, showed how not all the river stretches reached the "*Good*" status by 2015, according to the WFD.

The functional integrity evaluation of the Non-Treated Loads index underlined an improvement of the river water quality over the years, thanks also to the wide monitoring actions aimed at detecting illegal and incorrectly controlled discharge points, together with adjusting the treatment plants to the increasing population recorded in the last years. Only in some stretches, and in particular near Padova, the river still has a "*Bad*" status, which should lead to more actions and interventions addressing the pressure sources reduction.

The low values of the Total Pollutant Reduction index over the whole investigated period have instead demonstrated that the major causes of river deterioration are due to civil and industrial discharge points.

Therefore, the conclusions reached in this paper show how the integrity model, together with the existing classification systems for surface waters, can represent a useful and fast tool to support local technician and decision-makers when planning interventions and actions in order to mitigate the environmental pollution risk.

5. References

[1] Mirauda D, Erra U, Agatiello R and Cerverizzo M 2017 *Water* Applications of Mobile Augmented Reality to Water Resources Management **9(699)**

[2] Mirauda D, Erra U, Agatiello R and Cerverizzo M 2018 *Int. J. Sus. Dev. Plann.* Mobile augmented reality for flood events management **13(3)** pp 418-424

[3] Mirauda D and Ostoich M 2011 *Environ. Impact Assess. Rev.* Surface water vulnerability assessment applying the integrity model as a decision support system for quality improvement **31** pp 161-71

[4] Howard R A and Matheson J E 1984 *Readings on the Principles and Applications of Decision Analysis* vol 2, ed R A Howard and J E Matheson (Strategic Decisions Group, Menlo Park, CA) pp 719-62

[5] Schachter R 1988 *Oper. Res.* Probabilistic inference and influence diagrams **36** pp 589-604

[6] Shenoy P 1992 *Oper. Res.* Valuation-based systems for Bayesian Decision Analysis **40** pp 463-44

[7] Balbis L, Gaetani F, Minciardi R, Portella G, Sacile R and Trasforini E 2002 *Proc. 3rd International Conference on Computer Simulation in Risk Analysis and Hazard* (Mitigation, Sintra, Portugal) pp 243–52

[8] Mirauda D, Greco M and Moscarelli P 2011 *WIT Transactions on Ecology and the Environment* vol 146 (WIT Press: Southampton and Boston) pp. 355-67

[9] Greco M, Mirauda D and Plantamura Volpe A 2014 *Procedia Engineering* Manning's roughness through the entropy parameter for steady open channel flows in low submergence **70** pp. 773-80

[10] Greco M and Mirauda D 2015 *In Engineering Geology for Society and Territory, River Basins, Reservoir Sedimentation and Water Resources* vol 3 eds Lollino G et al. (Springer International Publishing, Switzerland) pp. 641-45

[11] Greco M and Mirauda D 2015 *J. Hydrol. Eng.* Entropy parameter estimation in large-scale roughness open channel **20(2)** pp. 04014047-1-6

[12] Mirauda D, De Vincenzo A and Pannone M 2017 *Spatial Statistics* Statistical characterization of flow field structure in evolving braided gravel beds DOI: 10.1016/j.spasta.2017.10.004

Acknowledgments

The study was developed within the project SMART BASILICATA, Call "Smart Cities and Communities and Social Innovation" (Advert MIUR n.84/Ric 2012, PON 2007 – 2013 of 2nd March 2012).

IOP Conf. Series: Earth and Environmental Science **120** (2018) 012007 doi:10.1088/1755-1315/120/1/012007

Indoor $PM_{2.5}$ and its Polycyclic Aromatic Hydrocarbons in Relation with Incense Burning

Susira Bootdee[2,4], Somporn Chantara[1,2] and Tippawan Prapamontol[3]

1 Environmental Chemistry Research Laboratory (ECRL), Chemistry Department, Faculty of Science, Chiang Mai University, Chiang Mai, Thailand, 50200
2 Environmental Science Research Center, Faculty of Science, Chiang Mai University, Chiang Mai, Thailand, 50200
3 Pollution and Environmental Health Research Program, Research Institute for Health Science (RIHES), Chiang Mai University, Chiang Mai, Thailand, 50200
4 Industrial Chemical Process and Environment Program, Faculty of Science Energy and Environment, King Mongkut's University of Technology North Bangkok (Rayong Campus), Rayong, Thailand, 21120
E-mail: somporn.chantara@gmail.com

Abstract. This study aims to determine fine particulate matter ($PM_{2.5}$) and polycyclic aromatic hydrocarbons (PAHs) emitted from incense burning to assess human health risk. $PM_{2.5}$ samples were collected for 8 hrs and 24 hrs during special occasions and normal period from two shrines in the city of Chiang Mai, Thailand. $PM_{2.5}$-bound PAHs were extracted and analyzed by GC-MS. The highest average $PM_{2.5}$ concentrations were found during Chinese New Year (625±147 $\mu g/m^3$ (8 hrs) and 406±159 $\mu g/m^3$ (24 hrs)). The highest total PAHs concentrations were also found during the same period (168±60 ng/m^3 for 8 hrs and 102±26 and for 24 hrs). Concentrations of $PM_{2.5}$ and carcinogenic-PAHs were highly correlated (r = 0.451-0.802) and were high during special occasions particularly during Chinese New Year due to high number of visitors and amount of incense being burned. The toxicity equivalent (TEQ) values were also relatively high during Chinese New Year (31-32 ng/m^3 (8 hrs) and 10-20 ng/m^3 (24 hrs)). It can be revealed that incense burning emits air pollutants and can increase degree of indoor air pollution and human health risk.

1. Introduction

The polycyclic aromatic hydrocarbons (PAHs) are comprised of a group of semi-volatile organic pollutants containing at least two fused aromatic rings. PAHs are toxic chemical group, which some of them are mutagenic or carcinogenic. The percentage of carcinogenic PAHs bound to PM_{10}, $PM_{2.5}$ and $PM_{1.0}$ emitted from vehicles and natural gas combustion in descending order were $PM_{1.0}$ (49-56%) > $PM_{2.5}$ (37-43%) > PM_{10} (35-41%). Therefore, PAH's concentrations associated with particulate matters are highly dependent on fine particles [1]-[3]. Furthermore, Hassanvand et al. [3] revealed the predominant PM-bound PAHs concentrations found in $PM_{2.5}$ were 83-88 %, which can penetrate deep into the alveolar regions of the lungs.

An epidemiological study found the relationship of traffic-related $PM_{2.5}$-bound PAHs and odd preterm birth was increased by 30 % per inter-quartile increase, which positively correlated and clustered together with the analysis [4]. Moreover, several studies assessed human exposure to airborne PMs and PAHs from heavy incense burning. The long-term exposure to incense burning in homes was associated with an increased risk of cardiovascular mortality [5], while Wang et al. [6] reported

incense burning at home was associated with an increased risk of asthma and wheezing. However, incense burning is a popular practice inside temples more than residential homes and the workplace. Chiang et al. [7] investigated the assessment of probabilistic human health risk exposed to carcinogenic PAHs in particles emitted from incense burning for temple goers/workers in a Taiwanese temple. A 95% probability total ILCR (9.87×10^{-4} to 1.13×10^{-3}) of workers extremely exposed to carcinogenic PAHs in the temple were indicated to be of a high potential health risk.
Incense burning is a serious issue concerning human health effect. The purpose of this study has been to provide reliable data and information concerning incense burning and its pollutant emissions. Consequently, the main objective of the research is to determine $PM_{2.5}$-bound PAHs emitted from incense burning to assess health risks.

2. Materials and methods

2.1 $PM_{2.5}$ sampling

The sampling shrines are located in Waroros market, which is one of the most crowded areas in the city of Chiang Mai (Figure 1). The first shrine is the oldest shrine. The size of the building is approximately 8x18 m^2 and 4 m height. The ventilation system is not well-equipped and the building was designed as almost a closed system with only one door open for visitors. The second shrine has an area of approximately 960 m^3. The ventilation system in the second shrine is better than the first one. It has 4 turbine ventilators on the roof of the worship room and its ceiling is open. It has quite a good ventilation system. At the first shrine, each visitor would purchase 8 large incense sticks (~ 9 mm diameter, ~ 48 cm long and ~ 34 cm coated combustible part). After each individual ceremony, the burned incense sticks are not allowed to be kept inside the worship area, but are kept outside. The second shrine would purchase a package of 28 smaller incense stick. Twenty – six of them were burned in the worship room and the rest were burned outdoors.
Indoor $PM_{2.5}$ samples were collected on Teflon fiber filters (2 µm PTFE, 46.2 mm diameter, Whatman's filter paper) using Minivol Air Samplers (Air metric, USA) at a flow rate of 5 L/min for 8 hrs (8 am – 4 pm) and 24 hrs (8 am – 8 am). The filters were pre-weighed by a microbalance in a controlled room (25.4±2.8 °C, 41.3±5.4% RH). After the sampling, the filters were kept in aluminium foil plates inside a desiccator for 48 h before being re-weighed and stored in a freezer (-4.0 °C) until analysis.
The sampling was carried out during special occasions and over normal periods (the background value). The special occasions were those associated with the Chinese lunar calendar, such as Chinese New Year, the Chinese Ghost festival, a vegetarian festival, the moon festival.

2.2 Extraction and analysis of PAHs

The samples were extracted in 25 ml dichloromethane (DCM) and n-hexane mixture (1:1, v/v) for 45 min by a ultrasonicator. The extracted solutions were than purified using a nylon syringe filter and were dried using a rotary evaporator. The solution was added with a mixture of internal standards (acenaphthene-d12 and pyrylene-d10) and was adjusted to 2 ml volumetric flask with solvent mixture. PAHs were analyzed by gas chromatography - mass spectrometer (GC-MS, Agilent, USA) equipped with a 30 m HP-5MS capillary column. The heating program was set for column oven from 60°C to 290°C at 6 °C/min, then hold for 20 min [8]. The MS was operated in selective ion monitoring mode (SIM). The 16-PAHs was identified and quantified including naphthalene (NAP), acenaphthylene (ACY), acenaphthene (ACE), fluorene (FLU), phenanthrene (PHE), anthracene (ANT), fluoranthene (FLA), pyrene(PYR), benz[a]anthracene (BaA), chrysene(CHR), benzo[b]fluoranthene (BbF), benzo[k]fluoranthene(BkF), benzo[a]pyrene (BaP), indeno[1,2,3-cd]pyrene (IND), dibenz[a,h]anthracene(DBA), benzo[g,h,i]perylene (BPER). The quality control of PAHs was studied using standard reference material (SRM 1649b; urban dust, NIST, USA) Recovery efficiency of PAHs (n=3) ranged 61 (CHR) to 126 (ANT) % and averaged 84 %. The detection limit in this study were 0.08 – 0.66 ng/m^3.

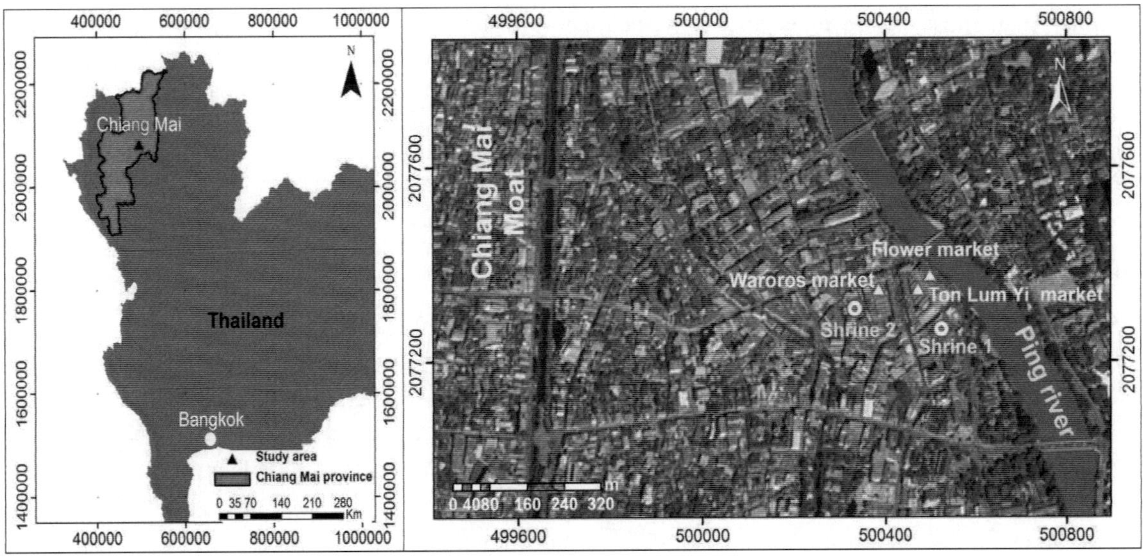

Figure 1 Map of sampling shrines in Chiang Mai City

3. Results and discussions

3.1 Concentrations of $PM_{2.5}$ and $PM_{2.5}$ – bound PAHs

Concentrations of indoor $PM_{2.5}$ measured on special occasions (with exclusion of Chinese New Year), Chinese New Year and the normal period at both shrines are shown in Table 1. The normal period of 8 hrs of indoor $PM_{2.5}$ ranged from 82±25 μg/m^3 (shrine 1) to 100±35 μg/m^3 (shrine 2), while those of the 24 hrs periods were 50±20 μg/m^3 (shrine 2) to 52±24 μg/m^3 (shrine 1). Concentrations of $PM_{2.5}$ at shrine 1 were slightly lower than those at shrine 2, but no significant difference ($p>0.05$). Based on the results, the different architecture of the two shrines might not play a significant role for PM concentrations. Otherwise, if the number of visitors is almost equal, the $PM_{2.5}$ concentrations in the shrine 1 should be higher than the shrine 2 due to its poorer ventilation system. In general at both shrines, the mean $PM_{2.5}$ concentrations of 8 and 24 hrs in normal period were significantly less than those of the Chinese New Year and other special occasions($p>0.05$). The highest value of indoor $PM_{2.5}$ concentrations was obtained in Chinese New Year for both of 8 and 24 hrs periods (524 ± 110 - 625 ± 147 μg/m^3 and 317 ± 109 - 406 ± 159 μg/m^3, respectively) at the two shrines. The $PM_{2.5}$ concentrations associated with number of visitor are highly dependent on the amount of incense being burned [9]. During the Chinese New Year and on other special occasions, the 24 hrs average values of $PM_{2.5}$ (35 μg/m^3) collected from both shrines were 1-11 times higher than standard value of the National Ambient Air Quality Standard (NAAQS), USA [10]. Moreover, it was found that the 24 hrs of incense burning in this study was 2-11 times higher than urban traffic site (28±6.3 - 38±13 μg/m^3) in Greece [1]. The main reason for high level of pollutants found indoors was clearly from number of visitors, which was related with the amount of incense being burned.

Table 2 shows the mean of total PAHs (t-PAHs) bounded with $PM_{2.5}$ concentrations. Mean t-PAHs of 8-hrs sampling time at shrine 1 were 168±60 ng/m^3 (Chinese New Year), 107±45 ng/m^3 (other special occasions) and 62±30 ng/m^3 (normal period), while those of 24-hrs in Chinese New Year, other special occasions and normal periods were 102±26 ng/m^3, 47±7 ng/m^3 and 32±19 ng/m^3, respectively. At shrine 2, the 8-hrs mean concentrations of total $PM_{2.5}$-bound PAHs were 90±41 ng/m^3 (Chinese New Year), 71±30 ng/m^3 (Other special occasions) and 25±15 ng/m^3 (Normal period). The 24-hrs mean PAHs concentrations in all periods at both shrines were lower than those of 8-hrs (46±29 ng/m^3, 30±12 ng/m^3 and 14±9 ng/m^3, respectively). However, the trend of PAHs concentrations found in each occasion was the same. The 8- and 24-hrs of total PAHs concentrations in Chinese New Year were

IOP Conf. Series: Earth and Environmental Science **120** (2018) 012007 doi:10.1088/1755-1315/120/1/012007

higher than those in the other special occasions and normal period. The values were significantly different between the other special occasions and normal period, while the values of the other special occasions were not significantly different from Chinese New Year. The dominant PAHs found in both shrines were BbF, BaP, IND and BPER, while the dominant particulate-bound PAHs collected at Thai temple in Samutprakarn, Chachengsao and Ayutthaya province were BaA, BbF, BaP and DbA [11]. Moreover, 8-hrs indoor BaP concentrations measured during Chinese New Year and other special occasions (1.61 – 10.0 ng/m^3) were higher than the values measured by personal samplers from workers in Thai temple (2.53±0.83 ng/m^3) [11]. This is probably due to the amount of incense being burnt during the sampling period.

Table 1 Mean of indoor $PM_{2.5}$ concentrations (µg/m^3) in both shrines in various occasions

Sampling periods	Shrine 1		Shrine 2	
	8 hrs	24 hrs	8 hrs	24 hrs
Chinese New Year	524 ± 110^a (n=2)	317 ± 109^a (n=6)	625 ± 147^a (n=3)	406 ± 159^a (n=6)
Other special occasions	228 ± 81^a (n=5)	178 ± 81^a (n=8)	184 ± 85^b (n=9)	133 ± 71^b (n=8)
Normal period	82 ± 25^b (n=7)	52 ± 24^b (n=8)	100 ± 35^b (n=9)	50 ± 20^c (n=13)

$^{a, b, c}$ = Significant differences ($p < 0.05$) among groups of sampling periods (vertical direction)

Table 2 Concentrations (ng/m^3) of t-PAHs, c-PAHs and nc-PAHs in both shrines in various occasions

Sampling sites	Sampling periods	Concentrations (Mean±SD)					
		8 hrs			24 hrs		
		t-PAHs	c-PAHs	nc-PAHs	t-PAHs	c-PAHs	nc-PAHs
Shrine 1	Chinese New Year	168±60^a (n=2)	83±35^a	85±25^a	102±26^a (n=6)	49±16^a	52±18^a
	Other special occasions	107±45^{ab} (n=5)	55±31^{ab}	52±16^{ab}	47±7^b (n=8)	26±3^b	20±6^b
	Normal periods	62±30^b (n=7)	24±15^b	38±12^a	32±19^b (n=8)	18±16^b	15±5^b
Shrine 2	Chinese New Year	90±41^a (n=3)	64±34^a	26±9^a	45±29^a (n=6)	27±14^a	18±16^a
	Other special occasions	71±30^a (n=9)	48±24^a	22±8^a	30±12^a (n=8)	20±10^{ab}	10±4^{ab}
	Normal periods	25±15^b (n=10)	15±12^b	10±7^b	14±9^b (n=13)	10±8^b	5±3^b

$^{a, b, c}$ = Significant differences ($p < 0.05$) among groups of sampling periods (vertical direction)

The average concentrations of carcinogenic PAHs (c-PAHs) and non-carcinogenic PAHs (nc-PAHs) are shown in Table 2. The c-PAHs were BaA, CHR, BkF, BbF, BaP, IND, and DBA while nc-PAHs were NAP, ACY, ACE, FLA, PHE, ANT, FLU, PYR and BPER [10]. The concentrations of c-PAHs and nc-PAHs were found to have the same trend as $PM_{2.5}$ concentrations. The 8- and 24-hrs average c-PAHs concentrations at both shrines in descending order were Chinese New Year > other special occasions > normal period. However, they were not significantly different ($p>0.05$) between Chinese New Year and other special occasions but significantly higher than normal period. Moreover, the c-PAHs concentrations of both 8- and 24-hrs samplings were about 1-2 times higher than those of nc-PAHs in every occasion including normal period. The c-PAHs were found in high concentrations because high molecular weight PAHs (>200) could be more easily adsorbed to particulate phase [12].

3.2 Correlations between $PM_{2.5}$ and PAHs

The Spearman's rank-correlation of $PM_{2.5}$, t-PAHs, c-PAHs and nc-PAHs concentrations are shown in Table 3. $PM_{2.5}$ concentrations were significantly correlated ($p<0.01$) with concentrations of t-PAHs ($r = 0.674$-0.822), c-PAHs ($r = 0.618$-0.802) and nc-PAHs ($r = 0.612$-0.914). Moreover, the strong positive correlations between $PM_{2.5}$ and all type of PAHs were found in 24 hrs sampling ($r = 0.674$-0.822 (t-PAHs), 0.618-0.700 (c-PAHs) and 0.738-0.914 (nc-PAHs)). The correlation between t-PAHs and c-PAHs ($r = 0.934$-0.982) was higher than that between t-PAHs and nc-PAHs ($r = 0.795$-0.956). The result obtained indicated that PAHs concentrations associated with fine particles are highly dependent on the number of incense being burned.

Table 3 Correlations of $PM_{2.5}$ and PAHs concentrations for 8-hrs and 24-hrs sampling at both shrines

	8 hrs (n = 14)				24 hrs (n = 22)			
	$PM_{2.5}$	t-PAHs	c-PAHs	nc-PAHs	$PM_{2.5}$	t-PAHs	c-PAHs	nc-PAHs
Shrine 1								
$PM_{2.5}$	1.000				1.000			
t-PAHs	0.688**	1.000			0.822**	1.000		
c-PAHs	0.802**	0.934**	1.000		0.700**	0.942**	1.000	
nc-PAHs	0.612**	0.956**	0.851**	1.000	0.914**	0.921 **	0.791**	1.000
	8 hrs (n = 21)				24 hrs (n = 27)			
Shrine 2								
$PM_{2.5}$	1.000				1.000			
t-PAHs	0.475*	1.000			0.674**	1.000		
c-PAHs	0.451*	0.982**	1.000		0.618**	0.980**	1.000	
nc-PAHs	0.448*	0.795**	0.700**	1.000	0.738**	0.850**	0.757**	1.000

** Correlation is significant at the 0.01 level (2- tailed)
*Correlation is significant at the 0.05 level (2- tailed)

3.3 Assessment of possible health effects based on PAHs concentrations

3.3.1 Toxicity equivalent (TEQ)

The toxicity equivalent concentration (TEQ) is widely used to assess risk of carcinogenic potency of each individual PAH. This parameter is calculated from total of individual concentration of each PAH multiply by their toxic equivalent factor (TEF) relative to the carcinogenic potency of BaP (Eq.1), which was used as a reference carcinogenic compound [13].

$$TEQ = \sum_i (PAHi \times TEF_i) \qquad (Eq\ 1)$$

Where PAH_i is concentration of an individual PAHs and TEF_i is the toxic equivalent factor.
Table 4 shows the values of TEQ of each sampling period. TEQ values obtained from both shrines in descending order were Chinese New Year > other special occasions > normal periods. The highest TEQ values were found during Chinese New Year. The mean TEQ values obtained in during Chinese New Year were 2-3 times higher than the normal periods, and 1-2 times higher than other special occasions. In comparison with the study in Taiwan [14], where PM_{10} samples were collected for 8 hours in a shrine, the TEQ value (36.6 ng/m^3) was almost the same with the value obtained during the Chinese New Year in this study (3.2 ± 1.8 to 31.8 ± 27.4 ng/m^3), while the 8-hrs TEQ values (0.29 ± 0.12

to 4.60±1.35 ng/m^3) from Thai temple were lower than in this study [10]. The TEQ values found in this study were much higher than the European guideline (1 ng/m^3).

Table 4 Average concentration (ng/m^3) of toxicity equivalent (TEQ) and the inhalation cancer risk of $PM_{2.5}$-bound PAHs at shrines in various occasions

Index	Sites	sampling hours	Chinese New Year	Other special occasions	Normal periods
TEQ	Shrine 1	8	32±12	20±14	14±12
		24	20±7	12±3	7±5
	Shrine 2	8	32±27	15±8	3±2
		24	10±4	7±4	3±2
ICR	Shrine 1	8	27×10^{-4}	18×10^{-4}	12×10^{-4}
		24	18×10^{-4}	10×10^{-4}	60×10^{-5}
	Shrine 2	8	28×10^{-4}	13×10^{-4}	28×10^{-5}
		24	85×10^{-5}	60×10^{-5}	23×10^{-5}
Risk ($ICR\times10^6$)	Shrine 1	8	2,700	1,800	1,200
		24	1,800	1,000	600
	Shrine 2	8	2,800	1,300	280
		24	850	600	230

The TEQ values of 8-hrs sampling were higher than those of 24-hrs sampling. Both values presented similar trends. Moreover, the TEQ values (8- and 24-hrs) at shrine 2 were lower than shrine 1. However, the obtained TEQ values emitted from motor vehicles and fuel burning for heating found in studied by [15], [16], were lower than those values Chinese New Year of this study. The main factors should be the ventilation system and the amount of incense being burned.

3.3.2 The inhalation cancer risk (ICR) assessment
The inhalation cancer risk (ICR) was used to estimate the value of cancer risk from PAHs exposure during different periods and can be calculated using Equations 2 [8], [15], [17], [18].

$$ICR = TEQ \times IUR_{BaP} \qquad (Eq\ 2)$$

Where, IUR_{BaP} is the inhalation unit risk

IUR_{BaP} values for lifetime (70 years) PAHs exposure were used in this study to estimate the inhalation cancer risk. The recommended IUR_{BaP} of the World Health Organization (WHO) is 8.7×10^{-5} m^3/µg [19]. The mean TEQ value was obtained from the calculation of each period (Eq. 1). The inhalation cancer risk was calculated and compared among different periods as shown in Table 4.
The 8- and 24-hrs of lifetime inhalation cancer risks (ICR) calculated based on WHO guideline obtained from both shrines in descending order were Chinese New Year > other special occasions > normal periods. Hence, if a million people were exposed to PAHs at the level of 10 ng/m^3 for 70 years, 850 persons may have a risk of cancer development. The ICR values between 10^{-6} to10^{-4} are potential risk and ICR of 10^{-6} represents a lower-bound zero risk, while the upper 10^{-4} of ICR indicates high potential health risk [17], [20]. The calculated ICR values were found to have a high potential health risk based on WHO, while those based on CalEPA were in a lower-bound zero risk. The societal inhalation cancer risk was obtained by multiplication of ICR values with a million people [15]. The societal inhalation cancer risk was obtained by multiplying the ICR by one million people. The

societal ICR (cases/million people) based on WHO for 8-hrs value calculation at shrines 1 and 2in descending order were Chinese New Year (2,700 and 2,800), other special occasions (1,800 and 1,300) and the normal periods (1,200 and 280). While those of the societal ICR for 24-hrs calculation in descending order were Chinese New Year (1,800 and 850), other special occasions (1,000 and 600) and the normal periods (600 and 230). The 24-hrs average value for the normal periods in this study was 600 or 8.6 cases/year (shrine 1) and 230 or 3.3 cases/year (shrine 2). In general, the values of societal ICR in all cases were found to be higher at shrine 1 than shrine 2. The main reason might be the poor ventilation system in the shrine 1. The societal ICR values of 8-hrs sampling calculated were higher than the 24-hr sampling due to the service period of the shrine. Generally, the shrines are ordinarily opened in the daytime (8 am – 5 pm) except only during Chinese New Year, when the shrines are opened all the time. However, it should be noted that the values are only a rough estimation of cancer risk from the $PM_{2.5}$-bound PAHs inhalation.

4. Conclusions

The highest indoor $PM_{2.5}$ and PAHs concentrations were found during Chinese New Year due to number of visitors and amount of incense being burned. $PM_{2.5}$ concentrations were highly correlated with c-PAHs indicated that carcinogenic compounds were dominant in particulate PAHs and generated from incense burning. It can be concluded that amount of incense being burned plays a significant role in amount of emitted pollutants and degrees of air pollution and human health risk. To improve indoor air quality, where a lot of incenses are used, the ventilation systems should be equipped to improve air circulation. In public places such as temples and shrines, the number of incenses to be burned by individual visitor should be strictly controlled, while the outdoor air circulation and indoor ventilation system should be ensured to be adequate for such practice.

5. References

[1] Manoli E, Kouras A, Karagkiozidou O, Argyropoulos G, Voutsa D and Samara C 2015 *Environ Sci Pollut Res.* Doi: 10.1007/s11356-015-5573-5.
[2] Yang TT, Lin ST, Hung HF, Shie RH and Wu JJ 2013 *AAQR.* **13** 662-671
[3] Hassanvand M S, Naddafi K, Faridi S, Nabizadeh R, Sowlat M H, Momeniha F, Gholampour A, Arhami M, Kashani H, Zare A, Niazi S, Rastkari N, Nazmara S, Ghani M and Yunesian M 2015 *Sci Total Environ* **527-528** 100-110
[4] Wilhelm M, Ghosh J K, Su J, Cockburn M, Jerrett M and Ritz B 2011 *Environ. Health.* Doi: 101186/1476-069X-10-89 1-12
[5] Pan A, Clark M L, Ang L W, Yu M C, Yuan J M, Koh W P 2014 *Enviro Health Perspect* **122** 1279-1284
[6] Wang I J, Tsai C H, Chen C H, Tung K Y and Lee Y L 2011 *Eur Respir J.* **37** 1371-1377
[7] Chiang K C, Chio C P, Chiang Y H and Liao C M 2009 *J. Hazard. Mater.* **166** 676-685
[8] Wiriya W, Prapamontol T and Chantara S, 2013 *Atmos. Res.* **124** 109-122
[9] Bootdee S and Chantara S 2014 *IJESD.* **5** 228-232
[10] U.S.-EPA. (U.S. Environmental protection Agency) 2008 *Polycyclic aromatic hydrocarbons (PAHs) – EPA fact sheet.* Washington, DC: National Center for Environmental Assessment, Office of Research and Development: 2008
[11] Navasumrit P, Arayasiri M, Tin Hiang O M, Leechawengwongs M, Promvijit J, Choonvisase S, Chantchaemsai S, Nakngam N, Mahidol C and Ruchirawat M 2008 *Chem. Biol. Interact.* **173** 19-31
[12] Orecchio S 2011 *Atmos. Environ.* **45** 1888-1895
[13] Nisbet I C T and LaGoy P K 1992 *Regul. Toxicol. Pharmacal.* **16** 290-300
[14] Kuo C Y, Yang Y H, Chao M R, Hu C W 2008 *Sci. Total Environ* **401** 44-50
[15] Jung K H, Yan B, Chillrud S H, Perera F P, Whyatt R, Camann D, Kinney P L and Miller R L 2010 *Int. J. Environ. Res. Public Health* **7** 1889-1900
[16] Krugly E, Martuzevicius D, Sidaraviciute R, Ciuzas D, Prasauskas T, Kauneliene V, Stasiulaitiene I and Kliucininkas L 2014 *Atmos. Environ.* **82** 298-306
[17] Sarkar S and Khillare P S 2012 *Environ Monit Assess* **185** 1199-1213

[18] Jia Y, Stone D, Wang W, Schrlau J, Tao S and Massey Simonich S L 2011 *Environ Health Perspect* **119** 815-820
[19] World Health Organization (WHO) 2000 *Air Quality Guideline for Europe*, second edition. WHO Regional Publications 2000: European Series, No.91. WHO, Copenhagen
[20] Chiang K C and Liao C M 2006 *Sci. Total Environ* 372 64-75

Acknowledgements
Financial support from the National Research University (NRU), Chiang Mai University is gratefully acknowledged.

IOP Conf. Series: Earth and Environmental Science **120** (2018) 012008 doi:10.1088/1755-1315/120/1/012008

Exposure Assessment of Polycyclic Aromatic Hydrocarbon (PAHs) in Childcare Centers of Muang, Nakhon Ratchasima

C Jitlada[1] and P Pentamwa[2]

1 Graduate Program in Environmental Pollution and Safety, Institute of Public Health, Suranaree University of Technology, Nakhon Ratchasima, Thailand.
2 Assistant Professor, School of Environmental Health, Institute of Public Health, Suranaree University of Technology, Nakhon Ratchasima, Thailand.
E-mail: jitlada141233@hotmail.com

Abstract. This study aims to characterize airborne polycyclic aromatic hydrocarbons (PAHs) as of particulate and vapor phases variation. The samples were collected from the childcare centers where divided into urban and rural areas in Nakhon Ratchasima Province of Thailand. The airborne samples were collected from five childcare centers during the dry season in the year 2017. The PAHs species were determined by the gas chromatography and mass spectroscopy (GS/MS) method. Results show that the total concentrations of PAHs were higher than vapor phase that both similar in urban area and rural area. The dominant PAHs compounds of both urban and rural areas were benzo(a)pyrene, benzo(a,h)anthracene and indeno(1,2,3-cd)pyrene, respectively. Furthermore, the concentrations of PAHs in municipality (urban) childcare centers were higher than rural childcare centers area of Nakhon Ratchasima province. The risks associated with exposure to PAHs were evaluated using the TEF approach. The estimated value of lifetime lung cancer risks children in urban were significantly ($p < 0.05$) 2 times of children in rural, thus demonstrating that exposure to PAHs at levels found at urban site may be cause potential health risks.

1. Introduction

The problem of air pollution that affects human health from exposure to various pollutants in the air. Particles of solid particulates are produced by the milling of both organic and inorganic materials which, when released into the atmosphere, can float in the air for a certain period of time and disperse [1]. At the same time, it also acts as a conduit for other substances that adversely affect the health of the particle, such as heavy metals, organic matter volatile Or Polycyclic Aromatic Hydrocarbon (PAHs). The United States Environmental Protection Agency (US EPA) has designated 16 PAHs as important carcinogens are Naphthalene, Acenaphthylene, Acenaphthene, Fluorene, Phenanthrene, Anthracene, Fluoranthene, Pyrene, Benzo (a) anthracene, Chrysene, Benzo (b) fluoranthene, Benzo (k) fluoranthene, Benzo (a) pyrene, Benzo (a,h) anthracene, Benzo (g,h,i) perylene and Indeno (1,2,3-cd) pyrene PAHs are caused by inadequate combustion of vehicle fuels [2]. People living in urban areas are therefore more likely to experience high levels of PAHs. The condition is stable in both gaseous and particulate state, depending on molecular size and atmospheric conditions. Normally, PAHs with low molecular mass are found in the gas phase due to the change in vapor pressure and high molecular weight PAHs are found in the particle phase [3]. PAHs enter the body through eating. Breathing and skin contact, which PAHs in the air. Dust particles are absorbed into the lungs. Tissue, fat, and fat accumulate in the liver, kidneys and fats. So long accumulated in the body., such as Benzo (a) Pyrene,

IOP Conf. Series: Earth and Environmental Science **120** (2018) 012008 doi:10.1088/1755-1315/120/1/012008

Benzo (b) fluoranthene, Benzo (g, h, i) perylene and Indeno (1, 2, 3-cd) Carcinogenicity in Groups 2B refers to a group that is most likely to cause cancer in humans. It can cause lung cancer and respiratory system [4]. When humans are exposed to PAHs. Indoor and outdoor environments can affect both acute and chronic [5]. Based on previous data, the health risk assessment of exposure to PAHs in the Bangkok metropolitan area by Toxic Equivalency Factor (TEF) was 1.35 ng / m^3, which is higher than the risk factor for carcinogenicity defined as 1.0 of Benzo (a) Pyrene. Norramit et al. [6] found that PAHs concentrations were equal to 1.86 and 1.54 ng / m^3 in urban and non-urban areas. 3.46 and 2.70 ng / m^3 in urban and non-urban areas respectively. Slezakova et al. [7] show that urban residents are more likely to experience PAHs than those who live in urban areas. Out of town According to data from the study of the concentration of PAHs inside and outside the school in Rome, Italy, the indoor PAHs amounted to 1.15 ng / m^3 and the outdoor area was worth At 0.74 ng / m^3 [8], it was shown that those living in the building were more likely to experience PAHs than those living outside the building. In Thailand, there are no studies on the health risk assessment of PAHs in young children classified as "Vulnerable groups" [9]. Have high health risks. This study investigated the levels of both PAHs and gaseous carcinogens. Both indoor and outdoor use, as well as the risk of exposure to PAHs in the child development center between the municipality and the area outside Nakhon Ratchasima municipality. This study evaluated the risk of exposure to PAHs. The data will be used as a basis to reduce the risk of exposure to PAHs.

2. Methods

2.1 Sampling location

The sampling was during the dry season of May 2017 on working days. Daily samples was collected for 6 h (from8 a.m. to 2 p.m.) that representing the spending time of children in childcare center. The particulate (PM) and vapor phase samples were both collected. The sampling included locations at ten childcare centers were divided into two groups (5 childcare centers in urban area and 5 childcare centers in rural area) in the Nakhon Ratchasima province of Thailand. The sampling locations were presented in Figure 1. All childcare centers environment had natural ventilation systems, the classrooms were often ventilated by opening windows. The study locations were generally located at a distance of 500 meters away from main roads. The study area of site1- site5 was found with high traffic volume and the ventilation through natural window and doors assumed to ventilation rate of more than 20 percent of the area. The average of ambient temperature was 31 degrees celsius (^oC) and relative humidity was 58%. For the location of site 6 – site 10, the traffic condition showed with low traffic volume. The average of ambient temperature was 32 ^oC and relative humidity was 65%. The predominant wind during the study was presented from south and southeast wind.

2.2. Extraction and analysis of PAHs

The analysis of PAHs, air samples were extracted for 6 h with 50 ml n-hexane in soxhlet by TO13A method [4]. The concentration of PAHs were analyzed by gas chromatograph mass spectrophotography (GC/MS) Agilent 6980. The used GC column was a 30 m DP-5 capillary column (0.25 mm id., 0.25 μm film thickness). The column temperature at 80°C hold for 5 min and increased to 290°C at 10 °C/min hold for 35 min . PAHs compounds were identified based on retention times.

2.3. Quality control

Target PAHs standard stock solution of 16 PAHs was used as a recovery standard for the measurement of recovery coefficients from 80% to 120% that recommend by TO13A method. Known volume of standard solutions was spike into the filters of the analysis procedure. This study spiked recoveries ranged from 85% to 107% that indicated acceptable for all 16 PAHs.

2.4 Statistical analysis

For all data sample t-test was applied to determine the statistical significance ($p<0.05$) of the difference between the mean in both group (childcare center in urban and rural in Nakhon Ratchasima Province).

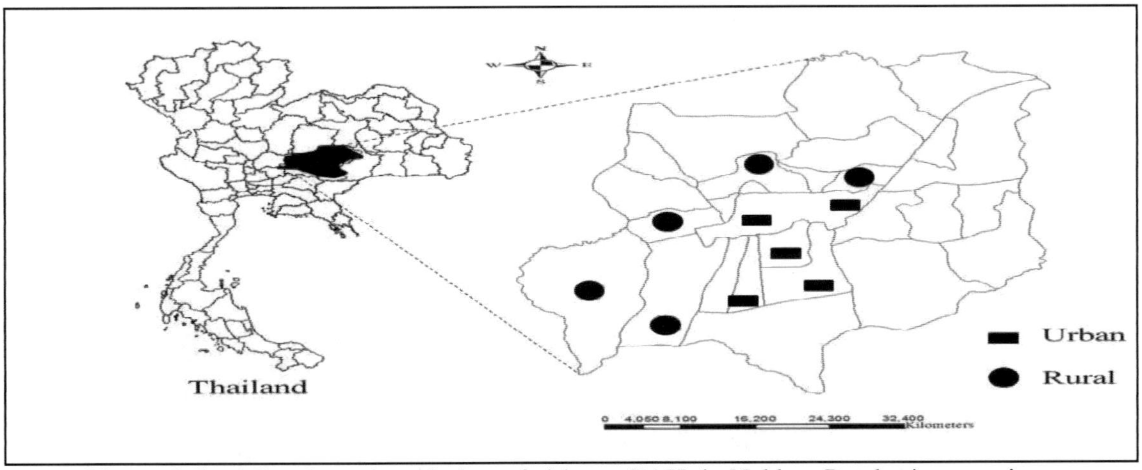

Figure 1. Sampling Location 10 sites of airborne PAHs in Nakhon Ratchasima province.

3. Result and discussion

3.1. Gas/particle distribution of PAHs

The gas-particle phase of 16 PAHs in this study is presented in Figure 2. The phase distribution show in the percentage of the concentration in both phases were naphthalene, acenaphthylene, acenaphthene, fluorene, phenanthrene, anthracene, fluoranthene, pyrene, benzo (a) anthracene, chrysene, benzo (b) fluoranthene, benzo (k) fluoranthene, benzo (a) pyrene, benzo (a, h) anthracene, benzo (g, h, i) perylene and indeno (1, 2, 3-cd) pyrene. PAHs found in particles phases larger than vapor phase. Amount of benzo (a) pyrene in particulate phase are 57% of total mass. For the Naphthalene, acenaphthylene, acenaphthene, benzo (a) anthracene and chrysene compound have distribution at all sites are same between vapor and particle phases. High molecule mass of PAHs were found in particle phase than vapor phase. The PAHs phase distributions in both sampling sites of urban and rural areas were similar [3]. This may be explained by the evaporation of PAHs in the room temperature distribution from the outdoor environment to the indoor air. The study of Wang et al. [5] the PAHs in air are divided between vapor and particulate phases depend on the ambient air temperature and humidity.

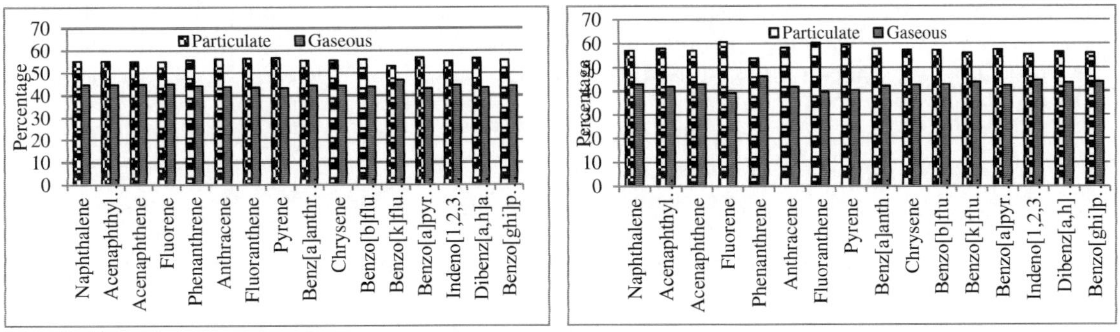

Figure 2. PAHs distribution in Particulate and vapor phase; concentrations of PAHs in urban area (Left), concentrations of PAHs in rural area (Right).

3.2. Concentration of PAHs in municipality (urban) and outside the municipality (rural) childcare centers of Nakhon Ratchasima province.

The concentration of 16 PAHs are showed in Figure 3. PAHs levels of urban childcare center ranged from 1.00 (minimum) to 2.02 ngm^{-3} (maximum), while rural areas ranged from 0.71 to 1.55 ngm^{3}. The results demonstrated that levels of 16 PAHs were significantly high in urban childcare centers area.

The total concentration of 16 PAHs for urban sites higher 1.3 times than rural sites. These findings concerned to children health impact from PAHs indicated to carcinogenic effects to body in benzo [a] pyrene, benzo [b] fluoranthene, benzo [k] fluoranthene and dibenz [a, h] anthracene compounds [10]. The highest concentration of PAHs is Indeno (1, 2, 3-cd) pyrene at mean of 2.02 ngm^{-3} for 8.6 % of ΣPAHs. The dibenz [a, h] anthracene were second most abundant group of urban childcare centers (8.5 % of ΣPAHs). 16 PAHs such as Benzo (a) pylene, benzo [ghi] perylene are possible human carcinogens. The least compounds of PAHs such as (naphthalene), at the urban childcare centers site less than 4.2 % of ΣPAHs.

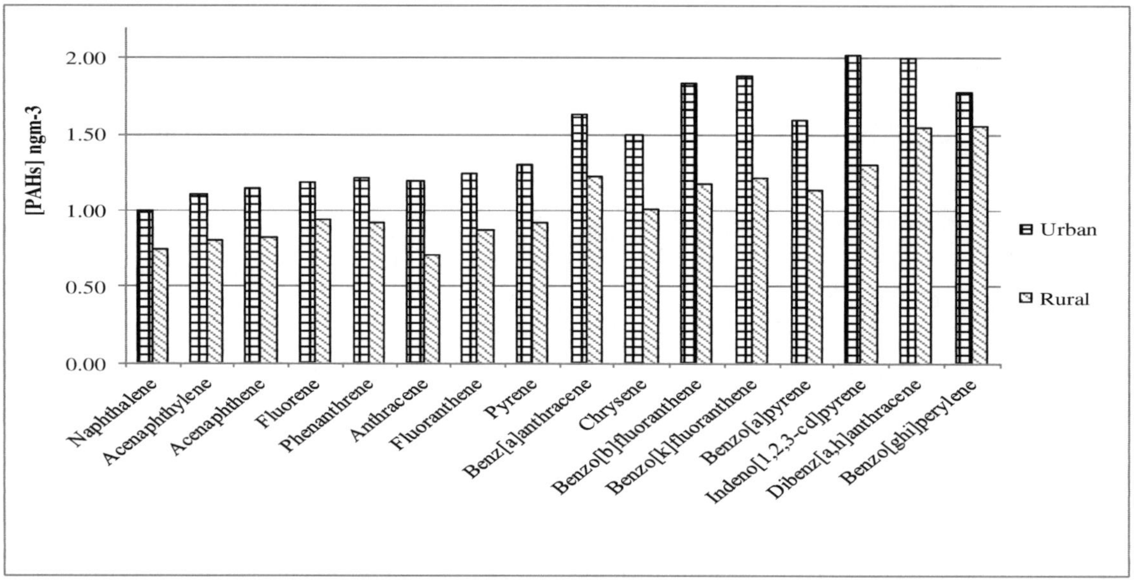

Figure 3. Concentration of PAHs in childcare centers municipality (urban) and outer municipality (rural) of Nakhon Ratchasima province.

3.3. Health risk assessment of PAHs in childcare center of municipality (urban) and outside the municipality (rural) of Nakhon Ratchasima province.

The 16 PAHs are carcinogenic, and their health risks in human can be calculate by BaP-equivalent ([BaP] eq) were calculate from PAHs concentration with toxicity equivalency factors (TEFs) of concentrations shown in equation (1) were multiply by UR [BaP] (unit risk) there are cancer risk shown in equation (2), while unit risk of life time cancer is 1.1×10^{-6} of 70 years [4].

$$\Sigma[BaP]eq = \Sigma(C_i \times TEF_i) \quad (1)$$

$$\text{Inhalation cancer risk} = \Sigma[BaP]eq \times UR[BaP] \quad (2)$$

The results of inhalation cancer risks for 16 PAHs, this study are urban and rural site and health risks of children in childcare center were found PAHs are presented table 1. The estimated of lifetime lung cancer risks based on Benzo[a]pyrene at the urban area were 9.5×10^{-6} for PAHs in urban area, and 1.8 times higher rural area is 5.2×10^{-6}. At the urban area, the values of lifetime lung inhalation cancer risks for childcare center in urban area were higher. The result demonstrates that urban area, higher expose PAHs, and traffic emissions might be the potentially major pollution source of PAHs to childcare center. The childcare center of municipality (urban) of Nakhon Ratchasima province close to major roads this study 500 meter from major road and where emissions from vehicular traffic might be significant effect to human [11]. This study indicated the influence of low temperature during sampling effect to high concentration of PAHs the risks associated with the urban area have significantly higher than rural area. Therefore a long distance from major road and traffic emissions to protect human health to reduce exposure PAHs in childcare center.

Table 1. Estimated inhalation cancer risk to measured concentrations of PAHs in childcare center.

PAHs	TEF	Cancer Risk	
		Urban	Rural
Naphthalene	0.001	6.38E-10 ±0.03	2.64E-10 ±0.10
Acenaphthylene	0.001	6.49E-10 ±0.05	3.08E-10 ±0.04
Acenaphthene	0.001	6.71E-10 ±0.04	3.30E-10 ±0.08
Fluorene	0.001	6.93E-10 ±0.02	2.42E-10 ±0.05
Phenanthrene	0.001	7.15E-10 ±0.03	2.53E-10 ±0.03
Anthracene	0.010	7.04E-09 ±0.01	3.41E-09 ±0.07
Fluoranthene	0.001	7.37E-10 ±0.03	3.19E-10 ±0.01
Pyrene	0.001	7.70E-10 ±0.07	3.30E-10 ±0.03
Benz[a]anthracene	0.100	9.57E-08 ±0.01	4.40E-08 ±0.12
Chrysene	0.010	8.80E-09 ±0.06	4.73E-09 ±0.06
Benzo[b]fluoranthene	0.050	5.39E-08 ±0.03	2.81E-08 ±0.02
Benzo[k]fluoranthene	0.050	5.50E-08 ±0.02	3.08E-08 ±0.04
Benzo[a]pyrene	1.000	9.50E-06 ±0.06	5.20E-06 ±0.01
Indeno[1,2,3-cd]pyrene	0.100	1.27E-07 ±0.05	7.48E-08 ±0.02
Dibenz[a,h]anthracene	0.100	1.18E-07 ±0.02	0.66E-07 ±0.01
Benzo[ghi]perylene	0.010	1.06E-08 ±0.01	7.04E-09 ±0.05

4. Conclusions

The concentrations of polycyclic aromatic hydrocarbons (PAHs) in childcare centers during dry seasons in Nakhon Ratchasima province of Thailand; the concentration of 16 PAHs in all sampling sites were followed from Benzo (a, h) anthracene, Benzo (g, h, i) perylene, Indeno (1, 2, 3-cd) pyrene, benzo [a] pyrene, Benzo (k) fluoranthene, Benzo (b) fluoranthene, Benzo (a) anthracene, Chrysene, Pyrene, Naphthalene, Acenaphthylene, respectively. The distribution of PAHs between particulate and vapor phases was similar to other studies, with particle phase higher than vapor phase. The major sources of PAHs are the traffic emissions and distance of childcare center from major road. Health risk assessment based on the carcinogenic that the concentrated in urban area were several times higher than rural area indicating a causing PAHs is adverse health effects for children live in childcare center in municipality (urban) Nakhon Ratchasima province.

5. References

[1] Mostert et al. (2011). Nitro-PAH in ambient particulate matter in the atmosphere of Athens. *Chemosphere. 40 (1)*: 533–537.

[2] Ruiz et al et al. (2009). Seasonally dependent size distributions of aliphatic and polycyclic aromatic hydrocarbons in urban aerosolsfrom densely populated areas. *Environ Science Technology 22 (2)*:896–908.

[3] Zheng et al. (2009). Research on distribution characteristics of polycyclic aromatic hydrocarbons in main traffic line crosses of Beijing urban district. *Atmospheric environment. 23 (1)*: 30–40.

[4] US.EPA. (1998). *Toxicological Review of PAHs. (CAS No. 91-20-3)***.** http://www.epa.gov./iris, March 1, 2015.

[5] Wang et al. (2013). Aggregate exposures of nine preschool children to persistent organic pollutants at day care and at home. *Exposure Analysis and Environmental Epidemiology. 13 (1)*: 187-202.

[6] Norramit et al. (2005). Characterization and carcinogenic risk assessment of polycyclic aromatic hydrocarbons in the respirable fraction of airborne particles in the Bangkok metropolitan area. *Journal of health science. 51(1)*: 437-446

[7] Slezakova et al. (2013). Impact of vehicular traffic emissions on particulate-bound PAHs: levels and associated health risks. *Atmospheric research. 127 (1)*: 141-147.
[8] Krugly et al. (2014). Characterization of particulate and vapor polycyclic aromatic hydrocarbons in indoor and outdoor air of primary schools. *Atmospheric environment 82 (1)*: 298-306.
[9] World Health Organization (WHO).(2002). *Environmental Health Criteria*: Selected Non–Heterocyclic Polycyclic Aromatic Hydrocarbons.World Health Organization Publication, Geneva.
[10] International Agency for Research on Cancer (IARC). (2010). Polynuclear aromatic compounds, Part 1. Chemical Environment and Experimental Data, 32.
[11] Naumova et al. (2013). Gas/particle distribution of polycyclic aromatic hydrocarbons in coupled outdoor/indoor atmospheres. *Atmospheric environment 37 (1)*: 703-719.

Acknowledgments
Authors are grateful to express my sincere thanks to my thesis advisor, Asst. Prof. Dr. Prapat Pentamwa for invaluable help and constant encouragement throughout the course of this research and Suranaree University of Technology for providing the financial support of this work.

IOP Conf. Series: Earth and Environmental Science **120** (2018) 012009 doi:10.1088/1755-1315/120/1/012009

Short-Term Introduction of Air Pollutants from Fireworks During Diwali in Rural Palwal, Haryana, India: A Case Study

S Gautam[1, 2], A Yadav[2], A Pillarisetti[3], K Smith[3] and N Arora[2]

1 Marwadi University, Rajkot, India
2 The INCLEN Trust International, Okhla Industrial Area, Phase-I, New Delhi, India
3 School of Public Health, University of California, Berkeley, Berkeley, United States
E-mail: sneha.gautam@marwadieducation.edu.in

Abstract. The contribution of firework-related air pollutants into the rural atmosphere was monitored by measuring ambient air concentrations of $PM_{2.5}$, CO, and metals over Mitrol–Aurangabad, Haryana, India, before, during, and after the 2015 Diwali celebration. $PM_{2.5}$ concentrations were observed to be approximately 5 times and 12 times higher than Indian and WHO 24-h standards, respectively. CO concentrations on the day of Diwali were found to be nearly 7.5 times and nearly 1.5 times higher than Indian standards and WHO 8-h standards, respectively. Increased concentrations of SO4, K, N3, Al, and Na were observed. SO4, K, N3, Al, and Na were found between approximately 2 and 5 times higher on festival days than on a normal, non-festival day in November. Use of firecrackers during Diwali and surrounding celebrations thus contribute to decreased air quality and elevated levels of air pollutants associated with adverse health impacts. Optimization or controlled use of firecrackers during Diwali is suggested in rural areas.

1. Introduction

In recent years, there has been concern that degradation of air quality and subsequent effects on human health has increased due to use of fireworks during festivals and celebrations [1]. Fireworks are closely linked to elevated levels of pollutants, including particulate matter, SO2, NOx, K, and ozone, amongst others [2]. In India, a number of festivals are celebrated throughout the year; Diwali, the Festival of Light, is one of the most important. Diwali is celebrated between the pre-winter season (October) and the post-monsoon season (November). As the festival of light, firecrackers are often used on the days before and during the festival, resulting in deteriorated air quality. Pathak et al. [3] and Wang et al. [4] reported that fireworks used during Diwali create high mass concentrations of particulate matter (PM) with a chemical composition that varies from normal atmospheric conditions and can be harmful to human health.

Pathak et al. [3] quantified the amount of PM exposure by air mass back trajectory analysis using the NOAA–HYSPLIT model in order to examine the existence of the transported aerosols which people in urban areas are exposed to over the short-term. PM emitted from firecrackers and its contribution to pollution in the urban atmosphere has also been measured [3], [5]. Further recent studies [3], [6] measured elevated local PM concentrations due to firecrackers during Diwali. Moreover, Hindustan Times [7] reported that the Delhi pollution Control Committee (DPCC)'s reported Air Quality Index (AQI), a combined measure of air pollution, peaks between 6 pm – 1am on Diwali. The AQI starts a downward trend after 2 am. Data from real-time monitoring stations operated by DPCC, Indian Institute of Tropical Meteorology (IITM), and Indian Meteorological Department (IMD) indicate that concentrations of PM_{10} and $PM_{2.5}$ exceed the national standards during festival periods, sometimes by greater than 10 times.

While these early studies evaluate concentrations in urban area of India, there is no information about the air quality of rural areas during Diwali. According to the recent report of The World Bank [8], 68% of the Indian population is rural. Despite growing evidence of adverse air quality and health effects attributable to the use of fireworks, no literature could be found measuring PM, its constituents, or CO in rural areas of India during Diwali. In the present study, the particle concentrations, including metals and ion analysis, and CO levels were recorded to describe the short-term impact of firecrackers on concentration due to Diwali in the rural area of Palwal District, Haryana state, India.

2. Material and methods

2.1. Study area

INCLEN has established SOMAARTH: a Demographic Development and Environment Surveillance Site (DDESS) in Palwal district, Hayana covering 51 villages (Figure 1). The demographic surveillance site includes 51 villages (approximately 200,000 population) from three Blocks of Palwal (Hathin, Hodal and Palwal) and is circumscribed by three major roads – Mathura National Highway (NH2) that forms eastern boundary of surveillance site; Palwal–Hathin State Highway, forming the western boundary; and intersecting roads that form the southern boundary (Figure 1).

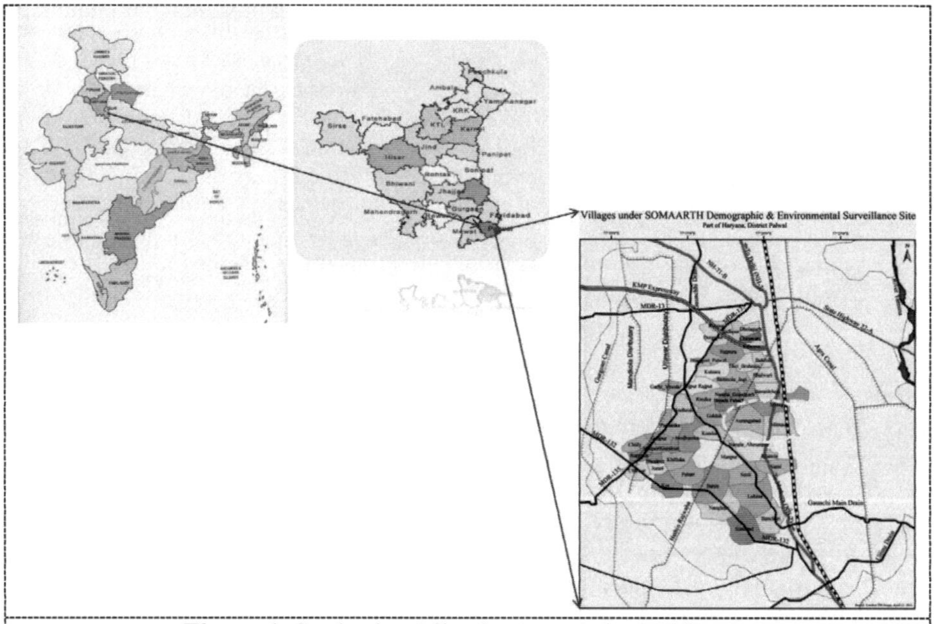

Figure 1. Study site: Mitrol– Aurangabad Villages.

2.2. Sampling protocol

In the year 2015, Diwali was celebrated on 11th November. Data were collected for 8 days, beginning 9th November 2015 and ending 16th November 2015. Firecracker activities were observed to begin in the evening (around 18:00) and continued until approximately 1:00. Sampling began nightly at 17:00. CO was measured for 24 hr, while PM was monitored in three 8 hr periods.

To measure $PM_{2.5}$, we utilized an URG-2000-30ED cyclone (URG Corporation, Chapel Hill, NC, USA) connected to an SKC PCXR8 pump (SKC Inc, Eighty-Four, PA, USA) running at a flow rate of 3 litters per minute. To ensure a stable power supply to the PCXR8, batteries were changed every 8 hours (daily at 6:00, 14:00, and 22:00). Flows were checked pre- and post- battery exchange using a Mesa DryCal Defender 530 (Mesa Labs, Butler, NJ, USA). $PM_{2.5}$ was collected on pre-weighed 47mm

PTFE filters. Filters were pre- and post-weighed at Kirk R. Smith's laboratory at University of California, Berkeley after being conditioned for 48 hours at 23-25ºC and a relative humidity of between 35 and 40%. Filters were weighed in triplicate on a Mettler-Toledo XP2U microbalance with a readability of 1 μgm^{-3}. To measure CO, we used a Lascar EL-CO-USB electrochemical monitor with a detectable range of 0 – 500 ppm. Batteries in the CO monitor were replaced every 48 hours.

2.3. Elemental, cation, and anion analysis

The sampled PTFE filters were analyzed for elements, anions, and cations by Desert Research Institute (Reno, NV, USA). Elemental analysis was performed by X-ray Fluorescence (XRF) using the PAN analytical Epsilon 5 XRF analyzer, which uses a side-window, dual-anode x-ray tube with Scandium and Tungsten anodes. Seven different condition sets are used during a single analysis run to enable detection of specific groups of elements. Anions and cations analysis was performed using the Dionex ICS 5000, a liquid chromatographic technique based on an ion exchange mechanism and suppressed conductivity detection for the separation and determination of cations and anions.

3. Results

3.1. CO concentration

Figure 2 depicts the CO concentration profile during the experiment. Mean CO concentrations (averaged over 8 hour periods) were higher during Diwali as compared to non-Diwali. The average concentration of CO varied from 0.62 – 15.4mg m^{-3} (Figure 2). The 8-hour CO standard for residential areas in India is 2 mg m^{-3}. In the present study, CO concentration varied between 0.62 – 15.1 mg m^{-3}.

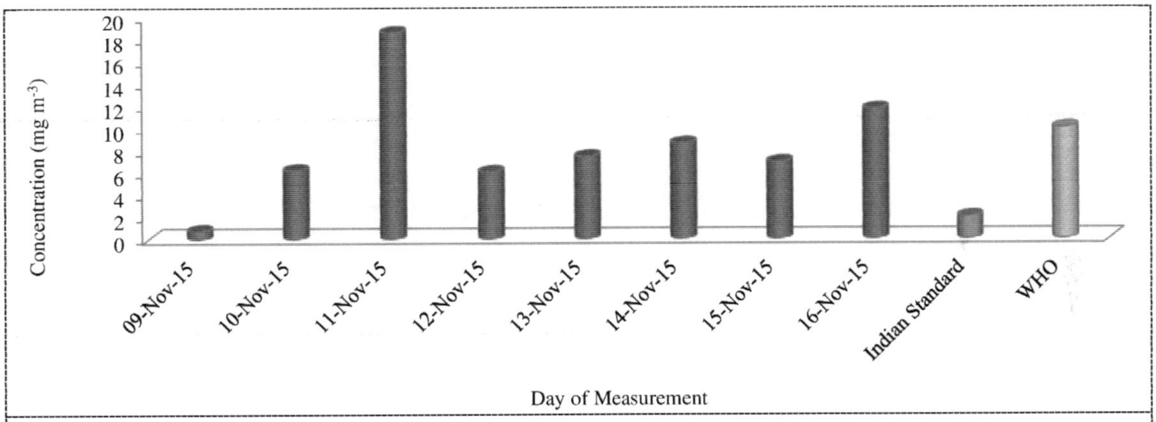

Figure 2. Average CO concentration profile with Indian standards and 8 hour WHO guideline at Mitrol – Aurangabad Village [9, 10].

3.2. PM concentration ($PM_{2.5}$)

The 24-hour average concentration of $PM_{2.5}$ during Diwali is depicted in Figure 3. The concentrations vary between 141 to 345 $\mu g\ m^{-3}$. The results indicate variability of fine particle concentration between the pre-Diwali (9th November) period, Diwali (11th November) itself, and the post-Diwali (12th November) period. Higher concentrations of $PM_{2.5}$ were recorded on 10th and 12th November, 2015. The 24-hour average concentration of $PM_{2.5}$ during pre-Diwali, Diwali, post-Diwali periods and a normal day are depicted in Figure 4.

The concentration of fine particle decrease as one moves further forward in time from Diwali. The festival day concentrations were 2.10 times, 1.31 times, and 1.34 times higher than a normal day, and the pre-Diwali and post-Diwali concentrations, respectively. The incremental difference in concentration from a normal day was 155 $\mu g\ m^{-3}$.

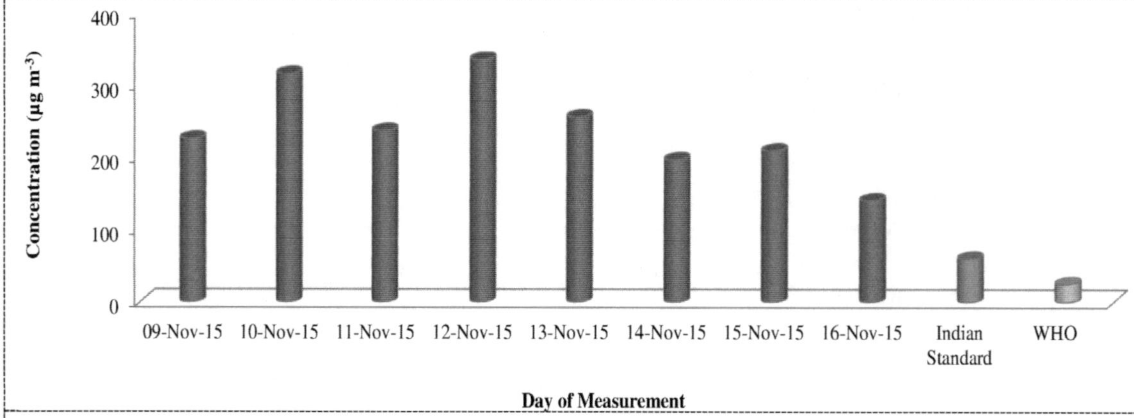

Figure 3. Daily average $PM_{2.5}$ concentration profile with Indian standards and WHO guidelines at Mitrol – Aurangabad Village [9], [11].

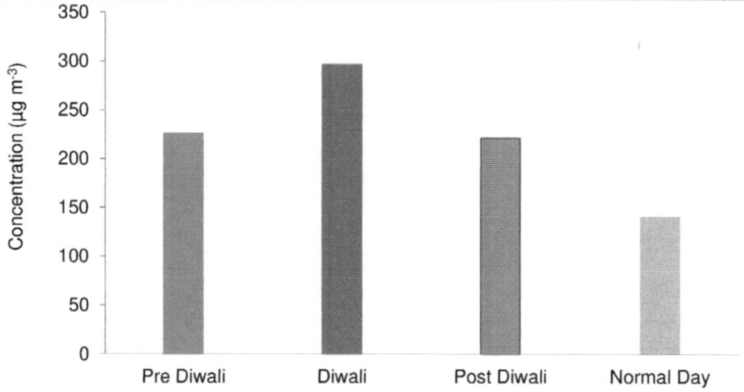

Figure 4. Variation of $PM_{2.5}$ concentrations on Pre- Diwali, Diwali, post-Diwali and Normal day at Mitrol – Aurangabad Village.

3.3. Variation of $PM_{2.5}$ and CO concentration at Mitrol – Aurangabad Village

The 24 hr and 8 hr average concentration of $PM_{2.5}$ and CO respectively in the study village during Diwali festival are depicted in Figure 5. For both $PM_{2.5}$ and CO, the average concentrations were observed higher during Diwali as compare to normal day of measurement. Figure 5 shows the higher concentration of CO during the night of Diwali, which could be due to the increased use of solid fuel to make sweets for the festival. On a normal day, the higher concentration of CO could be attributed to the maximum open pit fire observed near the sampling location. The higher concentration of $PM_{2.5}$ on the day following Diwali could be due to effect of "Goverdhan Puja," a separate celebration. The average concentration of $PM_{2.5}$ and CO during pre and post Diwali were observed to be almost same as on a normal day (Figure 5). Mg, Zn, Fe, Mn, Cu, Sr, and Cd have low concentrations pre-Diwali, post-Diwali, and on a normal day, but were elevated during Diwali. Similarly, SO4, K, N3, Al, Na, NH4, and Ba have been shown to have higher concentrations during Diwali than on a normal day. These increases potentially indicate the impact of firecrackers on ambient air quality on the night of Diwali, as evidenced by increased concentrations of $PM_{2.5}$ and CO.

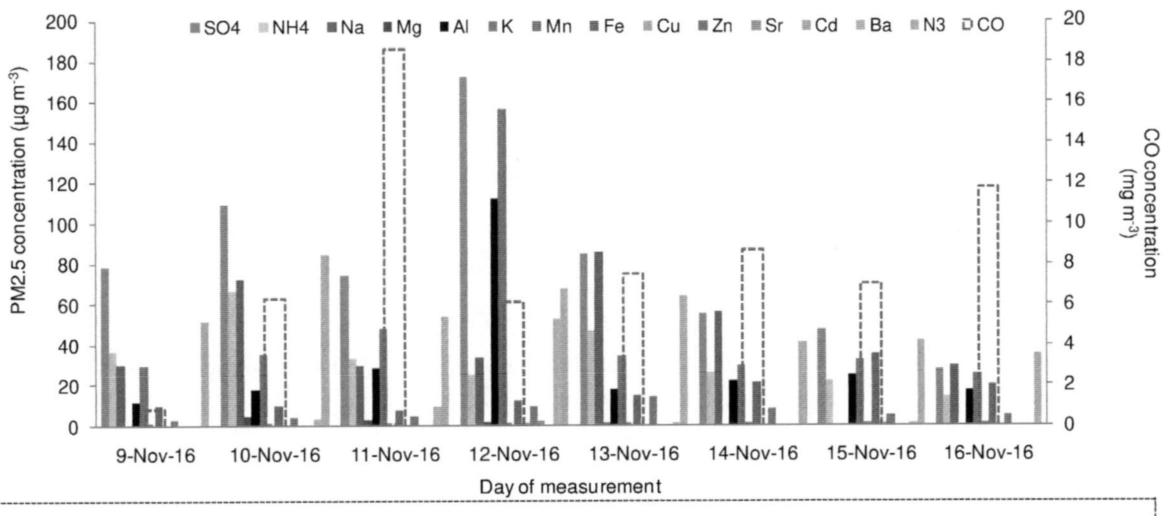

Figure 5. Variation of $PM_{2.5}$ and CO concentrations during Diwali festival at Mitrol – Aurangabad Village.

4. Discussion

The daily mean CO concentrations measured were as high as 7.5 times and 1.5 times the Indian National Ambient Air Quality Standards (NAQQS, 2 mg m^{-3} for 8 hr, [9]) and WHO standards (7 mg m-3 for 8 hr, [11]), respectively. The observed results are, however, comparable with CO concentrations of 0.21 – 26.1 mg m^{-3} [12] reported previously for rural area of developing country in a non-festival period.

The maximum average $PM_{2.5}$ concentration measured was 296 µg m^{-3} on Diwali, which were ~5 times higher than the Indian standard and ~12 times higher than the WHO 24-h standards [11]. The recorded concentrations during the festival were 1.32 times, 1.34 times and 2.1 times higher than the concentration of pre-Diwali, post-Diwali and normal day respectively. SO4, K, N3, Al, Na, NH4, and Ba were 1.52 times, 1.13 times, 1.51 times, 4.59 times, 2.74 times, 52.4 times, and 1.34 times higher than pre-Diwali concentration, 1.91 times, 1.33 times, 0.95 times, 2.47 times, 2.51 times, 35.9 times, 1.41 times higher than post-Diwali concentration, and 4.30 times, 3.03 times, 1.53 times, 3.11 times, 3.19 times, 86.6 times, and 1.93 times higher than normal day concentration. In our study setting, people generally celebrated "Choti Diwali" ("Small Diwali") on the 10th November and "Goverdhan Puja" on the 12th November with firecrackers, potentially explaining elevated PM concentrations on those days relative to non-festival days and Diwali itself.

Our findings are comparable with $PM_{2.5}$ concentrations 2-3 times higher than normal concentrations [13], 5.74 times higher than normal concentrations [3], and 34.3 µg m^{-3} greater than normal concentrations [14] reported across different urban areas of India during the Diwali. Beyond Diwali, our study indicates elevated levels of $PM_{2.5}$ in rural areas even during non-festival days, warranting further investigation.

SO4 and K show substantial elevations, presumably due to firework use on Diwali. $PM_{2.5}$ metals species concentration were recorded to follow a same trend at Mitrol – Aurangabad village during study period and which was, Diwali > post-Diwali > pre-Diwali > normal day. Our research indicates that an impact of Diwali firework uses on the next day (post-Diwali) concentrations.

5. Conclusion

We report on the first measurements of the impact of Diwali on air quality in rural India. 68% of the Indian population lives in rural areas; this percentage has been consistent for the past 4 years [8]. In India, all of the air quality monitoring stations is operated in urban areas; no such stations are being operated in rural areas. Our findings suggest non-trivial increases in $PM_{2.5}$ and CO concentrations, as well as changes in the elemental composition of $PM_{2.5}$, during this festival. The potential for exposure to the reported pollutants at elevated concentrations during Diwali can increase the potential for acute health effects, suggesting need for guidance in rural areas on ways to ensure that the most vulnerable populations are able to avoid additional exposure or rapidly seek treatment if needed. Stagnant air typical of winter months in North India prevents dispersion of pollutants. We suggest new strategies to reduce the emission from firework-related activities. Possible measures to reduce the emissions of or increase the dispersion of pollutants from fireworks include the following:

- Use of some plastic cracker (without chemical material and fire) with sound.
- Try to use firecrackers on flat open ground away from house or society to enhance proper pollutant dispersion.
- Ensuring that emergency medical treatment is available for vulnerable populations on festival and pre-festival days, which also see firecracker use.

6. References

[1] Vecchi R, Bernardoni V, Cricchio D, D'Alessandro A, Fermo P, Lucarelli F, Nava S, Piazzalunga A, Valli G 2008 *Atmos. Environ.* **42** 1121
[1] Moreno T, Querol X, Alastuey A, Minguillo´ n MC, Pey J, Rodriguez S, Miro,´JV, Felis C, Gibbons W 2007 *Atmos. Environ.* **41** 913.
[1] Pathak B, Biswas J, Bharali C, Bhuyan PK 2015 *Atmos. Pollut. Res.* **6** 220
[1] Wang Y, Zhuang GS, Xu C, An ZS 2007 *Atmos. Environ.* **41** 417
[5] Rao PS, Gajghate DG, Gavane AG, Suryawanshi A, Chauhan C, Mishra S, Gupta N, Rao CVC, Wate SR 2012 *Bull. Environ. Contam. Toxicol.* **89** 376
[6] Ambade B, Ghosh S 2013 *Nat. Hazards* **69** 589
[7] Hindustan Times; update on November 12 2015 22:00 IST. (http://www.hindustantimes.com/delhi/air-quality-deteriorates-on-diwali-night-pollution-levels-spiked/story-twUn7KQzhfDuzAcSKcVZGK.html)(Last accessed on 29January 2016).
[8] The World Bank 2014 (http://data.worldbank.org/indicator/SP.RUR.TOTL.ZS) (Last accessed on 29 January 2016).
[9] CPCB. 2009. National Ambient Air Quality Standards (NAAQS), Central Pollution Control Board, New Delhi. (http://cpcb.nic.in/National_Ambient_Air_Quality_Standards.php) (Last accessed on 8 November 2015).
[10] The World Bank 2014 (http://data.worldbank.org/indicator/SP.RUR.TOTL.ZS) (Last accessed on 29 January 2016).
[11] WHO, Ambient (outdoor) air quality and health. Fact sheet N°313, (2014), World Health Organization (http://www.who.int/mediacentre/factsheets/fs313/en/) (Last accessed on 8 November 2015).
[12] Deka D, Hoque RR 2014 *Aerosol Air Qual. Res.* **14** 1752
[13] Ravindra K, Mor S, Kaushik CP 2003 *J. Environ. Monit.* **5** 260
[14] Thakur B, Chakraborty S, Debsarkar A, Chakraborty S, Srivastava RC 2010 *Atmosfera* **23** 347

Acknowledgments
This work was carried out with support from the EPA-G2012-STAR-E1 (Measurements and Modeling for Quantifying Air Quality) project. We are thankful to INCLEN, New Delhi, India, University of California, Berkeley, USA and Marwadi University, Rajkot, India for providing us the required funding to complete this study.

Nutrient Supplying Potential of Different Spent Mushroom Substrate Preparations as Soil Amendment in a Potting Media

VU Ultra, Jr. [1], JME Ong Sotto[2] and MR Punzalan.[2]

1 Department of Earth and Environmental Sciences, Faculty of Science, Botswana International University of Science and Technology, Palapye, Botswana
2 College of Agriculture, Fisheries and Natural Resources, University of Eastern Philippines, Catarman, Northern Samar, Philippines
E-mail: mropunsalan_09@yahoo.com.ph

Abstract. A three consecutive cropping experiment was conducted to evaluate the nutrient supplying potential of different preparations of the spent mushroom substrate as an amendment of growing media for potted plants using pechay as test plant. There are 12 treatment combinations consisted 4 types of growing media containing soil alone and mixtures of soil with fresh SMS (FSMS), weathered SMS (WSMS) and carbonized SMS (CSMS) in combination with 0%, 50% or 100% recommended rate (RR) of nitrogen fertilizer. Succeeding two trials were conducted on the same pots and treatment assignments. The high yield of pechay during the first and second crop was observed on WSMS and CSMS treatments FSMS media produced high yields only during the 3rd crop. Yield was increased by N fertilizer in WSMS and CSMS treatments but not in FSMS. The growth differences is attributed to differences in available nutrients and C/N ratio between treatments. WSMS and CSMS increased the available N while FSMS immobilized N and other nutrients indicting that weathered SMS and carbonized SMS are more suitable as a component of potting media or as soil amendments without detrimental effect on immobilization and availability of nutrients.

1. Introduction

Nurseries and the emerging popularity of urban agriculture/containerized crop production resulted in high demands of growing medium for potted plants. Currently, peat is utilized as the main component of the growing media [1] and depleted the peatlands [2]. In agriculture, alternative sources of nutrients and soil amendments is needed to combat land degradation due to excessive chemical fertilizer application that utilization organic wastes into soil amendments or as a substrate component for nursery potting media instead of peat will provide solutions for waste disposal and conserve peat areas [3].

The boom of the mushroom industry is a global phenomenon with the market value of $35 billion in 2015 [4]. The production is 6,535,542 tons in 2009 and still increasing annually at 9.2% between 2002 to 2016. However, mushroom production system is accompanied with the generation of millions of tons of residue referred to as spent mushroom substrate (SMS) [5], [6]. The SMS by-product in mushroom cultivation amounts to 5 kg of SMS generated from the production of 1 kg of mushrooms [7]. It has high levels of residual nutrients which may result in environmental pollution if it is dumped as waste [6], [8]. Thus, if properly utilized, SMS could serve as an important agricultural resources that would increase farmer's productivity and at the same time protect the environment.

Current practices in the different production area are carried out on substrates formulated with lignocelluloses materials from different sources (such as wheat, rye or rice straw, sawdust and corncobs), either alone or mixed with supplements to overcome nutritional limitations and to provide suitable substrate structure and pH [7]. The SMS by-product from these process have a stabilized organic matter that make them a potential candidate as soil amendments or as a substitute for peat for horticultural growing media. Previous studies have shown some benefits of using SMS in several agricultural application such as soil amendment which increased the organic matter, improved soil structure and provided plant-available nutrients [9]; while as surface mulch controlled the weeds [10]. Therefore, this study evaluated the potential of different preparations of SMS as soil amendments or component of a growing medium in horticulture by evaluating the growth of pechay grown in the different combination of SMS preparation, nutrient availability and nutrient supplying/retention capacity in the presence of different N levels of fertilization.

2. Materials and methods

2.1. Treatment description and experimental design

Using pechay as test plant, a two-factorial experiment consisted of four kinds of growing media and three levels of N fertilization was laid out following complete randomized design with three replications. The 12 treatment combinations include: T1 - Garden soil (GS) only, T2 - GS + FSMS (1:2), T3 - GS + WSMS(1:2), T4 - GS + CSMS (1:2), T5 - GS only + 50% RR of N, T6 - GS + FSMS(1:2) + 50% RR of N, T7 - GS + WSMS(1:2) + 50% RR of N, T8 - GS + CSMS(1:2)+ 50% RR of N, T9 - GS only + 100% RR of N, T10 - GS + FSMS (1:2) + 100% RR of N, T11 - GS + WSMS(1:2) + 100%RR of N and T12 - GS + CSMS (1:2) + 100%RR of N. Pots designated 100% RR of N (160 - 0- 0) were fertilized with 7g of urea (46-0-0) while and 3.5 g for 50% RR of N.

2.2. SMS and growing media preparation

The garden soil in the experiment was collected from a surface soil of the UEP experimental station. The soil samples were sieved and heat sterilized prior to mixing. One (1) part of the soil was amended with two (2) parts of SMS (fresh, weathered or carbonized) by weight basis. The SMS was obtained from *Volvariella volvacea* grown in the Mushroom Production Project of UEP, Catarman, Northern Samar, Philippines. One hundred twenty (120) kilos each of weathered and FSMS was mixed separately with sixty (60) kilos of sterile garden soil. Another 120 kg of fresh SMS was subjected to carbonization based on the method of Orge & McHenry [11] and after cooling was mixed with sixty kilos sterile garden soil. The different growing medium preparations were potted at 4kg per pot using a 6 x 11 inches polyethylene bag. The CSMS amended medium which was potted at 3 kg per pot.

2.3. Seedling Preparation and plant establishment

Seedlings were germinated on a sterile mixture of garden soil, compost and carbonized rice hull (1:1:2) for 14 days before transplanting at the rate of one (1) seedling per pot, The plants were irrigated twice each day and maintained under screen house condition.

2.4. Plant yield measurements

Twenty-three days after transplanting, plants were harvested and the fresh weight (shoots and roots) was recorded after washing and subsequent blot drying.

2.5. Chemical analysis

Before and after the experiment during each cropping, soil and growing media samples were collected for chemical analysis. Samples were immediately frozen after sampling and maintained frozen while on transient for laboratory analysis. The samples were freeze-dried prior to analysis at the Laboratory of Environmental Science, Catholic University of Daegu, South Korea. The pH and electrical conductivity (EC) of the soil/growing media samples were measured on a 1:2.5 and 1:5 soil: water (w/v) ratio, respectively [12] while the oxidizable organic C (oxidizable OC) by the Walkley and Black modified method [13]. The total nitrogen by the Kjeldahl method while the available N (NH4-N and NO3-N) was determined by 1N KCl extraction (1:5 w/v, 2 hours shaking) followed by close-system steam distillation. Available P was determined colorimetrically [14]. Exchangeable Na, K, Ca and Mg in soil were determined in the ammonium acetate extract [15] by flame photometry (Na, K) and atomic absorption spectrometry (Ca, Mg). Available Fe, Cu, Mn and Zn in soil were measured in the DTPA extract [16] by atomic absorption spectrometry.

2.6. Statistical analysis

Fresh yield and chemical properties data were subjected to 2-way analysis of variance followed by treatment mean comparison using Tukey's Honesty Significant Difference (HSD) Test at 5% level of significance.

3. Results

3.1. Growth performance of pechay on different growing media preparations as affected by N fertilization

Table 1. Fresh weight, number of leaves and root length of pechay grown for 27 days after transplanting as affected by different growing media and N fertilization at different cropping period.[1]

Treatments	Fresh Weight (g/plant)		
	1st Crop	2nd Crop	3rd Crop
T1 –soil	14.96 d	20.12 e	47.83 c
T2 –soil + F SMS	2.00 d	2.84 f	110.00 b
T3 –soil + WSMS	83.80 c	63.84 c	93.6 0 b
T4 –soil + CSMS	20.20 d	35.27 e	91.63 b
T5 –soil + ½ N	26.70 d	25.26 e	81.19 b
T6 –soil +FSMS + ½N	15.40 d	49.8 d	128.13 a
T7 –soil + WSMS + ½N	182.30 a	116.27 b	117.70 b
T8 –soil + CSMS + ½N	167.30 a	147.34 a	143.20 a
T9–soil + 100% N	60.50 c	17.85 e	36.50 c
T10 –soil + FSMS + 100%N	20.80 d	44.4 d	144.97 a
T11 –soil + WSMS + 100%N	214.60 a	71.67 c	53.30 c
T12 –soil + CSMS + 100%N	140.00 b	57.16 c	51.40 c

[1]Means followed by the same letter(s) are not significantly different from each other based on Tukey's HSD test at 5% level of significance.

The fresh weight of pechay was significantly affected by different growing media and N fertilization at different cropping (Table 1). During the first crop, the average fresh weight of pechay from T11 was highest at 214.6 g while the lowest value of 2.0 g was from T2. Treatments with WSMS as soil amendments with 100% and 50% of urea added had higher yield compared to other treatments. Plants grown in CSMS + 100% RR of urea performed better than those with FSMS and pure soil but inferior to those with WSMS. N fertilization significantly enhanced growth in terms of average fresh weight since T3 (with WSMS) and T4 (with CSMS)which have 0% RR of urea added gave a significantly lower gross weight of 83.3 g and 20.2 g, respectively. Lowest yield was obtained in T2 which is composed of GS + FSMS only. During the second crop, significantly higher fresh yield was obtained in treatments with WSMS and CSMS at 1/2 RR of N fertilizer. Plants from treatments with FSMS are consistently inferior compared to other media regardless of the N fertilization. The full rate of N fertilization did not result in high yield as compared to 1/2 RR of N fertilizer. Lowest yields were observed from growing media in soil + FMS, soil alone at all levels of N fertilizer and soil + CSMS without N fertilizer. On the 3rd crop, plants grown in treatments with FSMS had higher or comparable yield to other treatments across different levels of N fertilizer application. A significantly lower yield was obtained from treatments with WSMS and CSMS with a full rate of N fertilizer and those from the control (soil alone) and soil + full RR of N fertilizer.

3.2. Chemical properties and nutrient availability of the growing media

The chemical properties of the different growing media were presented in Tables 2a, 2b, and 3a, 3b. Significance differences were observed on pH, EC, Total C, Total N, C/N ratio, available N and available P before and after cultivation of pechay for three cropping (Table 2). Before plant cultivation

Table 2a. The pH, electrical conductivity (EC), total C, total N, C/N ratio, available N and available P of growing media used in the experiments before planting and after the first cropping.[1]

Treatments	pH	EC uS/m	Total C (g/kg)	Total N (g/kg)	C/N ratio	Avail. N (mg/kg)	Avail. P (mg/kg)
Before planting							
Soil	5.92 c	126 e	42 c	1.44 c	28.82 c	12.00 e	62.00 a
Soil + FSMS	7.02 b	1021 b	68 a	1.8 b	37.78 b	5.20 f	54.00 b
Soil + WSMS	8.39 a	2640 a	62 b	2.7 a	22.96 c	14.85 d	65.00 a
Soil + CSMS	8.02 a	803 c	78 a	1.1 d	70.91 a	6.80 f	32.00 c
After First Crop							
T1 –soil	6.54 c	145 e	42 c	1.6 c	26.25 c	4.20 f	59.00 b
T2 –soil + F SMS	8.17 a	466 d	74 a	2.0 b	37.00 b	3.20 f	52.92 b
T3 –soil + WSMS	8.68 a	837 c	67 a	2.6 a	26.15 c	12.80 e	63.70 a
T4 –soil + CSMS	8.16 a	417 d	77 a	0.92	83.70 a	5.40 f	31.36 c
T5 –soil + ½ N	5.83 c	428 d	42 c	2.3 b	18.26 c	18.20 d	58.00 a
T6 –soil +FSMS + ½N	7.85 b	432 d	66 a	1.8 b	36.67 b	6.24 f	53.13 b
T7 –soil + WSMS + ½N	8.44 a	986 c	60 b	2.9 a	20.69 c	28.00 c	63.95 a
T8 –soil + CSMS + ½N	8.05 a	709 c	76 a	1.5 c	50.67 b	26.80 c	31.49 c
T9–soil + 100% N	6.00 c	426 d	41 c	2.1 b	19.52 c	42.00 a	42.00 c
T10 –soil + FSMS + 100%N	7.81 ab	473 d	70 a	2.2 b	31.82 c	8.42 ef	51.54 b
T11 –soil + WSMS + 100%N	8.29 a	1037 b	66 a	2.9 a	22.76 c	32.28 b	62.04 a
T12 –soil + CSMS + 100%N	7.88 ab	744 c	76 a	1.9 b	40.00 b	27.54 c	30.54 c

[1]Means followed by the same letter(s) are not significantly different from each other based on Tukey's HSD test at 5% level of significance.

mixing of different SMS preparation increased the pH, EC and total C of growing media compared to soil alone (Table 2a). The magnitude of pH increase was higher due to WSMS and CSMS addition than the FSMS. The EC was increased significantly by SMS addition to growing media with the highest increase due to WSMS. The total C also increased due to the mixing of SMS especially the CSMS. FSMS and WSMS increased slightly the total N content but CSMS decreased the total N of the

IOP Conf. Series: Earth and Environmental Science **120** (2018) 012010 doi:10.1088/1755-1315/120/1/012010

growing media. Consequently, highest C/N ratio was observed due to the addition of CSMS. The available N and P were significantly enhanced by WSMS addition but reduced due to FSMS and CSMS based on the control soil.

After the first (Table 2a) and second crop cultivation (data not shown), the pH, EC, and total C followed the same trend to that observed before crop cultivation regardless of the N fertilization. On the other hand, the total N, C/N ratio, available N, and P showed considerable differences and patterns. The total N was increased due to N application that consequently decreased the C/N ratio. However, the available N levels showed greater variability and behave differently compared to total N (Figure 1). Addition of N fertilizer increased the available N concentration in growing media with WSMS and CSMS but not in growing media with FSMS. The available P followed the same trend with available N but of lower magnitude. In table 2b, noticeable increase EC, available N and P and the decrease in total C and C/N in FSMS amended treatments as compared to other treatments and to its previous values obtained after the first and second crop. On the other hand, the available N and P of treatments with WSMS and CSMS supplied with 100% RR of N fertilizer declined and the values were even lower compared to WSMS and CSMS treatments with 1/2 N fertilizer.

Table 3a, showed a significant increased on the exchangeable K, Ca, Mg and Na and available Cu, Mn, and Zn due to mixing of different preparations of SMS before planting and after the first crop. Although the three preparations did not differ significantly, a higher concentration of exchangeable basis and available micronutrient was observed due to carbonized SMS. Only the available Fe were not altered significantly by mixing different SMS preparations with soil in the growing media. After the second crop (data not shown) and after the 3rd crop (Table 3b), a general decrease of the exchangeable K, Ca, Mg and Na and available Cu, Mn, and Zn were observed. However, after the 3rd

Table 2b. The pH, electrical conductivity (EC), total C, total N, C/N ratio, available N and available P of growing media used in the experiments.[1]

Treatment	pH	EC uS/m	Total C (g/kg)	Total N (g/kg)	C/N ratio	Availabl e N (mg/kg)	Availabl e P (mg/kg)
After 3rd Crop							
T1 –soil	6.12 c	146 d	42 c	1.5 b	28.00 c	4.38 e	32.28 b
T2 –soil + F SMS	7.94 a	824 a	52 b	1.7 b	30.23 b	38.28 b	64.22 a
T3 –soil + WSMS	8.04 a	423 b	61 a	1.6 b	38.13 b	16.28 d	36.12 b
T4 –soil + CSMS	7.72 a	386 c	76 a	0.8 c	90.48 a	12.18 d	21.48 c
T5 –soil + ½ N	5.82 c	468 b	41 c	1.8 b	22.78 c	19.22 c	32.18 b
T6 –soil +FSMS + ½N	7.41 ab	892 a	46 b	2.3 a	20.00 c	48.24 a	52.12 a
T7 –soil + WSMS + ½N	7.88 a	827 a	48 b	2.5 a	19.20 c	43.24 b	42.13 b
T8 –soil + CSMS + ½N	7.62 ab	726 a	74 a	2.4 a	30.83 b	39.48 b	24.18 c
T9–soil + 100% N	5.68 c	548 b	42 c	1.8 b	23.33 c	24.12 c	24.17 c
T10 –soil + FSMS + 100%N	7.54 ab	926 a	48 b	2.3 a	20.87 c	58.34 a	48.26 a
T11 –soil + WSMS + 100%N	8.00 a	724 a	49 b	2.7 a	18.15 c	19.24 c	24.00 c
T12 –soil + CSMS + 100%N	7.58 ab	634 b	72 a	2.1 a	34.29 b	18.26 c	16.21 d

[1]Means followed by the same letter(s) are not significantly different from each other based on Tukey's HSD test at 5% level of significance

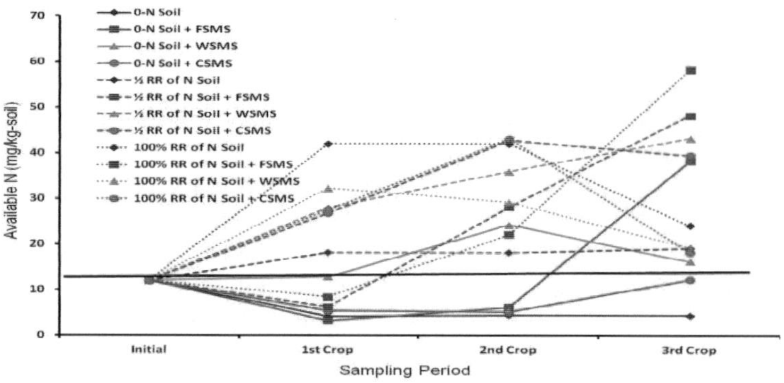

Figure 1. Amount of available N in the growing media before and after each crop of pechay for 3 consecutive cropping. Solid black line indicate the initial level of available N in soil.

crop, the amounts of exchangeable K and Ca in WSMS and CSMS in 1/2 RR and 100% RR N fertilized were significantly lower compared to other treatments. The exchangeable K and Ca in WSMS and CSMS in 1/2 RR and 100% RR N fertilized decreased in higher magnitude compared to others while those with FSMS had increased considerably.

4. Discussion

The variable growth based on the fresh yield of pechay grown in different growing media with different levels of N fertilization and at different successive cropping is an indication of the potential of different preparation of SMS as a component growing medium of potted plants. Our results showed how these different preparations would affect nutrient availability especially on N and P. The variable growth of pechay could be attributed to differences in the chemical properties of the different soil: SMS mixtures with different preparation as affected by time. Based on the initial chemical analysis

Table 3a. Exchangeable K, Ca, Mg and Na and the available Fe, Cu, Mn, and Zn of growing media used in the experiments after the first crop.[1]

Treatments and Soil Preparations	Exchangeable Bases				Available Micronutrients			
	K (g/kg)	Ca (g/kg)	Mg (g/kg)	Na (g/kg)	Fe (mg/kg)	Cu (mg/kg)	Mn (mg/kg)	Zn (mg/kg)
Before planting								
Soil	0.38 c	3.67 b	0.58 b	0.35 b	1.85 a	1.38 b	6.40 b	2.12 b
Soil + FSMS	0.43 b	4.11 a	0.65 a	0.39 a	2.07 a	1.55 a	7.47 a	2.37 a
Soil + WSMS	0.45 a	4.32 a	0.68 a	0.41 a	2.18 a	1.62 a	7.53 a	2.49 a
Soil + CSMS	0.46 a	4.80 a	0.70 a	0.42 a	2.22 a	1.66 a	7.68 a	2.54 a
After First Crop								
T1 –soil	0.39 c	3.62 b	0.58 b	0.34 b	1.70 a	1.43 b	6.24 b	2.18 b
T2 –soil + F SMS	0.42 c	4.03 a	0.64 a	0.38 a	2.03 a	1.51 a	7.02 a	2.33 a
T3 –soil + WSMS	0.44 a	4.23 a	0.67 a	0.40 a	2.13 a	1.59 a	7.38 a	2.44 a
T4 –soil + CSMS	0.45 a	4.70 a	0.68 a	0.41 a	2.17 a	1.62 a	7.52 a	2.49 a
T5 –soil + ½ N	0.32 d	3.58 b	0.59 b	0.36 b	1.78 a	1.39 b	6.45 b	2.01 b
T6 –soil +FSMS + ½N	0.42 b	4.04 a	0.64 a	0.39 a	2.04 a	1.52 a	7.05 a	2.34 a
T7 –soil + WSMS + ½N	0.44 a	4.25 a	0.67 a	0.40 a	2.14 a	1.60 a	7.41 a	2.45 a
T8 –soil + CSMS + ½N	0.45 a	4.72 a	0.68 a	0.41 a	2.18 a	1.63 a	7.55 a	2.50 a
T9–soil + 100% N	0.28 d	3.64 b	0.52 b	0.42 a	1.68 a	1.32 b	6.14 b	2.21 ab
T10 –soil + FSMS + 100%N	0.41 c	3.92 ab	0.62 a	0.39 b	1.98 a	1.48 a	6.84 a	2.27 ab
T11 –soil + WSMS +	0.43b	4.12 a	0.65 a	0.39 a	2.08 a	1.55 a	7.18 a	2.38 a

100%N								
T12 –soil + CSMS + 100%N	0.44 a	4.58 a	0.66 a	0.40 a	2.12 a	1.58 a	7.33 a	2.43 a

[1]Means followed by the same letter(s) are not significantly different from each other based on Tukey's HSD test at 5% level of significance.

conducted before cropping, SMS mixture would increase the pH of soil which could be attributed to the liming contents of SMS during its preparation. Calcium sulfate is added on the mushroom substrate during preparation to neutralize the pH of the composting mixture of residues and to supply the suitable level of Ca for mushroom [17]. The pH observed in the growing media is quite high than the normal pH for crops which is about 6.5 [18]. Increase in pH will affect nutrient dynamics especially the availability of P and other micronutrients (Table 2a and Table 3a). Similarly, the EC values in SMS amended growing media was increased due to SMS preparations indicating the capacity of SMS preparation to increase the soluble nutrients in the growing media. This is supported by a higher concentration of exchangeable basis and available micronutrients determined after crop cultivation. However, it should be noted that before crop cultivation, the EC of the growing media with different SMS exceed the suggested reference level (605 uS m^{-1}) [17].

Based on the data presented, the different preparations of SMS had greatly affected the N availability in the growing media. It appeared that FSMS will result to high N immobilization throughout the duration of the first and 2nd cropping period resulted in an inferior growth of pechay even in the presence of full rate of N fertilizer. This could be attributed to high C/N ratio and the nature of organic C present in the growing media containing FSMS (Figure 1). It should be noted that the N-immobilization in FSMS containing growing media was occurring at a very high rate because the added N fertilizer did not improve the growth of pechay (low fresh weight) and that the available N levels after crop cultivation remained low (Table 1, 2a, 2b). In contrast, WSMS and CSMS showed higher N availability (Table 2a and 2b) which implied that net N mineralization takes place. Weathering of SMS prior to mixing in the growing media resulted to low C/N ratio that favored mineralization [19], [20]. In fact, the addition of WSMS slightly increased the available N in the growing media indicating that WSMS released some available N ready for plant uptake. Consequently, additional N fertilizer increased the available N for plant use resulted in enhanced growth of pechay during the first and second crop. On the other hand, carbonization process of SMS transformed the SMS-C into a non-bioavailable form that is not reactive to microbial attack [21], [22]. Even if the C/N ratio of the growing media containing CSMS is high, this did not result to biological N immobilization. Probably, CSMS had increase available N absorption into the C matrix that resulted to N-fixation (physical N-immobilization) as can be observed by low growth of pechay receiving half rate of fertilizer N [20]. During the 3rd crop, a considerable increased in the fresh yield of pechay in growing media containing FSMS while those plants grown in media containing WSMS and CSMS had low fresh yield regardless of the N fertilizer level (Table 1). The increase in yield in FSMS containing growing media could be attributed to an increase in available N and P, and other nutrients as indicated in the residual nutrients after the 3rd crop. This would imply that at this stage, FSMS was able to reach a condition that promotes nutrient mineralization rather immobilization as observed during the 1st and 2nd cropping, hence an increase of nutrients occurred (Table 2c and 3c). In contrast, the decline in fresh yield of pechay from WSMS and CSMS containing growing media despite the supply of N fertilizer would indicate that some other nutrient became limiting for optimal growth as indicated in lower K and Ca contents in these treatments. It appeared that at 3rd-crop, nutrient imbalance occurred in WSMS and CSMS containing media causing slow growth of test plant. This nutrient imbalance could be due to previous nutrient uptake depleting some nutrients.

Table 3b. Exchangeable K, Ca, Mg and Na and the available Fe, Cu, Mn, and Zn of growing media used in the experiments after the third crop.[1]

Treatments	Exchangeable Bases				Available Micronutrients			
	K	Ca	Mg	Na	Fe	Cu	Mn	Zn
	(g/kg)				(mg/kg)			
After Third Crop								
T1 –soil	0.35 c	3.56 b	0.50 c	0.32	1.68	1.42	6.02	2.10 b

				b	a	b	b	
T2 –soil + F SMS	0.48 a	4.28 a	0.62 b	0.43 a	2.16 a	1.56 a	6.52 b	2.46 a
T3 –soil + WSMS	0.42 b	4.12a	0.63 b	0.38 a	2.06 a	1.32 b	7.06 a	2.42a
T4 –soil + CSMS	0.40 b	4.24 a	0.62 b	0.39 a	2.01 a	1.46 b	7.04 a	2.48 a
T5 –soil + ½ N	0.32 d	3.46 b	0.58 b	0.34 b	1.84 a	1.36 b	6.33 b	2.15 b
T6 –soil +FSMS + ½N	0.46 a	4.38 a	0.68 a	0.46 a	2.13 a	1.65 a	6.76 a	2.53 a
T7 –soil + WSMS + ½N	0.36 c	3.08 b	0.60 b	0.41 a	2.02 a	1.56 a	7.16 a	2.38 a
T8 –soil + CSMS + ½N	0.38 c	3.34 b	0.55 c	0.42 a	2.00 a	1.52 a	7.24 a	2.42 a
T9–soil + 100% N	0.31 d	3.64 b	0.51 c	0.40 a	1.79 a	1.34 b	6.14 b	2.22 a
T10 –soil + FSMS + 100%N	0.48 a	4.53 a	0.69 a	0.44 a	2.56 a	1.64 a	6.96 a	2.48 a
T11 –soil + WSMS + 100%N	0.32 d	3.20 b	0.57 c	0.37 a	2.19 a	1.48 b	6.88 a	2.30 a
T12 –soil + CSMS + 100%N	0.31 d	3.14 b	0.62 b	0.40 a	2.15 a	1.54 a	7.02 a	2.26 a

[1]Means followed by the same letter(s) are not significantly different from each other based on Tukey's HSD test at 5% level of significance.

5. Conclusions

Overall, our study has shown that the different preparation of SMS as a component in growing media of potted plants resulted to varying biochemical properties affecting nutrient dynamics and availability. Based on our results, WSMS is the best preparation for the component in growing media as it increases the N and P availability, favorable C/N ratio and increased micronutrient contents. CSMS could be favorably used as a component of growing media as it also improves nutrient availability especially increasing the exchangeable basis and available micronutrient essential for plant growth. This would also follow that WSMS and CSMS will be appropriate as soil amendments in the field compared to the addition of FSMS which could be detrimental to N availability immediately after application. Application of FSMS may cause nitrogen immobilization for a prolonged period which may result in nitrogen deficiency of existing crops or newly planted crops when applied directly to the soil. Therefore, WSMS and CSMS could be used in the preparation of growing media or as soil amendments that will contribute to their disposal in an environment-friendly way and reduces the need for peat simultaneously.

6. References

[1] Raviv M, Chen Y and Inbar Y 1986 Peat and peat substitutes as growth media for container-grown plants. In *The role of organic matter in modern agriculture.* 257-87

[2] Page S E and Baird A J 2016 Peatlands and global change: Response and resilience. *Annu. Rev. Env. Resour.* **41** 35-57

[3] Abad M, Noguera P and Bures S 2001 National inventory of organic wastes for use as growing media for ornamental potted plant production: case study in Spain. *Bioresource Technol.* **77** 197–200

[4] FAO 2013 *FAO STATISTICAL YEARBOOK 2013 World Food and Agriculture* Food and Agriculture Organization of the United Nations Rome 2013 (http://www.fao.org/docrep/018/i3107e/i3107e.PDF)

[5] Philippoussis A N 2009 Production of mushrooms using agro-industrial residues as substrates In *Biotechnology for Agro-Industrial Residues Utilisation* Springer Netherlands 163-96

[6] Rinker D L 2002 Handling and using "spent" mushroom substrate around the world. *Mushroom Biology and Mushroom Products, UAEM, Cuernavaca.*

[7] Semple K T, Reid B J and Fermor T R 2001 Impact of composting strategies on the treatment of soils contaminated with organic pollutants. *Environ. Pollut.* **112(2)** 269-83

[8] Sanchez J E and Royse D J 2009. *Scytalidium thermophilum*- colonized grain, corncobs and chopped wheat straw substrates for the production of *Agaricus bisporus. Bioresource Technol.* **100** 1670–74

[9] Paula F S, Tatti E, Abram F, Wilson J and O'Flaherty V 2017 Stabilisation of spent mushroom substrate for application as a plant growth-promoting organic amendment *J. Environ. Manage.* **196** 476-86

[10] Nguyen V T and Wang C H 2016 Effects of Organic Materials on Growth, Yield, and Fruit Quality of Honeydew Melon. *Commun. Soil Sci. Plant Anal.* **47**(4) 495-504

[11] Orge, R. F., & McHenry, M. P. (2013). Biowaste conversion technology for household food and energy security in the Philippines: appropriate on–site small–scale rice husk waste carbonisation. *International Journal of Innovation and Sustainable Development*, *7*(4), 387-399.

[12] Allison LE, Moodie CD 1965 Methods of soil analysis. In: Black CA, Evans DD, Ensminger LE, White JL, Clark FE, Dinauer RC (Eds.) *Part 2. Chemical and Microbiological Properties. Agronomy No. 9.* American Society of Agronomy, Madison WI 1379–96

[13] Yeomans JC, Bremner JM 1989 A rapid and precise method for routine determination of organic carbon in soil *Commun. Soil Sci. Plant Anal.* **19** 1467–75

[14] Olsen SR, Xole, CV, Watanabe FS and Dean LA 1954 Estimation of Available Phosphorous in Soils by Extraction with Sodium Bicarbonate. US Department of Agriculture, *Circular* 939

[15] Knudsen D, Peterson GA and Pratt PF 1982 In: ASA-SSSA (Ed.), *Lithium, Sodium and Potassium: Methods of Soil Analysis* Madison **2** 225–46

[16] Lindsay E L and Norvell W A 1978 Development of a DTPA soil test for zinc, iron, manganese, and copper. *Soil Sci. Soc. Am. J.* **42** 421–28

[17] Stamets P and Chilton JS 1983 *The mushroom cultivator*. Agarikon Press, Olympia

[18] Abad M, Noguera P and Bures S 2001 National inventory of organic wastes for use as growing media for ornamental potted plant production: case study in Spain *Bioresource Technol.* **77** 197–200

[19] Ultra Jr VU, Nunez J P and Lee S C 2016 Influence of charcoal-based soil amendments on growth and nutrient uptake of rice (Oryza sativa) in Cadmium contaminated soil *Emir. J. Food Agric.* **28**(12) 872

[20] Nunez JP, Ultra VJr, Park JS, Seo PD and Lee SC 2012 Soil nutrient dynamics, supplying capacity and agronomic effects of charcoal-based amendments to rice. Proc. The 50th Anniversary of the Korean Society of Crop Science: 2012 International Symposium on Current Status and Prospects of Environment-Friendly Agriculture in Asian Region. Oct.11-12, 2012. Naju-si, Jeollanam-do, Korea. *Korean J.Crop Sci.* **52** (2) 44

[21] Asai H, Samson BK, Stephan HM, Songyikhangsuthor K, Homma K, Kiyono Y, Inoue Y, Shiraiwa T & Horie T 2009 Biochar amendment techniques for upland rice production in Northern Laos 1. Soil physical properties, leaf SPAD and grain yield. *Field Crops Res.* **111** 81–84

[22] Masulili A, Utomo WH and Syechfani MS 2010 Rice husk biochar for rice based cropping system in acid soil 1. The characteristics of rice husk biochar and its influence on the properties of acid sulfate soils and rice growth in West Kalimantan, Indonesia. *J. Agric. Sci.* **2**(1) 39.

Acknowledgement

Gratitude was extended to the Environmental Science Laboratory, Catholic University of Daegu, South Korea for the chemical analysis of soil and growing media samples.

IOP Conf. Series: Earth and Environmental Science **120** (2018) 012011 doi:10.1088/1755-1315/120/1/012011

Algae-Based Carbon Sequestration

Cai Haoyang[1]
1 Hangzhou Foreign Language School, China
E-mail: caipei1970@163.com

Abstract: Our civilization is facing a series of environmental problems, including global warming and climate change, which are caused by the accumulation of green house gases in the atmosphere. This article will briefly analyze the current global warming problem and propose a method that we apply algae cultivation to absorb carbon and use shellfish to sequestrate it. Despite the importance of decreasing CO2 emissions or developing carbon-free energy sources, carbon sequestration should be a key issue, since the amount of carbon dioxide that already exists in the atmosphere is great enough to cause global warming. Algae cultivation would be a good choice because they have high metabolism rates and provides shellfish with abundant food that contains carbon. Shellfish's shells, which are difficult to be decomposed, are reliable storage of carbon, compared to dead organisms like trees and algae. The amount of carbon that can be sequestrated by shellfish is considerable. However, the sequestrating rate of algae and shellfish is not high enough to affect the global climate. Research on algae and shellfish cultivation, including gene technology that aims to create "super plants" and "super shellfish", is decisive to the solution. Perhaps the baton of history will shift to gene technology, from nuclear physics that has lost appropriate international environment after the end of the Cold War. Gene technology is vital to human survival.

1. Introduction

Our civilization is facing a series of environmental problems, including global warming and climate change, which are caused by the accumulation of green house gases in the atmosphere. Various possible methods have been proposed including industrial recycling, compressing it into the ocean floor, replacing fossil fuels with solar panels and land-based plant cultivation. However, the key way is carbon sequestration, which may not be accomplished by simply planting trees because carbon will be released again as the remnants of tree decay. Currently, the only way to permanently store carbon on a large scale is carbonates formation, which can be achieved by using shellfish to transform the carbon in the atmosphere into calcium carbonates.

2. Current environmental problems

2.1 Global warming and sea-level rise

2.1.1 Data from NASA

Currently, we are facing a severe global-warming problem. Based on research data from various research institutions, it can be concluded that global warming is indeed occurring, and that the increasing accumulation of greenhouse gases such as carbon dioxide in the atmosphere is causing the green-house effect to be more and more notable.

The global temperature data provided by NASA[1] (as shown in Figure 1) over the past few decades shows a growth of temperature of about 0.94 degrees Celsius, compared with that in 1880. Its global temperature map also shows the global average temperature with a significant increase, especially the annual average temperature in the Northern Hemisphere.

The diagrams below show that global warming is indeed occurring; it is not a hoax. This dramatic warming process has been happening over the past one hundred years, indicating the seriousness of the global warming problem that the earth is facing.

The sea-level rise, which is at least partially caused by global warming, can be proved by the diagram provided by NASA [2] (as shown in Figure 2):

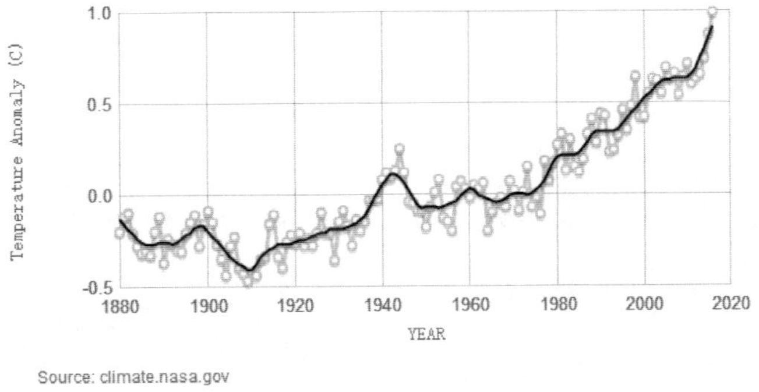

Figure 1, Temperature Anomaly.

Figure 2, Sea Level Change.

2.1.2 Data collection in Zhoushan

However, the data provided by NASA and IPCC is too far away from where I live--the melting icecaps are in Antarctica, thousands of miles away. In order to search more local sea-level data and make further proof of sea-level rises, I went to the Zhoushan monitoring center of National Maritime Bureau and asked for assistance (Figure 4). By looking up local sea -level measurements, which accumulated for decades, and comprehensive analysis of data sampling of observation points, a chart was drawn (Figure 3):

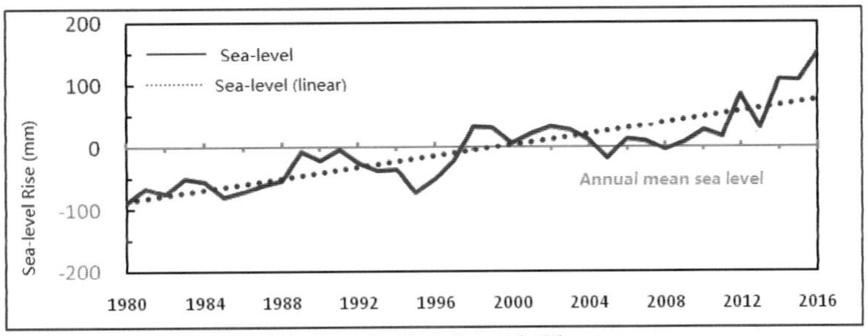

Figure 3, Sea Level Change

IOP Conf. Series: Earth and Environmental Science **120** (2018) 012011 doi:10.1088/1755-1315/120/1/012011

From 1980 to 2016, the average rising speed of the sea-level near Zhoushan was 4.5mm/year, which was close to the data provided by the fourth assessment of the IPCC. An increasing rate of 4.5mm per year seems very little, but with the accelerating melting speed of the glaciers in Antarctica and Greenland, the actual sea-level rise by 2050 will be considerable.

According to the data, the sea level rise is consistent with global warming. Although the theory that the sea level rising should be completely attributed to global-warming is not proofed, there is no doubt that global warming caused the melting and disintegration of glaciers and icecaps, and forced the sea-level to rise.

Figure 4, Zhoushan Data Collection

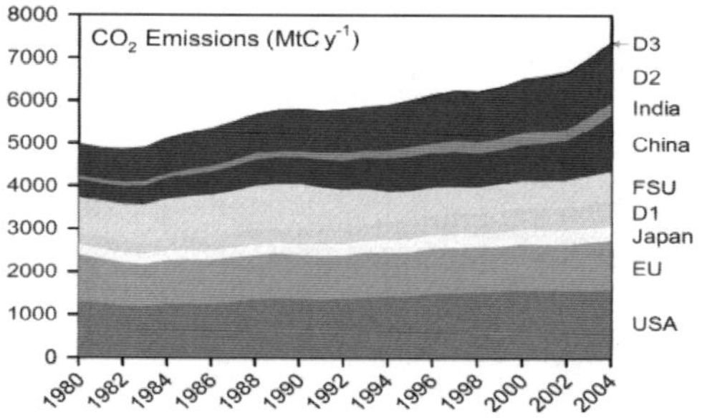

Figure 5, CO_2 Emissions

2.2 Carbon emissions

2.2.1 Industrial carbon-dioxide emissions

Except for a tiny decline in carbon-dioxide emissions in the early 1980s, which was partially caused by the energy crisis, the world's carbon-dioxide emission has increased by 46 percent in total. [3] (as shown in Figure 5)

The amount of carbon-dioxide in the atmosphere is increasing because the carbon cycle was broken by the emission of greenhouse gases, creating more carbon-dioxide than plants and the formation of calcium carbonate can absorb.

2.2.2 CO_2 emissions from forest loss

G. R. van der Werf mentioned in his research report that "Within the science and policy communities, carbon emissions from deforestation and forest degradation have been estimated to account for about 20% of global anthropogenic CO2 emissions."[4]

Forest loss greatly decreased the amount of carbon-dioxide that is absorbed and contributed 20% to the total carbon-dioxide increase. Though 20% is a relatively small percentage, compared with that caused by industrial emissions, this carbon emission illustrates the fact that plants are vital in controlling the green-house effect because they are one of the most important parts of the earth that transfers carbon-dioxide into organisms. Forest loss creates a greater gap between the production and the absorption of carbon-dioxide. Thus, based on current technology levels, the use of plants is a vital way to sequestrate carbon-dioxide in the atmosphere and fix the carbon cycle.

2.3 Brief analysis of global warming problem

In brief, the global-warming problem is caused by the inefficiency of modern day carbon-free energy sources like solar energy, wind energy, hydro-power and nuclear energy.

Solar energy is a prospective source, but the producing process of solar panels is not only costly, but also pollutes the environment with chemicals, including lead and cadmium.

The suitable areas for wind farms and hydropower stations are also limited.

Bio-fuels and methane are regarded as "clean energy", but they emit greenhouse gases, like CO2 and CH4. In some situations, using bio-fuels may be an irrational measure because a lot of grains would be consumed, which are needed by people who suffer from famine in South-Sahara regions.

Although nuclear energy comparatively produces little green-house gases and nuclear power stations can be built in many places, it causes political problems, in which no country can guarantee that those nuclear fuels are not used for military purposes. It is a difficult task to landfill nuclear waste without polluting local environments. Moreover, there are many legitimate objections from the public because of concerns that the consequence of a nuclear leak would be devastating.
Thus, whether the global-warming problem would be successfully and completely solved depends on the technology developments of clean energy.
However, based on technology, there are lots we can do in order to slow down the warming trend. In addition to expecting technological breakthroughs in nuclear fusion, we can sequestrate the carbon by using plants. Simply planting trees may not be an effective method because the carbon absorbed by trees will be released again after the trees die. There is not much land to grow extra plants on the continent, while there are huge amounts of water areas that remain uncultivated (Figure 6). So, it is possible to grow algae in some areas in which the growing algae is relatively easy to control (to prevent it from growing all around the world). In order to solve the problem of decay, we can let shellfish to sequestrate the carbon that is in the form of organic compounds into calcium carbonates.

Figure 6, Uncultivated Water Areas [5]

3. Algae Cultivation

3.1 Brief introduction

Algae cultivation aims to reduce the green-house effect in the atmosphere by using algae to absorb carbon-dioxide and letting algae sink into deep oceans. Carbon, together with the remnants of algae, would be stored in deep oceans, on the seafloor, for several centuries.
Since algae is a fast-growing species, the effort we would need to spend to cultivate them is very little. It is widely known that the ability to grow fast is not a trouble, since some algae reproduce quickly and pollute water areas, causing "red tides", but fast-growing is also a quality that can be used to absorb carbon-dioxide in a large scale.
In addition to the fast-growing quality, algae that live on the ocean surface (such as plankton) can be controlled by adding and reducing the amount of nutrients in the ocean.

3.2 Effectiveness of algae

Since each solution listed above has its insufficiencies, algae is outstanding, compared with the drawbacks of the other solutions. Carbon Recycling Process can only absorb a little amount of carbon-dioxide, while algae have a large potential to reduce carbon-dioxide from all around the atmosphere. Compressing carbon-dioxide into the seafloor is risky if we put exceeding amounts of green-house gases into it. Algae, however, can sink onto the seafloor as remnants of organisms, being prevented from being transformed into carbon-dioxide. Also, the costs of Tung trees' cultivation on coastal deserts are also high, preventing this solution from coming true. In contrast, cultivating algae is cheap.
Land-based plants contribute 52% of the total carbon-dioxide absorbed by the earth's biosphere, while ocean-based algae contributed 45% to 50% of that, which means that despite their small size, algae can absorb carbon-dioxide efficiently because of their comparatively short life cycles. Algae can even reduce the amount of carbon-dioxide in the atmosphere more if we can support them with the appropriate environment, including temperature, nutrients (iron is one of the most important and

effective ones), sunlight and the number of animals that eat them. Thus, algae cultivation is a promising solution to reduce greenhouse effects.

The effectiveness of algae can be well illustrated by an experiment. A group of scientists from Alfred Wigner Polar Oceanographic Institute, claim that if we put extra iron into some water areas, algae is able to grow much more rapidly and absorb more carbon-dioxide. In February 13th, 2004, they sprayed 7 tons of $FeSO_4$ into acidic sea water, covering approximately 167 square kilometers of water area, and monitored the change of algae growth in the following five weeks. As they predicted, six species in diatoms grew rapidly. After that, researchers stopped adding iron and saw that the majority of diatoms died and sunk onto the seafloor, which proved that the growth of algae can be partially controlled. Its low cost is attractive because this method could be used extensively, in oceans all around the world, to absorb considerable amounts of greenhouse gases.

It is true that this method may show many drawbacks after being applied, but at least, the potential of algae is fully displayed.

John Martin, a marine scientist of the Marine Observation Station in California, said, "If you give me half a tanker of iron, I can return you an ice age."

3.3 Cultivation in restricted areas and marine ranches

3.3.1 Brief introduction

Despite the low price, problems, including the difficulty to control and the potential emission that will happen centuries later, will be caused if we apply iron fertilizers to open oceans. One way to avoid them is to cultivate algae in restricted areas, which means letting algae grow on ropes that extend into the ocean. Constructing pipes under the ropes that transport carbon-dioxide and releasing it for algae to use could also be a method. Algae can grow 1-2 centimeters per day, while they can grow much faster if they absorb extra carbon-dioxide. When the algae is bigger and longer, storing more carbon from the atmosphere and the gas we apply, we should harvest them. While in open sea areas, algae is difficult to collect and store, we just need to pull up the ropes and harvest them if algae is cultivated on ropes, which means it is restricted. After harvesting, we can either use it to generate electricity or sequestrate it.

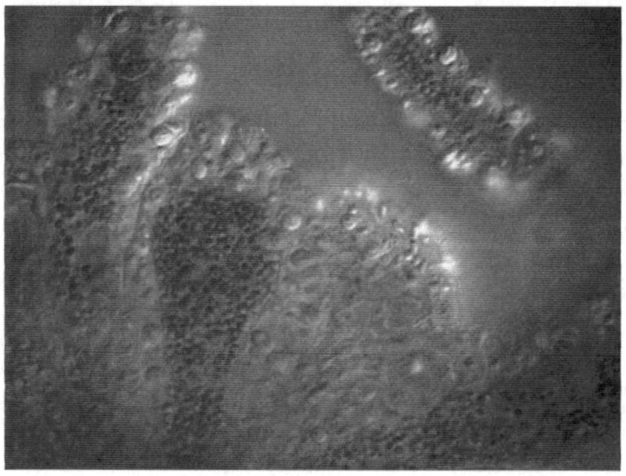

Figure 7, Red Tides

Figure 8, Chlorella

While fertilizer could improve the productivity of farmlands, it can also increase the amount of harmful algae and form "red tides". If we cultivate kelp in these areas, nutrients in the sea water would be largely absorbed and would not promote the growth of "red tides"(as shown in Figure 7 [6]) and harm fisheries . Thus, the bio-diversity in coastal would also be protected.

For a marine ranch that aims to sequestrate carbon, there are two possible choices of algae.

If we need to harvest algae for subsequent treatments, large algae, like kelp would be a better idea. Though small algae, such as chlorella (Figure 8 [7]), can grow rapidly, they are difficult to harvest and process because they distribute all around the surface of the water, forming a very thin layer of organisms and sinking quickly. In contrast, large algae can be controlled to grow on ropes.

However, if there is no need to harvest algae, single-cell algae would be a better choice because of its fast metabolism rate. They can also be easily absorbed by shellfish.
Through previous experiments, Martin J.H. and Fitzwater S.E. [8] proposed that in ocean areas with high nutrition content and low chlorophyll content, the limiting factor for algal growth is Fe, which means if enough amount of Fe is provided, algae growth can be greatly improved. Other fertilizers are needed to cultivate either single-celled algae or multicellular algae, which will not be introduced in this article.

3.3.2 Subsequent treatment
Subsequent Treatment is the key process to algae cultivation because the way we process the remnants of algae directly determines whether the whole process will make sense for carbon sequestration. The target is that we should not let the carbon kelp have an absorbed flow into the atmosphere, in other words, to store the carbon. Three sequestrating methods will be introduced in this passage.

3.3.2.1 Generating electricity
The most fundamental way to generate electricity is to burn combustibles, mainly fossil fuels, to generate steam, to rotate the steam turbine and to generate electric power (as shown in Figure 9[9]). However, the fundamental problem is that burning fossil fuels releases the carbon that was previously stored in the crust, and thus increases the amount of carbon-dioxide in the atmosphere. If what we burn is not fossil fuels and we burn plants instead, the carbon-cycle will be reestablished. The amount of carbon we emit by generating electricity will be the exact number those carbon plants absorbed. Then the question is what kind of plant should we burn? It is obvious that we cannot burn the trees, which was rejected by the United Kingdom in the 19th century because of the consumption of forest resources.

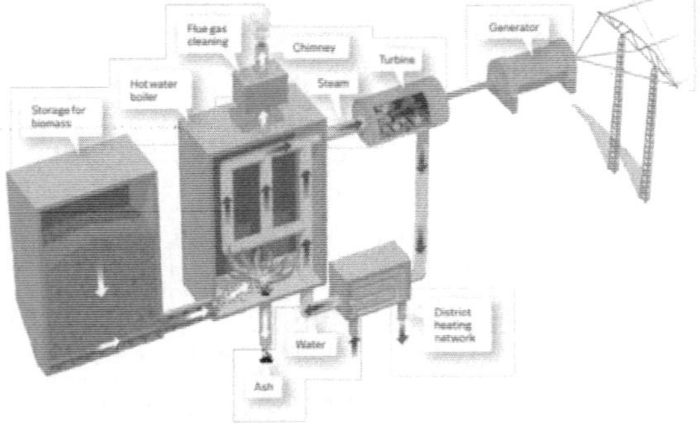

Figure 9, Generating Electricity

Compared with the disadvantages of burning trees, kelp seems a feasible idea. After being harvested and dehydrated, kelp can be used as fuel. The calorific value of standard coal is 29.27MJ/kg, while the average calorific value of dehydrated algae is 12.16MJ/kg, which means the potential heat that can be generated by kelp will be considerable.
However, if we want to completely fix the carbon cycle by simply using the remnants of kelp to generate electricity, the use of fossil fuels must be totally abandoned, which is currently impractical . The cycle itself, cultivating kelp and burning them, does not absorb any extra carbon in the atmosphere but consumes energy that needs to be used to manufacture fertilizers. It is clear that how much carbon it can sequestrate directly will depend on the amount of fossil fuels we are going to save. Unfortunately, the amount of fossil fuel that is going to be saved is limited because the cost of burning algae is apparently higher than excavating and using fossil fuels. Currently, using dehydrated kelp is not a practical idea.

3.3.2.2 Sequestration by shellfish
The only way that the earth itself permanently decreases carbon-dioxide in the atmosphere is to sequestrate it into calcium carbonates. The carbon absorbed by other methods, including photosynthesis and seawater dissolution will eventually be emitted into the atmosphere again. To

permanently store carbon, we not only need to use photosynthesis to change it into organic tissue, but must find a reliable way to store the organisms-- and storing it in shellfish's shell, which mainly consists of calcium carbonates, is a good idea. Shellfish can transform the carbon from organisms into calcium carbonates by using this process:

$$Ca^{2+}+2HCO_3^{-}{=}CaCO_3\downarrow+CO_2+H_2O$$

Since chlorella cannot be controlled by binding it on a rope, it needs to be restricted in an area by physical measures. Otherwise, the chlorella might float everywhere around the ocean with the tides and ocean currents. Dams are a possible choice for restriction, for the mature technology and good effect, which is greater than any of the chemical or biological restrictions. For the whole locked water area, the water inlet and the water outlet need to be designed in the middle of the dam, which looks like a tidal power station (as shown in Figure 10 [10]). However, the inlet and outlet water speed, especially outlet water speed needs to be small enough to not discharge too many organisms, or we cannot achieve the best results and will pollute nearby sea areas.

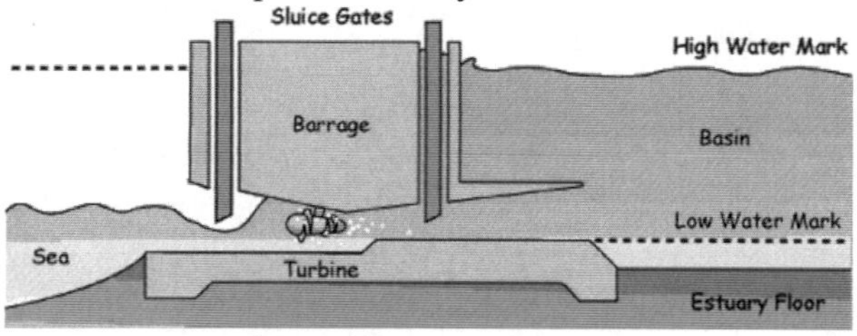

Figure 10, Tidal Power Station

The most significant drawback of dams which damages local bio-diversity, needs not to be of concern because the chlorella cultivation has already destroyed local ecosystems. Actually, if we want to keep local environments constant everywhere on the earth, not changing them in some way, it is impossible to achieve the sequestration of carbon-dioxide. Usually when we need to improve the environment in areas, we have to sacrifice others, though at a different intensity. Global Warming is more severe than local ecosystems. Thus, building dams is an acceptable method.

After the small algae that live on the surface of the water die, they sink into the water and can be absorbed by shellfish as nutrition. Carbon would be sequestrated in its shell as the shellfish grows.

Other than the permanent storage of carbon, the most significant advantage is that shellfish has a high filtration efficiency, which enables them to use nearly all the organisms that distribute in seawater nearby and transform them into a part of their shells quickly. It has been proved that breeding shellfish on large scales has significant effects on regulating organisms in local water areas.

To absorb the remnants of algae that lives on the surface of the water and store them permanently, large amounts of shellfish need to be bred on the seafloor near coastlines. If it succeeds, unlike burying the remnants of kelp into ground, we do not need to do anything because calcium carbonates are extremely difficult to dissolve in seawater, creating a nearly permanent storage on the time scale of human civilization.

Some people may not desire to use biological methods because these methods do not appear as efficient as drilling a hole and compressing carbon-dioxide into seafloor. However, if the amount of shellfish is great enough, it would be the most efficient solution because of its low cost and high productivity. Zhang Jihong, a researcher in aquiculture ecology, announced that 1hm^2 of chlamys farreri (1125000 per hm^2) sequestrates 3.36 tons of carbon [11]. Though the chlamys farreri was not provided by intensive organisms as nutrition, it can still sequestrate 336tons of carbon per square kilometer. Thus, the carbon that can be sequestrated by shellfish is considerable.

It is believed that the white remnants on the Dover Cliff contribute to the climate significantly by storing carbon in the atmosphere. Those white cliffs display natural carbon sequestration process that happens near coastlines all around the world. As a species that is trying to make up for the environmental loss created by us, we can learn from nature and apply the process on a more concentrated scale to store the carbon.

4. More ambitious plans

4.1 Enlargement of cultivating scales

Since our civilization extremely relies on fossil fuels, it is impractical to suddenly quit using it or to greatly decrease the use. Otherwise, the global economy will collapse. In order to sequestrate the carbon on a much larger scale and cancel out the emission of green-house gases, we have to plant algae and shellfish on broad water areas, such as the Hudson Bay.

To cultivate algae on the surface of bay areas, the length of dams needed would be relatively short, compared to the broadness of water areas. It is true that the ecosystem in the bay would be destroyed by algae that cover the whole water surface, but the beneficial aspects would be more important than the drawbacks, which is the remission of the global warming problem.

4.2 Visions of "super plants" and "super shellfish"

Despite many of the advantages that algae cultivation may have, algae, including kelp, are still not the perfect species of plants that reduce the amount of carbon-dioxide in the atmosphere. There might be potential diffusion problems, and the cost will be high, too. All plants on earth have evolved in the way we see today because they need to compete with other organisms, rather than absorbing carbon-dioxide. In order to effectively mitigate the greenhouse effect, we must genetically modify the plant's genes and make "super plants" for our use.

An ideal "super plant" would meet these following conditions.

First, a "super plant" must have a high rate of photosynthesis to capture the energy in sunlight efficiently and grow quickly. This is the main reason why we expect to have a "super plant"-the photosynthesis efficiency of current plants, even including algae, is far below the demand, which is using them to change the amount of carbon-dioxide significantly.

Second, its Light Saturation Point(LSP) should be high enough to make full use of abundant tropical sunlight. Typically, a plant species with high photosynthesis efficiency would have relatively low LSP, being appropriate to high-latitude environments but unsuitable for tropical regions.

Third, although we could provide extra nutrients via fertilizers, the "super plant's" demand of nutrients should not be too high. Otherwise, producing these fertilizers may cost a lot of energy (the total consumption of manufacturing 1 ton of $CO(NH_2)_2$ is 1555.49kg standard coal, including the raw material and fuel [12]). Plus, too much nutrients will allow plants and bacteria in this region to grow rapidly and pollute the local environment.

Since the subsequent treatment is the key issue of carbon sequestration in plant cultivation, there would be a huge demand for "super shellfish". An eligible "super shellfish" may meet the following requirements.

The first is that "super shellfish" need to have high metabolism rates and good filtration ability to absorb organisms in the water quickly, which is the major goal of "super shellfish". The shellfish we have needs tremendous amounts of seafloor to produce a notable decrease of carbon-dioxide in the atmosphere. The whole process will make little sense if the shellfish's metabolism rate is low, even after "super algae" is developed because there would be much more organisms being produced than being absorbed.

Second, the "super shellfish" must be relatively easy to breed, in order to save costs. To cover vast marine areas, we need a tremendous amount of "super shellfish", which leads to the demand that "super shellfish" need to be easy to breed and can reproduce quickly.

The third is that the shell much be thick and big, to increase the portion of weight that will not decay and can permanently store carbon.

It used to be supposed in the 1950s and 1960s, that the 21st century would be an age of atomic energy and space colonization, so the environmental problems and energy crisis would not exist on Earth. Such estimations were reasonable, since the development of atomic technology and space technology were the fastest at that time. However, the energy problem has not been solved, and the development of atomic energy and space exploration is much slower than we once estimated. Until now, one of the major problems, which is the difficulty of using nuclear fusion to generate electricity, it is still several decades away from being solved. It is a shame that time estimations have remained constant for fifty years. Instead of having breakthroughs in energy technology and spaceships, computer science and bio-technology are developing at an amazing speed now (Figure 11 [13]). Perhaps, the mission that remained incomplete in the 20th century, causing an energy crisis, global warming and sea-level rise, will be accomplished through "super plants", which can be developed by computer technology and gene technology. Now, the baton of history has shifted to computer science and bio-technology, from nuclear physics and space technology that lost appropriate international environments after the Cold War. The consequence of "Transition Studies" depends on their performances.

Figure 11, Gene Technology

5. References

[1] Climate.nasa.gov. (2017). *Global Climate Change*. [online] Available at: https://climate.nasa.gov/ [Accessed 20 Aug. 2017].

[2] Climate.nasa.gov. (2017). *Global Climate Change*. [online] Available at: https://climate.nasa.gov/ [Accessed 20 Aug. 2017].

[3] Princiotta, F. (2017). *Global Climate Change-the Power Generation Mitigation Challenge (Powerpoint)*. Yantai, China, p.3.

[4] G. R. van der Werf, D. C. Morton, R. S. DeFries, J. G. J. Olivier, P. S. Kasibhatla, R. B. Jackson, G. J. Collatz and J. T. Randerson. (2009). CO2 emissions from forest loss. *Nature Geoscience*, 2(737–738; 2009), p.737.

[5] Picture provided by Baidu Images, from *http://image.baidu.com/*

[6] Picture provided by Baidu Images, from *http://image.baidu.com/*

[7] Picture provided by Baidu Images, from *http://image.baidu.com/*

[8] Martin J.H, Fitzwater S.E, Nature, 1988, 331(6154):341

[9] Picture provided by Google Images, from *https://www.google.co.jp/*

[10] Retrieved 20 August 2017, from *http://theearthproject.com/wp-content/uploads/2016/01/Tidal-Power-Diagram.gif*

[11] Zhang Jihong, Fang Jianguang, Tang Qisheng, Ren Lihua, Carbon sequestration rate of the scallop Chlamys Farreri cultivated in different areas of Sanggou Bay. *Progress In Fishery Sciences*, *34*(1), 14, from *http://www.doc88.com/p-7018374947275.html*

[12] Analysis *of energy consumption and consumption of chemical fertilizer production in China*. (2017). *Wenku.baidu.com*. Retrieved 20 August 2017, from https://wenku.baidu.com/view/d3d9da8f8762caaedd33d45e.html

[13] Picture provided by Baidu Images, from *http://image.baidu.com/*

IOP Conf. Series: Earth and Environmental Science **120** (2018) 012012 doi:10.1088/1755-1315/120/1/012012

Affecting Factors on Local Waste Management in Penyangkringan Village, Weleri: an Identification

Nadia Puspita Adriyanti[1], Ova Candra Dewi[2], Ahmad Gamal[3], Mohammad Joko Romadhon[4], and Raditya[5]

1 Master Degree Student Dept. of architecture, Universitas Indonesia
2 College Lecturer, Universitas Indonesia
3. Assistant Professor, Universitas Indonesia
4. Master Degree Student, Universitas Indonesia
5. Master Degree Student, Universitas Indonesia
E-mail: nadia.puspita61@ui.ac.id

Abstract. Villages in Indonesia usually does not have proper waste management and it is affecting the environmental and social condition in those places. Local governments have been trying to implement many kinds of solid waste management systems and yet many of them does not bear fruit. We argue that the failure of the waste management implementation in Indonesian villages is due to several aspects: the geographic condition of the villages, the social conditions, and the availability of facilities and infrastructures in those villages. Waste management should be modeled in accordance to those three aspects.

1. Introduction

Solid waste is a recurring environmental problems which affects community livelihood. Penyangkringan, which is located in sub-district Weleri, Central Java, is also one of many local communities that have been affected by the absence of suitable solid waste management program for many years. About 39% of drainage networks in Penyangkringan is in bad condition because of the solid wasted was disposed carelessly in to the drainage (RPLP Desa Penyangkringan 2017). Not only it will affect people's health who live there but it will also affect the environment by indirectly causing flood and it is because the clogged drainage cannot keep the rain water. The local government has been trying to solve the solid waste problems by implementing various solid waste management plan on the Penyangkringan, such as implementing solid waste recycling program and giving free garbage transports. But, the programs cannot be able to solve, even reduce, the solid waste management problem in Penyangkringan.

Penyangkringan is characteristically a typical Indonesian village. First, Penyangkringan covers agricultural area (rice fields), commercial area, and residence area. On the downtown area, Penyangkringan has not only one but two traditional markets that act as the center of economy activities in sub-district Weleri. They sell not only daily necessities but also other things like livestock products and electronics. Second, Penyangkringan has different type of road networks. The road that connects houses and the main road have big difference in size and characteristics. Even the narrower roads that connect houses have different size in each neighborhood. Because of these characteristics, local government is trying to implement different kind of solid waste transportations, in accordance with the type of the road.

IOP Conf. Series: Earth and Environmental Science **120** (2018) 012012 doi:10.1088/1755-1315/120/1/012012

We identify factors hampering local waste management in Penyangkringan based on three analytical dimensions: road network, gender roles in local social institution, and the availability of facilities and infrastructures.

2. Literature Review

Waste management in developing countries such as Indonesia has many difficulties compared to more developed countries. Furthermore, the solid waste generation rate in developing countries is higher compared to the developed countries and it is because of the rapid development of urban population and industries (Boadi, 2005). According to Ogawa (2005), there are several constraints that affecting waste management in developing countries: technical constraint which is about the lack of human resources, especially technical expertise, in national and local level and also the lack of research and development activities regarding waste management. Developing countries usually do not have stable economic and industrial development so there is no sufficient fund for development of sustainable solid waste management system. Social aspect also takes part in constraining waste management. Collaborative projects from the government to encourage community to take part in developing waste management often fail because usually there are not enough people to take part in the project and the incentive is too small.

Informal sector in solid waste management prevails because of those constraints. The lack of service provided by the governments is seen by people who have no skill nor proper education as chances to get money for a living. The informal recycling sector can also bring economic benefits to developing countries as they minimize capital expenditures and maximize man power (Haan et al. in Wilson, 2006). However, the informal sector in solid management can affect health of the people working in that sector such as exposed to toxic, infectious components, and gaseous emissions. (Cointreau in Wilson, 2006). To decrease the negative effect, there's a need to integrate informal sector to formal solid waste management sector by helping organize the informal sector workers and adding value to the recycled materials before selling them (Wilson, 2006).

Aside from integrating informal sector to formal sector, another strategy for successful implementation of solid waste management is to increase degree of youth participation and including their ideas, skill, and experience (Srivastava, 2005). Women participation in solid waste management is also important due to the nature of solid waste management as house chores and it should be taken care of by women (Abadi, 2013). Woman's activities that related to solid waste management are: managing traditional environment, rehabilitators of the domestic environment, innovators in the use of new more appropriate technology, and protectors and caretakers of domestic environment (Mahapatra in Davies, 2016)

Solid waste management studies need to look at the integration between several aspects. Social structure and behavior can affect the availability of solid waste management infrastructure and how a transportation path for waste transport is determined. It is also can affect the type of waste generated in household or commercial area. Lack of research about solid waste management in Indonesia villages can lead to inappropriate technology and model of solid waste management applied in those area. The technology and models of solid waste management that are taken from developed countries might not work if applied to local areas, that is the reason why local studies is needed to make wise decision making in solid waste management.

3. Methodology

3.1. Literature review

Review of relevant literatures related to waste management in developing countries and informal waste management.

3.2. Field survey

Field survey consists of two kind of data collecting:

- Observation

Observation was done to confirm solid waste treatment in Desa Penyangkringan. Observing several locations in Penyangkringan and taking photos of solid waste dumped on the side-road, on the market, and on the drainage. Observation took 2 days.

- Interview

To gain information about social conditions in Penyangkringan, there is interview with several important figures in Penyangkringan. Those important figures include KOTAKU facilitators, BKM Barokah's chief, Penyangkringan's village-chief, and Weleri sub-district's head-chief.

3.3. Data analysis

Analyzing data from field survey and compare the data with the literature review.

4. Result and discussion

4.1. Geographic conditions

Penyangkringan has an administrative area of 1.78km2. It has a large agricultural land, almost 55.49% of land in Penyangkringan Village is land for rice fields (RPLP Desa Penyangkringan 2017). The soil in the percolation village is partly a sedimentary soil, also called alluvial soil, which is the soil formed from the finer material resulting from the deposition of river flows in the lowlands. Penyangkringan Village is located in the lowlands. The average temperature in the village is 20^0C. (Central Bureau of Statistics Kendal 2011).

Penyangkringan and other neighbor villages are separated by highways. Penyangkringan also has small river that stretches from South part of Penyangkringan to the East part. On the West side, Penyangkringan adjacents to rice fields.

4.2. Social conditions

According to 2017 survey data at Penyangkringan, this area has decent population of people in productive age group (ranging from 15-64 years old, according to Organization for Economic Co-operation and Development). Around 70% population of Penyangkringan is in productive age group (KOTAKU, 2017). The ratio between male and female population in Desa Penyangkringan is almost balance. According to KOTAKU's data, 29% of the female population in Desa Penyangkringan works as housewife, which means they are the one who usually responsible for taking care of household matter, including solid waste management. Empowering housewives to become part in the decision making in solid waste management can help the system to grow in accordance with the social condition in Desa Penyangkringan.

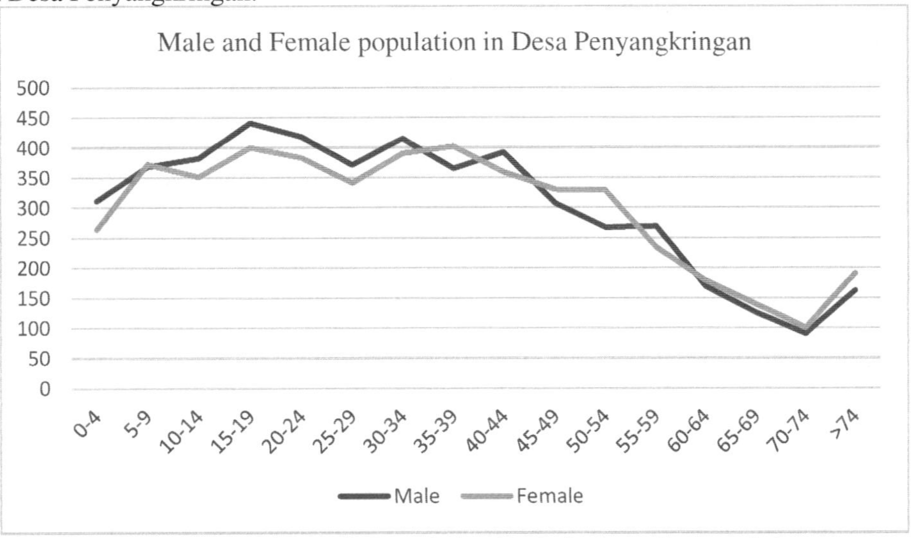

Table 1 Male and Female Population Ratio in Desa Penyangkringan
data source: KOTAKU Facilitator (2017)

IOP Conf. Series: Earth and Environmental Science **120** (2018) 012012 doi:10.1088/1755-1315/120/1/012012

* KOTAKU is national government program to develop suburban areas

Penyangkringan has self-help community organization called BKM Barokah. BKM is the abbreviation of *Badan Keswadayaan Masyarakat*. BKM Barokah helps people in Penyangkringan by channeling their aspirations to local government through discussions and proposals. However, BKM Barokah do not gather together very often to have internal discussion and coordination with regard to village planning. Other than BKM Barokah, Penyangkringan people can still gathered together through other social activities like local social gathering (*arisan*) and Moslems routine activities in learning *Quran* together. The community and its social programs can boost people awareness about solid waste management and how to do it.

4.3. Environment conditions

Most of the Penyangkringan neighborhoods have not yet undertaken waste management as environmental threat. Most of the waste is dumped into the rivers and vacant lots, and some of them are partially burned in the backyard of their homes. Penyangkringan Village has an informal temporary garbage dump located in *Pasar Kidul* area (also called Weleri II Market). This area has been used as a temporary disposal for more than 10 years, but at the same time, it is still used for buying and selling activities on certain Javanese days, such as on *Kliwon* and *Legi* days—part of Javanese calender days's name that consists of five days, *Kliwon, Legi, Pahing, Pon*, and *Wage.*

Penyangkringan has a waste recycling building located in *Pasar Kidul* area. This building has a waste recycling facility for composting and has been operating for 1.5 years since 2014 and it also has one garbage transport to pick up and deliver garbage from several Penyangkringan neighborhoods. However, the waste recycling facilities was deserted by the employee because the income from selling the fertilizer and the salary were deemed very low and not enough for a living. *Pasar Kidul* also has temporary waste collection bin which is usually transported to waste landfills located on the other village by local government's garbage trucks every 2-3 days.

4.4. Road networks

There are different kind of roads in Penyangkringan, some roads can be accessible to trucks and bus, others can only be accessible to motorcycle. Because of that, Penyangkringan needs different kind of garbage transportations to be able to pick up and deliver garbage from houses or from temporary dumpsters from each neighborhood.

There are two kind of garbage transportations used to take solid waste from houses, garbage carts and garbage motorcycle. Both has their own advantages and disadvantages.

4.5. Social conditions

Penyangkringan people understood the effect of poor solid waste management to their environment and health but they keep littering the road and river. Sometimes they burn the garbage at their backyard. It happens because for them, it is easier to just throw away the garbages and burn the rest rather than actually managing them. There is also no community garbage cleaning activity that is usually held weekly or monthly by villages. Looking at the total population of productive age group, it is possible to organize weekly community garbage cleaning activity in each neighborhood in Penyangkringan.

4.6. Facilities and infrastructures

There are some problems faced by Penyangkringan associated with waste management, the first is the lack of trash bin in the roadside area. The area along the main road does not have enough trash bins, so it indirectly encourages people to litter. As a result, not only solid waste is accumulating at certain areas, but also clogging the drainage channels. In addition to blockage of drainage channels, the solid waste is also piling up on *Pasar Kidul*, where the place is not supposed to be used as a waste disposal site while Penyangkringan. The solid waste has piled up high enough that it makes the cost of moving the waste to the landfill area becomes expensive.

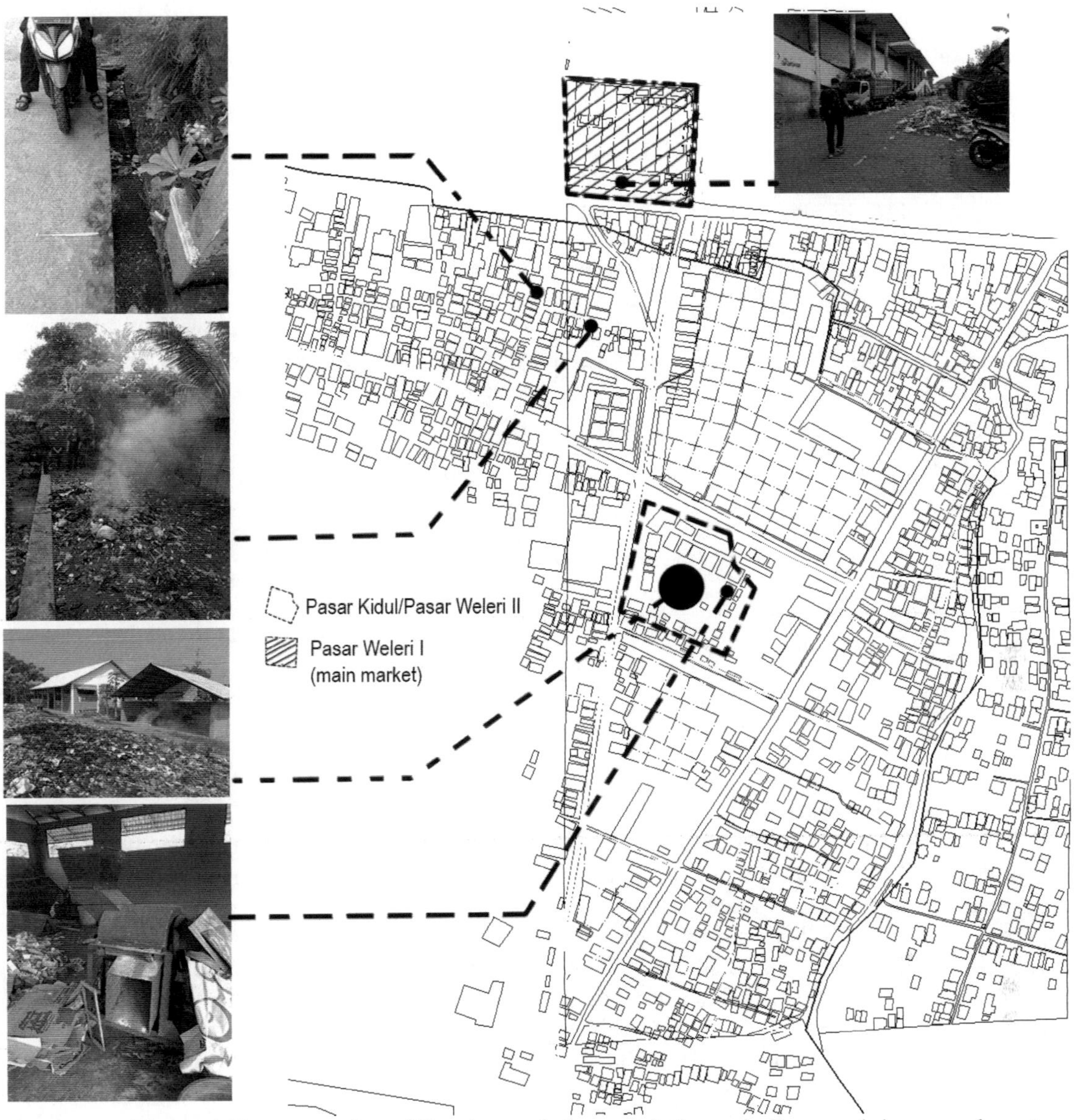

Figure 1 Few examples of littering and clogged drainage in Penyangkringan and waste recycling facility in Pasar Kidul
source: KOTAKU RPLP map 2017, with changes

The waste recycling facility provided by the government to reduce solid waste and to empower local community is stopped because the amount of salary and the income from selling the fertilizer were deemed not enough for a living. There were supposed to be five employee working at the waste recycling facility but all of them resigned after working for around 1.5 years, from 2014 to 2015. The salary for the employee was Rp 800.000, which was below the standard salary in Kendal. The standard salary for employee in Kendal in 2015 was Rp 1.383.000. One of the ex-employee is now working as scavenger as he sees it to be more profitable compared to his old job. It is assumed that the waste recycling facility is not supported by other local industries. Producing fertilizer from solid waste will be profitable if the location of the seller or other local industries are near the waste recycling facility so that the transportation cost can be reduced.

IOP Conf. Series: Earth and Environmental Science **120** (2018) 012012 doi:10.1088/1755-1315/120/1/012012

Garbage Cart	Garbage Motorcycle
Garbage cart is the cheapest garbage transportation since it only uses man-power. The smallest garbage cart's capacity is 1m3. However, since it is operated only by man power, it is not time-effective. It is slow and the operator might need to take a break for a few times especially if they have to cover wide area. It is more effective to use garbage cart to serve in smaller neighborhood, taking the trashes and the put it in the designated temporary dumpster, from there the trashes can be picked by bigger vehicles such as truck.	Garbage motorcycle can serve more houses in one day because it uses engine power which is faster than garbage cart. But It should be noted that garbage motorcycle needs gasoline per 100 km, the average use of gasoline for motorcycles is 3L / 100km. The cost will be more expensive because the need to put gasoline price into the budget and the impact on the environment is greater compared garbage cart because the gasoline will emit gas emission out into the air. Also, the garbage motorcycle operator should be someone with skill to operate special type of motorcycle. Local government in Sub-District Weleri has provided one garbage motorcycle to each villages to help them taking trashes from each houses. But since Penyangkringan area is big, it needs more than one garbage motorcycle to take trashes from each house. On the main road, garbage truck is used to take trashes from temporary dumpster and then transport them to the landfill area, which is located on Pagergunung village.

5. Conclusion and suggestion

In conclusion, the connection between road networks, social conditions/gender roles, and facilities and infrastructures affected solid waste management in Penyangkringan. It is important to see solid waste management as a whole system that includes many aspects, starting from the generation, treatment and disposal. To implement solid waste management system to Penyangkringan, it is not only by providing facilities and infrastructures but it is more urgent to manage the local communities in Penyangkringan, encouraging them to understand the importance of managing solid waste in their neighborhood (the capacity building). With the help of self-help community organization, BKM Barokah, the neighborhoods in Penyangkringan can organize weekly community garbage cleaning activity in each neighborhood. Monthly social gatherings such as *arisan* and *pengajian* can also be used as media to educate people how to manage their household waste.

6. References

[1] Abadi, Ronny Setiawan. 2013. Biro Penerbit Planologi UNDIP. *Keberlanjutan Pengelolaan Sampah Domestik di Kampung Menoreh, Kelurahan, Sampangan, Semarang*. Vol 9. Pg 87-96.

[2] Bernache, Gerardo. 2003. Resources, Conservation and Recycling. *The Environmental Impact Of Municipal Waste Management: The Case Of Guadalajara Metro Area*. Vol 39. Pg 223-237.

[3] Boadi, Owusu Kwadi and Markku Kuitunen. 2005. Journal of Environmental Health. *Environmental and Health Impacts of Household Solid Waste Handling and Disposal Practices in Third World Cities: The Case of the Accra Metropolitan Area, Ghana*. 68,4, ProQuest. Pg 32.

[4] Davies, Nyatsanza Taurai and Ndebele Sharon Kudzai. 2016. IOSR Journal of Humanities and Social Science. *The Usefulness of Including Women in Household Solid Waste Management: A Case Study of Dzivaresekwa High Density Suburb, Harare*. Vol. 21. Pg 92-108

[5] Ogawa, Hisashi. 2005. *Sustainable Solid Waste Management in Developing Countries*. http://www.gdrc.org/uem/waste/swm-fogawa1.htm

[6] *Rencana Penataan Lingkungan Permukiman Desa Penyangkringan Kota Tanpa Kumuh* (RPLP KOTAKU). 2017.

[7] Srivastava, P.K. et al. 2005. Waste Management. *Stakeholder-based SWOT analysis for successful municipal solid waste management in Lucknow, India.* Vol. 25. Pg. 531-537
[8] Wilson, David C., Costas Velis, Chris Cheeseman. 2006. Habitat International. *Role of Informal Sector Recycling in Waste Management In Developing Countries.* Vol 30. Pg 797–808.

IOP Conf. Series: Earth and Environmental Science **120** (2018) 012013 doi:10.1088/1755-1315/120/1/012013

Multiple Statistical Models Based Analysis of Causative Factors and Loess Landslides in Tianshui City, China

Xing Su[1, 2], Xingmin Meng[1], Weilin Ye[1, 2], Weijiang Wu[2], Xingrong Liu[2], and Wanhong Wei[2]

1 Key Laboratory of Western China's Environmental Systems(Ministry of Education), College of Earth and Environmental Sciences, Lanzhou University, Lanzhou 730000, China
2 Geological Hazards Prevention Institute, Gansu Academy of Sciences, Lanzhou 73000, China
E-mail: geocity@163.com

Abstract. Tianshui City is one of the mountainous cities that are threatened by severe geo-hazards in Gansu Province, China. Statistical probability models have been widely used in analyzing and evaluating geo-hazards such as landslide. In this research, three approaches (Certainty Factor Method, Weight of Evidence Method and Information Quantity Method) were adopted to quantitively analyze the relationship between the causative factors and the landslides, respectively. The source data used in this study are including the SRTM DEM and local geological maps in the scale of 1:200,000. 12 causative factors (i.e., altitude, slope, aspect, curvature, plan curvature, profile curvature, roughness, relief amplitude, and distance to rivers, distance to faults, distance to roads, and the stratum lithology) were selected to do correlation analysis after thorough investigation of geological conditions and historical landslides. The results indicate that the outcomes of the three models are fairly consistent.

1. Introduction

Tianshui, a city in the southeastern part of Gansu Province, located in the south of the Longzhong Loess Plateau and north of the Qinling Mountains. The Weihe River and its tributaries cross through the Tianshui City from west to east. Thus, Tianshui City has a typical valley basin landform. Special geological conditions in the region lead to frequent geological disasters. Historically, geological disasters have caused serious casualties and economic losses. For example, on July 12, 1978, the rainstorm in Bo Yang caused a number of loess landslides, debris flow. Not only killed 7 people, but also buried Baiyang railway station and interrupted Baotian railway traffic 360h [1]. In the autumn of 1984, rainfall triggered 88 landslides in the Tianshui City and its surrounding areas, railways and highways were seriously threatened [2]. On August 11, 1990, rainstorm in Tianshui City induced dozens of landslides, 22 people were killed, and 7 of whom were killed by one landslide. Besides, in Qinzhou District, Jiaoshuwan landslide, Taishanmiao landslide deformed intensively threated lives and property of 7,780 citizens [3]. From June 19 to July 26, 2013, a total of four heavy rainstorms in Tianshui area led to extensive geological disasters which resulted in 24 deaths [4]. The landslides in Dagou on July 21 [5], Yushu Village on July 29 [4], and Caowang village on July 29 [6] are all typical loess landslide caused by this round of heavy rainstorms.

Due to the lack of detailed landslide cataloging and high accuracy of the data, studies regarding spatial correlation of the loess landslide and the disaster factors of Tianshui City are rare. With the development of 3S (GIS, GPS and RS) technology, spatial correlation of landslide and causative factors based on statistics and probability analysis is becoming well developed. Various approaches and models have been proposed and popularized, such as certainty factor (CF) [7], weights-of-evidence (WOE) method [8], information quantity (I) method [9], conditional probability method [10] and so forth. Therefore, investigation of loess landslide in Tianshui City area (Qinzhou District and Maiji District) using statistical models (i.e., certainty factor (CF) method, weights-of-evidence (WOE) method, and information quantity method (I)) is of great theoretical and practical significance to further understand the formation mechanism of landslides and carry out effective risk assessment in this area. It will complement and improve the current relevant studies in the Tianshui City area, as well as providing scientific basis and technical support on landslide hazard prevention and management decision-making.

2. Method

From the spatial scale, research on landslide hazard could be divided into regional studies and monomer studies. Regional studies mainly focus on the analysis of causative factors, susceptibility, risk, and risk assessment of the landslide hazard in specific areas. All these analysis were actually based on the correlation of landslide in the study area and causative factors. Therefore, analysis of potential causative factors is the key to evaluate the success or failure of landslide susceptibility and hazard risk.

2.1. Certainty factor method

Certainty Factor method [7] used as a probability function for landslide hazard analysis. The basic assumption is that the landslide will occur in the future if the geological conditions of a certain area are the same or similar to those of the landslide in the past, which could be shown as:

$$CF = \left\{ \begin{array}{ll} \dfrac{PP_a - PP_s}{PP_a(1-PP_s)} & if \quad PP_a \geq PP_s \\ \dfrac{PP_a - PP_s}{PP_s(1-PP_a)} & if \quad PP_a \leq PP_s \end{array} \right\} \qquad (1)$$

Where, in the formula (1), PP_a is the conditional probability of occurrence of event (landslide) in data class a, which can be expressed as the ratio of landslide area to unit area in the unit representing data class a when the landslide is applied. PP_s is the prior probability of occurrence of the event throughout the study area A and can be expressed as the ratio of landslide area to study area in the whole study area.

CF value change in the interval of [-1, 1]. If CF value is positive, representing the event of deterministic growth, that is, the landslide occurrence of certainty. In other words, this unit is prone to be a landslide area. While negative CF value represents the certainty of reduction, that is, the certainty of landslide occurrence is low (not prone to landslides). If CF value is close to 0, which means the prior probability and the conditional probability are very close (the certainty of event occurrence cannot be determined). In short, the unit cannot determine whether or not it is a landslide prone area.

Using the spatial analysis module of ArcGIS 10, the number of landslide units and the number of factor units in all the factors were calculated, and the CF values of the factor layers were calculated, subsequently.

According to Eq. (2), the influence and susceptibility of each factor to landslide are calculated.

$$S_{CF} = CF_{(i,\max)} - CF_{(i,\min)} \qquad (2)$$

Where, S_{CF} is overall susceptibility of the landslide index for each factor. $CF_{(i,\max)}$ is the maximum value of the certainty factor of landslide for each factor i; $CF_{(i,\min)}$ is the minimum value of the certainty factor of landslide for each factor i. The higher the S_{CF}, the higher probability of the response and susceptibility of the factor to the landslide.

2.2. Weights-of-evidence method

Based on the Bayesian statistical model, Weights-of-evidence method is originally applied in the evaluation of mineral resources reserves in earth science [11]. In recent years, this method has been widely used in landslide susceptibility evaluation [12]. In Landslide hazard susceptibility and risk

mapping, weight of evidence method is be used to calculate the weight for each landslide causative factors based on the two cases of landslides (i.e., existence and absence). The basic assumptions of this approach are: 1) the conditions of future landslides are similar to those that favor landslide occurrence in the past; and 2) the causative factors for landslide mapping do not change with time.
According to the Bayesian rule, the probability of occurrence of landslide and the conditional probability function expressed by equation (3) under the condition of evidence factor F:

$$P\{L|F\}=\frac{P\{L\cap F\}}{P\{F\}}=\frac{N\{L\cap F\}}{N\{F\}} \quad \ldots\ldots\ldots\ldots\ldots\ldots\ldots(3)$$

Namely,

$$P\{F|L\}=\frac{P\{L\cap F\}}{P\{L\}}=\frac{N\{L\cap F\}}{N\{L\}} \quad \ldots\ldots\ldots\ldots\ldots\ldots\ldots(4)$$

Formula (5), (6) and (7) are used to calculate the evidence of landslide occurrence. $P\{F|L\}/P\{F|\overline{L}\}$ represents the adequacy of landslide occurrence; $P\{\overline{F}|L\}/P\{\overline{F}|\overline{L}\}$ represents the necessary rate of landslide occurrence. W^+ indicates the probability of a landslide at the current causative factor level, and the size indicates the positive correlation between the causative factor level and the landslide. W^- represents the probability of a landslide occurring outside of the causative factor level, which represents the negative correlation coefficient of the landslide within the causative factor level. The difference (W_c) represents the weight of the landslide within the causative factor level.

$$\mathbf{W}_i^+=\ln\left(\frac{P\{F|L\}}{P\{F|\overline{L}\}}\right) \quad (5) \qquad \mathbf{W}_i^-=\ln\left(\frac{P\{\overline{F}|L\}}{P\{\overline{F}|\overline{L}\}}\right) \quad (6) \qquad \mathbf{W}_c=\mathbf{W}_i^+-\mathbf{W}_i^- \quad (7)$$

The following formulas (8) and (9) are given by the formulas (5) and (6):

$$\mathbf{W}_i^+=\ln\left(\frac{\frac{N_1}{N_1+N_2}}{\frac{N_3}{N_3+N_4}}\right) \quad \ldots\ldots\ldots\ldots(8) \qquad \mathbf{W}_i^-=\ln\left(\frac{\frac{N_2}{N_1+N_2}}{\frac{N_4}{N_3+N_4}}\right) \quad \ldots\ldots\ldots\ldots(9)$$

Where N_1 represents the number of grids in which the landslide occurred within the stage of the factors; N_2 represents the number of grids that landslide occurred outside the stage of the factors; N_3 represents the number of grids that landslide did not occurred within the stage of the factors; and N_4 represents the number of grids that landslide did not occurred outside the stage of the factors.
The weight (W_c) reflects the importance of the factor to the landslide. Positive total weight indicates that the level is favorable for the occurrence of the landslide, while negative total weight indicates that it is not conducive to the occurrence of the landslide. The correlation of landslides is very small if total weight is close to 0.
Using the spatial analysis module of ArcGIS 10, the number of landslide units and the number of factor units of all grades in each factor as well as W^+, W^- and W_c were calculated.
According to Eq. (10), the influence and susceptibility of each factor to landslide are calculated.

$$S_W=W_{(i,\max)}-W_{(i,\min)} \qquad (10)$$

Where, S_W is the overall susceptibility index of the landslide for each factor; $W_{(i,\max)}$ is the maximum value of weights-of-evidence for each category of factor i; $W_{(i,\min)}$ is the minimum value of weights-of-evidence for each factor i. The higher the S_W, the higher probability of the response and susceptibility of the factor to the landslide.

2.3. Information quantity method

The information quantity method [9] assumed that the occurrence of landslide hazards correlated with the amount of information obtained during forecast. Equation of the amount of information $I(H,\ x_i)$ provided by each factor x_i to the landslide occurrence event (H) can be seen below:

$$I(H,x_i)=\ln\frac{P(H|x_i)}{P(H)}=\ln\frac{N_i/S_i}{N/S}=\ln\frac{N_i/N}{S_i/S} \quad (11)$$

Where $I(H, x_i)$ is the information value of the unit, $P(H|x_i)$ is the probability of occurrence of landslides in the condition x_i, $P(H)$ is the probability of landslides in the study area, S is the total number of landslides in the study area, N is the total number of landslide units, S_i is the evaluation area Factor x_i, N_i is the number of landslide units distributed within the factor x_i.

Using the spatial analysis module of ArcGIS 10, the number of landslide units, the number of factor units of all grades in each factor, and the I value of each factor were calculated.

The influence and susceptibility of each factor to landslide were calculated using equation (12) below:

$$S_I = I_{(i,\max)} - I_{(i,\min)} \quad (12)$$

Where S_I is the overall susceptibility index of the landslide for each factor; $I_{(i,\max)}$ is the maximum value of the landslide information value for each category of factor i; $I_{(i,\min)}$ is the minimum value of landslide information for each factor i. The higher the S_I, the higher probability of the response and susceptibility of the factor to the landslide.

3. Study area

The study area, Tianshui City, is located in southeastern Gansu Province. The geographical location of Tianshui City is: longitude 105° 13'15 "~ 106° 42'58" E, latitude 34 ° 5'5 "~ 34 ° 49'40" N with a total population of 1.3 million and a total area of about 5,833 km^2. The area of the two major districts of Tianshui City (i.e., Qinzhou and Maiji District) is about 2349 km^2 and 3484 km^2, respectively (Figure 1). The population density is around 223 person / km2. Tianshui is a typical populous city with fragile geological environment. Tianshui City is a high incidence area of landslide disaster, which seriously threatens people's life and property safety.

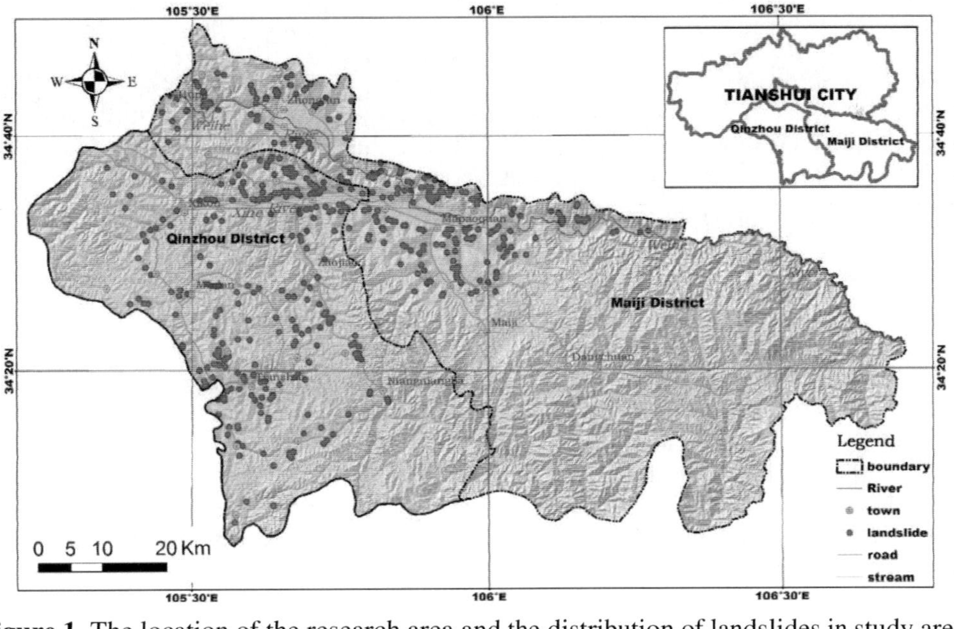

Figure 1. The location of the research area and the distribution of landslides in study area.

4. Data preparation

The data used in this study including detailed landslide inventory and high-accuracy causative factors data. Landslide inventory were obtained from existing documents, remote sensing images and field surveys. High-accuracy causative factors data were derived from SRTM DEM data at 30m resolution,

1: 200,000 regional geological map and google earth. The characteristics and causes of loess landslide disasters in the study area were identified after systematic summary and investigation. Causative factors of landslide development were selected based on identified characteristics and origin. The above typical middle - scale landslides with area greater than $10,000m^2$ and landslide length greater than 100 m in the study area is emphasized.

4.1. Landslide inventory

A detailed and reliable database of landslide spatial distribution in the study area was established by means of analysis and discrimination of existing landslide hazard research data, interpretation of remote sensing images and verification of field investigation. The data of 475 landslide hazard points were acquired, including 257 in Qinzhou District and 218 in Maiji District (Figure 2). In the study area, the area of 470 landslides are larger than 10,000 m^2, accounting for 99% of all landslides. In general, the selected landslide hazards meet the requirements of sampling. The total landslide hazard area is 89 km2, and the landslide vector map is converted to raster at a grid size of 30m × 30m. A total of 6,481,583 grid cells were obtained after rasterization with 98,968 landslide grid cells which took around 1.5% of the total area. The density of the landslide is 0.1/ km^2.

Figure 2. Pictures of the typical landslides in study area.

4.2. Data of environmental factors

A total of 12 causative factors were categorized into three classes (i.e., geomorphological factors, geological factors, and human activity factors) selected in this study. Geomorphological factors are altitude, slope, aspect, curvature, plan curvature, profile curvature, roughness, relief amplitude and distance to rivers (Figure 3a-3h). The geological factors include distance to faults, and the stratum lithology; and human activity factor is the distance to roads (Figure 4a-4d). The geomorphological factors were obtained from DEM data of 30m resolution in the study area by using ArcGIS software. The geological factors are based on 1: 200,000 regional geological map, using GIS tools to transform the raster data. Human activity factors were subsequently obtained through the google earth map, and then use ArcGIS 10 software to do more processing and statistical analysis (Table 1).

5. Results and discussion

5.1. Susceptibility analysis of landslide and classification factors

According to the data of the landslide and the cause of disaster in the study area, the CF, Wc, and I of the grading factor layers in the 12 factors were calculated, respectively. From the results in Table 2, among the 12 landslide causative factors in the region, 1200m ~ 1400m above sea level, 10 ° ~ 15 °

slope, southwest slope, -0.5 ~ 0.5 curvature, -0.5 ~ 0.5 plan curvature, -0.5 ~ 0.5 profile curvature, 1 ~ 1.05 roughness, 60 ~ 120 m relief amplitude, Quaternary Upper Pleistocene (Q_3) Malan loess and Neogene (N) mudstones in the stratum lithology, 4000m ~ 5000m away from the fault, 1200m ~ 1600m away from the river and 500m ~ 1500m away from the road are the advantage factor interval of the loess landslide hazard.

Table 1. Loess landslides 12 causative factors of Tianshui City

Data source	Factors	Classification of factors
SRTM DEM in the resolution of 30m	Altitude	<1000 m、1000 ~ 1200 m、1200 ~ 1400 m、1400 ~ 1600 m、1600 ~ 1800 m、1800 ~ 2000 m、2000 ~ 2200 m、2200 ~ 2400 m、>2400 m
	Slope	0 ~ 5°、5 ~ 10°、10 ~ 15°、15 ~ 20°、20 ~ 25°、25 ~ 30°、30 ~ 35°、35 ~ 40、>40°
	Aspect	Flat, north、Northeast, Southeast, South, South West, West, North West
	Curvature	<-4、-4 ~ -2、-2 ~ -0.5、-0.5 ~ 0、0 ~ 0.5、0.5 ~ 2、2 ~ 4、>4
	Plan curvature	<-2、-2 ~ -1、-1 ~ -0.5、-0.5 ~ 0、0 ~ 0.5、0.5 ~ 1、1 ~ 2、>2
	Profile curvature	<-2、-2 ~ -1、-1 ~ -0.5、-0.5 ~ 0、0 ~ 0.5、0.5 ~ 1、1 ~ 2、>2
	Roughness	1 ~ 1.05、1.05 ~ 1.1、1.1 ~ 1.15、1.15 ~ 1.2、1.2 ~ 1.25、1.25 ~ 1.3、1.3 ~ 1.35、1.35 ~ 1.4、>1.4
	Relief amplitude	<30 m、30 ~ 60 m、60 ~ 90 m、90 ~ 120 m、120 ~ 150 m、150 ~ 180 m、180 ~ 210 m、210 ~ 240 m、>240 m
	Distance to rivers	<400 m、400 ~ 800 m、800 ~ 1200 m、1200 ~ 1600 m、1600 ~ 2000 m
1: 200,000 regional geological map	Stratum lithology	Quaternary Holocene (Q_4), Quaternary Upper Pleistocene (Q_3), Neogene (N), Paleogene (E), Jurassic (J), Carboniferous (C), Devonian (D), Paleozoic (P_{Z1}), Sinian (Z), anterior Sinian (AnZ), granite, diorite (γ)
	Distance to faults	<1000 m、1000 ~ 2000 m、2000 ~ 3000 m、3000 ~ 4000 m、4000 ~ 5000 m、5000 ~ 6000 m、6000 ~ 7000 m、7000 ~ 8000 m、>8000 m
Google earth images	Distance to roads	<500 m、500 ~ 1000 m、1000 ~ 1500 m、1500 ~ 2000 m、2000 ~ 2500 m、2500 ~ 3000 m、>3000 m

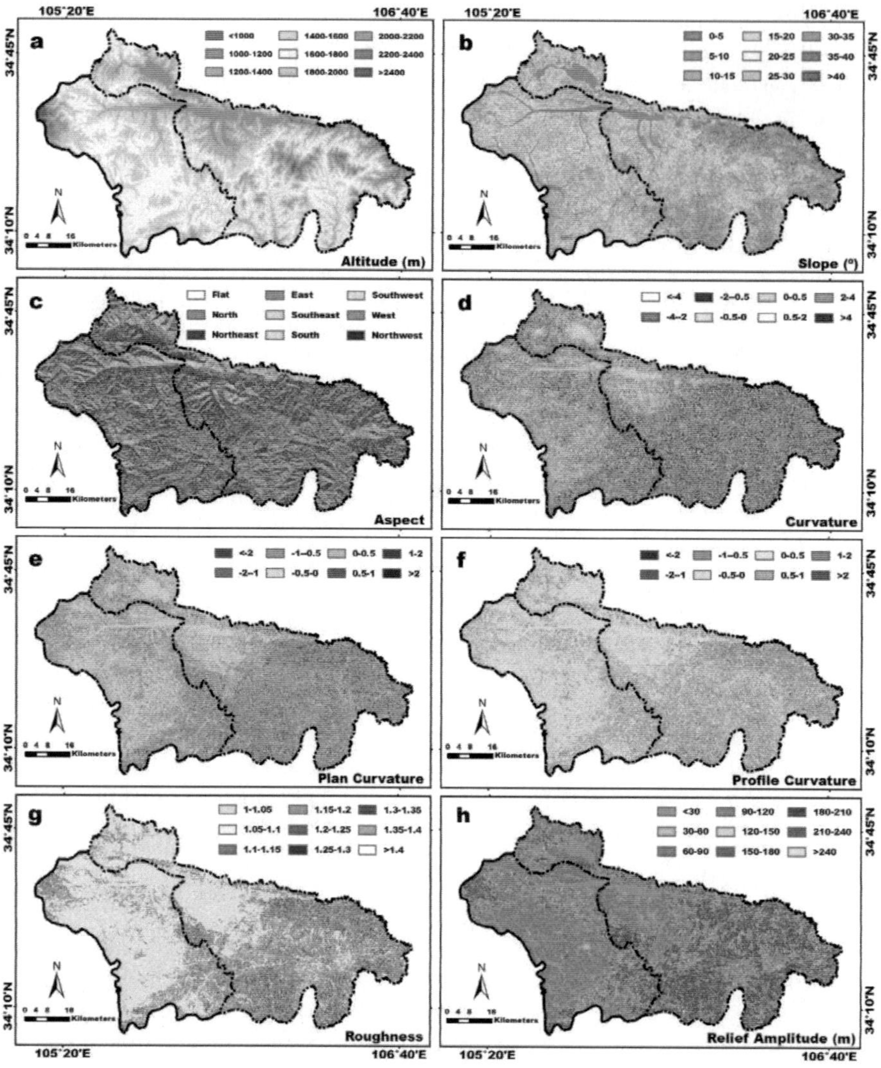

Figure 3. The grading map of various geomorphological factors in the study area.

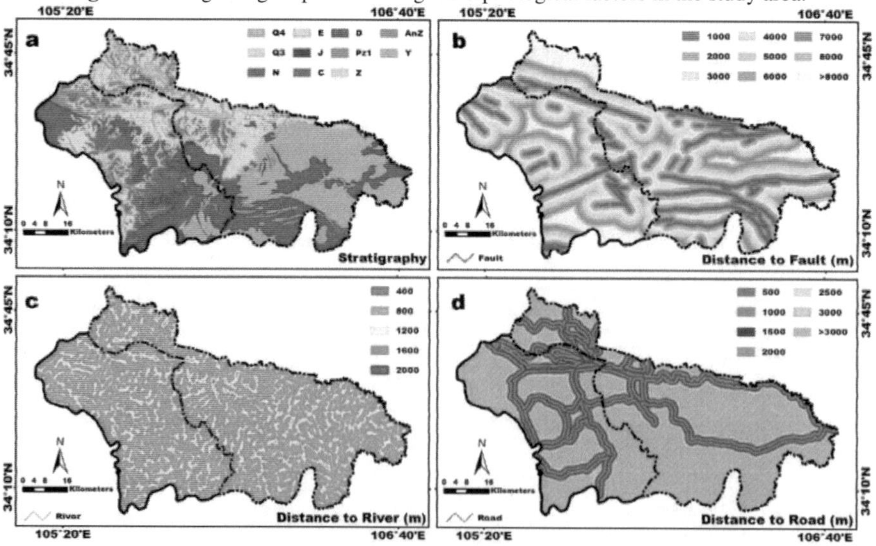

Figure 4. The grading map of various geological factors in the study area.

Table 2. Statistics of 12 causative factors of loess landslide in Tianshui City

Factors and its grades	CF	W_c	I	Factors and its grades	CF	W_c	I
Altitude				**Slope**			
1: <1000	-0.97	-3.59	-3.57	1: 0-5	-0.26	-0.32	-0.30
2: 1000-1200	0.40	0.56	0.50	2: 5-10	0.48	0.76	0.64
3: 1200-1400	0.69	1.56	1.13	3: 10-15	0.51	0.96	0.71
4: 1400-1600	0.16	0.22	0.17	4: 15-20	0.31	0.46	0.36
5: 1600-1800	-0.40	-0.69	-0.50	5: 20-25	-0.34	-0.47	-0.41
6: 1800-2000	-0.94	-2.97	-2.78	6: 25-30	-0.71	-1.35	-1.23
7: 2000-2200	-1			7: 30-35	-0.82	-1.79	-1.68
8: 2200-2400	-1			8: 35-40	-0.86	-2.00	-1.93
9: >2400	-1			9: >40	-0.90	-2.38	-2.34
Aspect				**Roughness**			
1: Flat	-0.70	-1.21	-1.20	1: 1-1.05	0.42	1.31	0.53
2: North	0.10	0.12	0.10	2: 1.05-1.1	-0.14	-0.19	-0.14
3: Northeast	-0.16	-0.20	-0.18	3: 1.1-1.15	-0.70	-1.30	-1.18
4: East	-0.31	-0.41	-0.36	4: 1.15-1.2	-0.80	-1.71	-1.62
5: Southeast	-0.29	-0.38	-0.34	5: 1.2-1.25	-0.84	-1.89	-1.83
6: South	0.15	0.18	0.16	6: 1.25-1.3	-0.87	-2.08	-2.04
7: Southwest	0.20	0.25	0.21	7: 1.3-1.35	-0.90	-2.27	-2.25
8: West	0.09	0.10	0.09	8: 1.35-1.4	-0.90	-2.27	-2.25
9: Northwest	0.16	0.20	0.17	9: >1.4	-0.92	-2.58	-2.56
Curvature				**Plan curvature**			
1: <-4	-0.89	-2.17	-2.15	1: <-2	-0.91	-2.43	-2.41
2: -4--2	-0.82	-1.73	-1.69	2: -2--1	-0.78	-1.57	-1.52
3: -2--0.5	-0.15	-0.21	-0.16	3: -1--0.5	-0.34	-0.46	-0.41
4: -0.5-0	0.29	0.49	0.34	4: -0.5-0	0.24	0.44	0.26
5: 0-0.5	0.26	0.39	0.30	5: 0-0.5	0.20	0.34	0.21
6: 0.5-2	-0.28	-0.41	-0.32	6: 0.5-1	-0.45	-0.67	-0.60
7: 2-4	-0.82	-1.73	-1.69	7: 1-2	-0.78	-1.57	-1.52
8: >4	-0.91	-2.37	-2.35	8: >2	-0.92	-2.56	-2.54
Profile curvature				**Distance to roads**			
1: <-2	-0.86	-2.00	-1.99	1: <500	0.40	0.59	0.50
2: -2--1	-0.75	-1.41	-1.36	2: 500-1000	0.62	1.15	0.94
3: -1--0.5	-0.45	-0.65	-0.59	3: 1000-1500	0.60	1.07	0.90
4: -0.5-0	0.20	0.35	0.22	4: 1500-2000	0.49	0.75	0.66
5: 0-0.5	0.23	0.43	0.26	5: 2000-2500	-0.06	-0.06	-0.06
6: 0.5-1	-0.36	-0.50	-0.45	6: 2500-3000	-0.11	-0.12	-0.11
7: 1-2	-0.75	-1.42	-1.37	7: >3000	-0.79	-2.17	-1.54
8: >2	-0.86	-1.97	-1.95				
Relief amplitude				**Distance to faults**			
1: <30	-0.55	-0.82	-0.79	1: <1000	-0.07	-0.09	-0.07
2: 30-60	0.44	0.63	0.58	2: 1000-2000	-0.10	-0.12	-0.10
3: 60-90	0.50	0.94	0.68	3: 2000-3000	0.12	0.16	0.13
4: 90-120	0.45	0.82	0.58	4: 3000-4000	0.16	0.19	0.17
5: 120-150	-0.31	-0.44	-0.37	5: 4000-5000	0.34	0.46	0.40

6: 150-180	-0.73	-1.42	-1.29	6: 5000-6000	0.01	0.02	0.01
7: 180-210	-0.86	-2.06	-1.94	7: 6000-7000	-0.17	-0.20	-0.18
8: 210-240	-0.92	-2.68	-2.60	8: 7000-8000	-0.74	-1.40	-1.35
9: >240	-0.98	-3.79	-3.73	9: >8000	-0.16	-0.18	-0.17
Stratum lithology				**Distance to rivers**			
1: Q_3	0.59	1.31	0.87	1: <400	-0.08	-0.16	-0.08
2: E	-0.08	-0.09	-0.08	2: 400-800	-0.00	-0.00	-0.00
3: N	0.67	1.47	1.08	3: 800-1200	0.11	0.13	0.11
4: D	-0.95	-3.42	-3.08	4: 1200-1600	0.71	1.26	1.19
5: P_{z1}	-0.28	-0.34	-0.33	5: 1600-2000	-0.92	-2.47	-2.45
6: Y	-1.00	-6.11	-5.80				
7: AnZ	0.58	0.88	0.85				
8: J	-1						
9: Q_4	-0.10	-0.11	-0.10				
10: C	-1						
11: Z	-0.65	-1.04	-1.03				

In the altitude layer, the value which the interval between 1200m and 1400m is the highest, that is obviously higher than other grades, which implies that the factor level between 1200m and 1400m is the advantage interval of landslide development. The bedrock areas for hard rock distributed that the elevation is higher, where are not prone to landslides. The results are in accordance with the results of [12].

In the slope gradient factor layer, the value of the interval between 5° and 20° are all greater than 0, which indicates that this slope range is sensitive to the landslide, which is in accordance with the research of (Wu et al. 2006; 2014) [1], [6]. Besides, the value of 10° ~ 15° in the factor layer is the highest, which is significantly higher than other grades. It shows that the slope of 10° ~ 15° is extremely sensitive to the development of medium-scale landslides.

In the aspect layer, the value of southwest direction is the highest, while the northwest direction is the second highest. The two factor layers are the prone layer of the landslide. The slope aspect determines the hours of receiving sunshine and the intensity of solar radiation, which influences the soil moisture content, the degree of weathering and fragmentation of rock, soil, and the vegetation coverage, thus affecting the formation of landslide.

Regarding all the curvature, plan curvature and profile curvature of the three curvature factors, -0.5 ~ 0.5 factor intervals are prone to factor layer. Further analysis showed that the -0.5 ~ 0 interval of the curvature -0.5 ~ 0 of the plan curvature and the 0 ~ 0.5 interval of the profile curvature are the most sensitive intervals. The curvature of the three groups unanimously reveals the upward concave. The susceptibility of the element to the landslide, i.e., the concave type slope, greatly dominated the development of the landslide above medium scale.

In the roughness layer, the range of 1 to 1.05 is the only positive value interval, which revealed it's the control effect on the landslide in study area. But 1 ~ 1.05 is the lowest in the roughness layer, which indicates that the variation of surface fluctuation is relatively low.

In the relief amplitude factor layer, the value of the 60 ~ 120 m interval is the highest, which becomes the landslide prone area. Analyzing the most susceptible grading layer of the five kinds of factors, such as curvature, plan curvature, profile curvature, roughness and relief amplitude, the landslide prone areas are the lower relief amplitude and lower slope erosion which is closely related to the low-angle loess-mudstone interface landslide developed extensively in the study area.

In the stratum lithology factor layer, the values of the Quaternary Upper Pleistocene (Q_3) Malan loess and the Neogene (N) mudstone are both greater than 1. This factor layer is most sensitive to the development of the landslide in the area, and is well consistent with the common development of the sliding-prone stratum such as the Quaternary Upper Pleistocene Malan loess and the Neogene mudstone of the in the past research.

IOP Conf. Series: Earth and Environmental Science **120** (2018) 012013 doi:10.1088/1755-1315/120/1/012013

Apparently, the fault structure controls the development of landslides, especially the deep and active faults. Fracture structure leads to the concentration of ground stress, rock fragmentation, weathering enhanced, mechanical strength decreased, and active groundwater. From the buffer of fault layer, the total weight value of the buffer zone between 4000m and 5000m is the highest, which is the most sensitive interval of landslide development. Mohammady et al. (2012) [13] revealed that the closer to the fault, the more favorable is the occurrence of the landslide. But in this research, the total weight value of distance to faults between 4000m and 5000m is the most sensitive interval of landslide development. This result is similar to the results of recent researches [8], [12].
The influence of water system on the landslide is mainly caused by the strong erosion including the erosion at the slope feet, at the bank slope, changing the stress state of the slope body, softening the rock mass and so on. In the buffer of river layer, the buffer interval value between 1200m ~ 1600m is the largest, which is the dominant interval range of the landslide. In the past, Kayastha et al. (2012) [14] showed that the closer to the water system, the more favorable the landslide occurred. However, in this research, the results show that the distance from 1200m to 1600m in the buffer zone is the most sensitive to the landslide response, and not the closer to the river, the more sensitive to the landslide. This result is consistent with some other researches [8], [12].
The influence of road on landslide is mainly caused by cutting slope in highway construction, irrational drainage and frequent vibration of passing vehicles. The value of the distance between 500m and 1500m in the distance layer of the road is the largest, which is the easy-to-hit interval of the landslide. Previous studies [13], [15] suggest that the closer to the road, the more favorable is the occurrence of landslide. The results of the study show that the distance from 500m to 1500m is the most sensitive to the landslide response, and Pourghasemi et al. (2013) [8] and Regmi et al. (2010) [12] have got similar results.

5.2. Susceptibility analysis of landslide and integral environmental factors

It could be found from the analysis of above section that the causative factors and susceptibilities of the grading factor layers of 12 factors are different, and their respective interval of advantages can be calculated. For a certain type of factor, the influence on landslide is also different. In this paper, we try to use the S value of landslide susceptibility index to represent the extent of influence of 12 factors on landslide development. According to Eq. (2), Eq. (10) and Eq. (12), the susceptibility index S of the whole 12 factors are calculated, and the results are normalized (Table 3 and Figure 5).

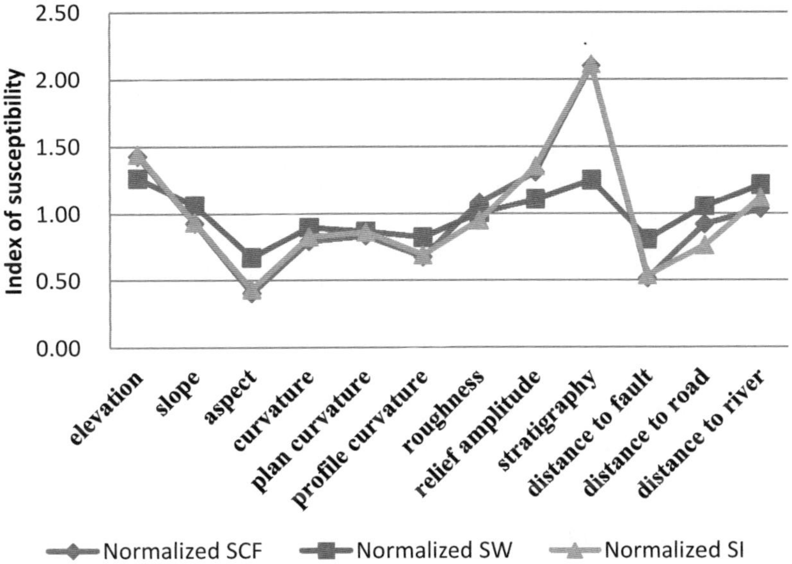

Figure 5. The normalized S value of each factor in the study area.

Table 3. The 12 susceptibility indexes of the loess landslide in the study area

Factors	S_{CF}	Normalized S_{CF}	S_W	Normalized S_W	S_I	Normalized S_I
Altitude	1.688155641	1.43	5.150495705	1.26	4.697856841	1.44
Slope	1.41891446	0.92	3.336565972	1.06	3.042007717	0.93
Aspect	0.898600963	0.41	1.465036763	0.67	1.416765423	0.43
Curvature	1.196644714	0.79	2.855810308	0.89	2.685647148	0.82
Plan curvature	1.158014129	0.83	3.0020889	0.87	2.805750813	0.86
Profile curvature	1.09886579	0.67	2.435052377	0.82	2.247890036	0.69
Roughness	1.340586274	1.08	3.895171347	1.00	3.088135409	0.95
Relief amplitude	1.478893993	1.31	4.722414083	1.11	4.413450141	1.35
Stratum lithology	1.670676325	2.10	7.580035321	1.25	6.8816904	2.11
Distance to faults	1.081167547	0.52	1.860457967	0.81	1.755280914	0.54
Distance to roads	1.40670078	0.92	3.313498213	1.05	2.479135575	0.76
Distance to rivers	1.622626683	1.03	3.721115156	1.21	3.644729969	1.12

Among the 12 factors, lithology, altitude, and relief amplitude are the key control factors of landslide development. The distance to rivers, roughness, slope, and distance to roads are moderate sensitive to landslide. The slope aspect, curvature and distance to faults have relatively weak control effect on landslide.

6. Conclusion

(1) According to the Certainty Factor (CF) method, the Weights-of-evidence method (WOE) method and the amount of information (I) analysis of the loess landslide in Tianshui City area, the most favorable factors for the occurrence of landslide in the 12 factors are as follows: altitude between: 1200m to 1400 m, slope between 10° to 15°, aspect of southwest, curvature between -0.5 to 0.5, plan curvature between -0.5 to 0.5, profile curvature between -0.5 to 0.5, roughness between 1 to 1.05, relief amplitude between 60 m to 120 m, stratum lithology of the Quaternary Upper Pleistocene (Q_3) Malan loess and the Neogene (N) mudstone distribution areas, distance to faults between 4000m to 5000m, distance to rivers between 1200m to 1600m, distance to roads between 500m to 1500m. The above 12 factors are the susceptible factors of loess landslide hazard. Their responses and susceptibility to landslide is the highest, which makes them the key factors of landslide development.
(2) Among the 12 causative factors, the three factors of lithology, altitude and relief amplitude are the key factors controlling the landslide development according to the susceptibility index analysis. The susceptibility of buffer zone of rivers/roads, roughness, and slope to the development of landslide is moderate. Aspect, curvature, and fault buffer have relatively weak control effects on landslide.

7. References

[1] Wu WJ and Wang NQ 2006 Landslide Hazards in Gansu. Lanzhou University Press, Lanzhou, China: 1-287 (In Chinese)
[2] Wu WJ 2003 Landslide and debris flow hazards in city of Tianshui. Hydrology and Engineering Geology 30(5): 75-78 (In Chinese)

[3] Meng XM, Xu YH, Guo T, and Zhang SW 1991 Research of Jiaoshuwan and Taishanmiao landslides in Tianshui City. Journal of Gansu Science 3(2): 36-44 (In Chinese)

[4] Guo FY, Meng XM, Li ZH, Xie ZT, Chen G, and He YF 2015 Characteristics and causes of assembled geo-hazards induced by the rainstorm on 25th July 2013 in Tianshui City, Gansu, China. Mountain Research 33(1): 100-107 (In Chinese)

[5] Peng JB, Fan ZJ, Wu D, Zhuang JQ, Dai FC, Chen WW, and Zhao C 2015 Heavy rainfall triggered loess-mudstone landslide and subsequent debris flow in Tianshui, China. Engineering Geology 186: 79-90.

[6] Wu WJ, Su X, Liu W, Wei WH, Feng LT, and Yang T 2014 Loess-mudstone interface landslides: Characteristics and causes. Journal of Glaciology and Geocryology 36(5): 1167-1175 (In Chinese)

[7] Lan HX, Zhou CH, Wang LJ, Zhang HY, and Li RH 2004 Landslide hazard spatial analysis and prediction using GIS in the Xiaojiang watershed, Yunnan, China. Engineering geology 76(1): 109-128.

[8] Pourghasemi HR, Pradhan B, Gokceoglu C, Mohammadi M, and Moradi HR 2013 Application of weights-of-evidence and certainty factor models and their comparison in landslide susceptibility mapping at Haraz watershed, Iran. Arabian Journal of Geosciences 6(7): 2351-2365.

[9] Gao KC, Cui P, Zhao CY, and Wei FQ 2006 Landslide hazard evaluation of wanzhou based on GIS Information value method in the three gorges reservoir. Chinese Journal of Rock Mechanics and Engineering 25(5): 991-996 (In Chinese)

[10] Chen BL 2012 Sensitivity analysis of topographic factors of landslide disaster in Qingjiang River Basin. Yangtze River 43(15): 39-42 (In Chinese)

[11] Bonham-Carter GF, Agterberg FP, and Wright DF 1989 Weights of evidence modeling: a new approach to mapping mineral potential. Statistical Applications in the Earth Science, Geological Survey of Canada Paper 89-9:171-183.

[12] Regmi NR, Giardino JR, and Vitek JD 2010 Modeling susceptibility to landslides using the weight of evidence approach: Western Colorado, USA. Geomorphology 115: 172-187.

[13] Mohammady M, Pourghasemi HR, and Pradhan B 2012 Landslide susceptibility mapping at Golestan Province, Iran: A comparison between frequency ratio, Dempster-Shafer, and weights-of-evidence models. Journal of Asian Earth Sciences 61: 221-236.

[14] Kayastha P, Dhital MR, and Smedt DF 2012 Landslide susceptibility mapping using the weight of evidence method in the Tinau watershed, Nepal. Natural Hazards 63: 479-498.

[15] Chen W, Ding X, Zhao RX, and Shi SL 2016 Application of frequency ratio and weights of evidence models in landslide susceptibility mapping for the Shangzhou District of Shangluo City, China. Environmental Earth Sciences 75: 64.

Acknowledgements

This study was funded by the National Natural Science Foundation of China (Grant No. 41362014); National Major State Basic Research Development Program (973 Program, Grant No.2014CB744703); Scientific Research Project of Gansu Academy of Sciences (Grant No. 2013JK-07, 2015JK-01, 2017JK-01, 2017HZ-01).

Green Open Space: Awareness for Health or Sustainability?

O C Dewi[1], I Chairunnisa[2], T Hidayat[2], M Anggraini[3], A Napitupulu[4]

1 Lecturer, Architecture Dept. Faculty Engineering, Universitas Indonesia, Indonesia
2 Master Student, Architecture Dept. Faculty Eng., Universitas Indonesia, Indonesia
3 Field Coordinator, Indonesian Red Cross, Indonesia
4 Program Manager, American Red Cross, Indonesia
E-mail: ova.candra@gmail.com

Abstract. Universitas Indonesia in cooperation with American Red Cross and Indonesian Red Cross have been assisting green open space revitalisation program in 7 locations in Bogor Regency (2016-2017). The program was held under The Urban Disaster Risk Reduction Greater Jakarta Project; an initiative program from American Red Cross Indonesia. This project was not only improving the existing green open space quality, but also creating one adapted from public land. The revitalization project figures what happened on daily basis on the existing land, proposing new programming facilities, community-based construction, monitoring and handing over. This paper discovers the meaning of a green space for the community, whether the community aware of its benefit on human health or environmental sustainability. The research question is does the community aware of green open space benefit for human health or environmental sustainability? Or both? The original data from the community was gathered and grouped based on its relevance with environmental quality and public health.

1. Introduction

The open green spaces are becoming more difficult to find in urban area. Many green open spaces were transferred in to a concrete and brick blocks. Such tendency can also be seen in big cities in Indonesia, nevertheless in Bogor city, which is marked by the decreasing of green open space and the dominance of physical buildings.

Studies have found that when people have access to green open space such as park, they exercise more [1]. Moreover evidence also suggest that people who live close to park and recreation facilities have lower rates of obesity and engage in more physical activity than those who do not [2].

Green open space is a space that is planned to meet the need of community interaction and joint activities. It gives space that serves as a means of container of human life, both individually and groups as well as other living creatures that live and grow sustainably [3]. The space also serves as an active playground for children and adults, a passive lounge for adults and as a green conservation area. Republic Indonesian Act No.26 Year 2007 about green open space stated that the proportion of forest should be minimum 30% from water shade to maintain environmental sustainability.

Universitas Indonesia, American Red Cross and Indonesian Red Cross implemented the Urban Disaster Risk Reduction (DRR) Greater Jakarta Project, specifically on Green Open Space Revitalization Program. The project involves Department of Architecture Faculty of Engineering, Department of Anthropology Faculty of Social and Political Science and Faculty of Psychology. Green open space in this paper refers to human built environment such as urban park, garden or any green space that is located in the middle of housing settlement. It exists to serve the community needs of community interaction and joint activities.

This paper emphasizes the pre-design phase of the green open space revitalization program in Bogor Regency [4]. It determines the community recognition about the green open space elements. The recognition was done by analyzing input from the community and observing the actual activities. The three stages involved in this phase are (1) to discover what is the meaning of green open space for the community, (2) to observe the existing daily basis activities on the current green open space, and (3) to find out what kind of green open space elements do the community needs or wants to have on the revitalized green open space. The community opinion about green open space elements will determine whether the community aware of its benefit on health human or environmental sustainability.

2. Theoretical review of green open space for health and sustainability

Penataanruang.com describes some of the backgrounds behind the green open space preservations and those are aiming to create the microclimate and to reduce the air pollution potential from the surrounding activities, to conserve the natural resources and to serve as evacuation areas during disasters [5]. The green open space is dominated by natural environment outside and within the urban area and can be along with recreation area and green line.

2.1. Human health

Based on medical dictionary, human health or health is a relative state in which one is able to function well physically, mentally, socially and spiritually in order to express the full range of one's unique potentialities within the environment in which one is living.

2.2. Environmental sustainability

Based on medical dictionary, environmental sustainability is long-term maintenance of ecosystem components and functions for future generations.

2.3. Benefit of urban green space

Haq (2011) underlined the 3 types of benefit from urban green space: (1) environmental benefit which consist of ecological benefits, pollution control, biodiversity and nature conservation (2) economic and aesthetic benefits such as energy savings and property values and (3) social and physical benefits which include recreation and wellbeing and human health [6]. Referring to the benefit of green open space, there is an opportunity to differentiate two direct benefits for the community; those are for environmental sustainability (benefit no.1) and human health (benefit no.3). The benefit of economic sustainability (benefit no.2) is out of discussion in this paper as these benefits were observed during the pre-design phase and based on the actual input from the community.

3. Methodology

3.1. Locations

This paper analyzes the result based on community participation and in depth interview method during the pre-design phase. The seven local communities from seven locations in Bogor Regency, West Java, were involved during the revitalization program between 2016-2017. The number of household from each location is around 300 Households. Community participation is one of key features of Healthy City based on World Health Organization (WHO). The involvement of the local community in this revitalization program is a must in order to identify the activity program on the green open space. It is a bottom-up approach and therefore these needs have to come from the community's side. It is expected that the involvement will increase community belonging sense and willingness to spend more time there, which will directly affect the sustainability of the green open space.

The 7 locations of green open space revitalization and the year of the projects conducted are:

- Location 1: Local Community in Green Open Space Karadenan, 2016
- Location 2: Local Community in Green Open Space Pondok Rajeg, 2016
- Location 3: Local Community in Green Open Space Sukahati, 2016

- Location 4: Local Community in Green Open Space Waringin Jaya, 2016
- Location 5: Local Community in Green Open Space Kedung Waringin, 2017
- Location 6: Local Community in Green Open Space Sukahati, 2017
- Location 7: Local Community in Green Open Space Waringin Jaya, 2017

Table 1. Green Open Space Element Definition

Elements	Definition*
Playground	An outdoor area featuring equipment designed for recreation and play, such as seesaws and swings
Gathering Place	Place for assembly of person; a meeting place
Sport Area	Area for activity involving physical exertion and skill that is governed by set of rules or customs and often undertaken competitively
Green Environment	Grassy environment area located usually at the centre of a city or town and set aside for common use
Trees plantation	An area under cultivated trees or plants
Open space for safety area	An area which is free obstacles
Water Catchment Area	A reservoir, esp. for catching and retaining surface drainage over a large area, in which sediment may settle
Educating Area (depending on kind of activities)	Area to educate
Therapy Area	Area for treatment
3R Facility	Facility for Reduce-Reuse-Recycle
Climbing Wall	Mountaineering a specially constructed wall with recessed and projecting holds to give practice in rock climbing, a feature of many sport centres
Drainage	Systemic withdrawal of fluids and discharge from a wound, sore of cavity
Toilet	An apparatus into which defecation and urination can be done

*) Relevant definition taken from the website [7] [8]

3.2. Data collection method

The data was collected during the Charrette process and held directly on the green open space, within different time and locations (Figure 1). The Charrette is an intensive and multi-disciplinary workshop design [9]. This method is commonly used to facilitate discussion between stakeholders on a project or program. A charrette, a group of experts met with the community groups and related stakeholders. The stakeholders for this program is the community, The Indonesian Red Cross Volunteer (SIBAT), The American Red Cross and the Universitas Indonesia team.

Four questions were raised during the charrette process and the local community was expected to answer. The questions were: (1) what is the definition of the green open space for me (the local people)? (2) What are the activities in this green open space during weekends, weekdays, days and night (3) What do they wish to see or have in a green open space, and (4) what are the priorities between the wish list from answer of question number 3?

To make sure the originalities of the data, only question number 3 is taken in to account as relevant question to this study. All together the wish list from the local community consists of playground, sport are, therapy area, educating area, green environment, trees plantation, 3R facility, catchment area, open space (safety area), toilet, climbing wall, gathering place, and drainage. In this study, we named this wish list as “program activity” list.

IOP Conf. Series: Earth and Environmental Science **120** (2018) 012014 doi:10.1088/1755-1315/120/1/012014

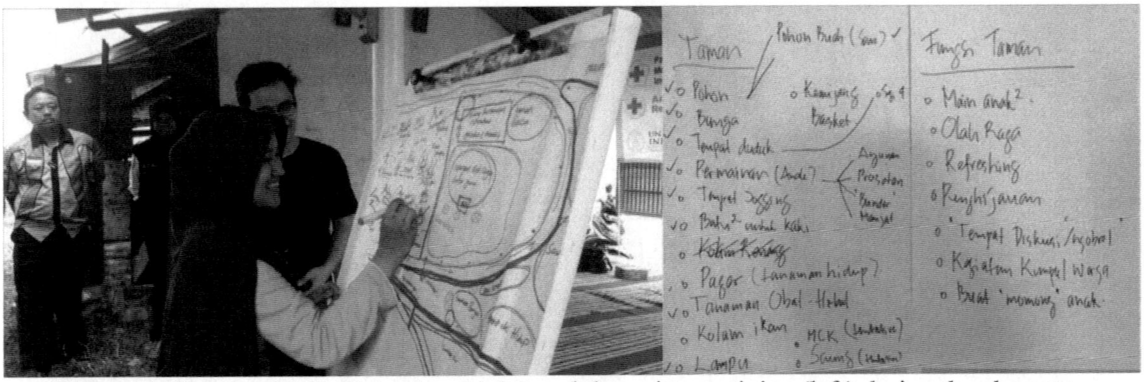

Figure 1. Community Participation (right) and the written opinion (left) during the charrette

3.3. Type of benefit from urban green space for the community

This study only involves 2 out of 3 types of benefit from an urban green space for the community and those are human health benefit and environmental sustainability benefit. However from the discussion the study found some whishes of the program activities from questions number 3 could also be part of both human health and environmental sustainability benefits. In this way the group of program community will be (1) program activities which include as part of human health only, (2) environmental sustainability only and (3) the last group, which are considered to be part of both human health and environmental sustainability.

4. Result and discussion

This paper summarizes the pre-design phase results of the green open space revitalization. This study emphasises the awareness of green open space for the community, whether it is benefiting for health, for sustainability or for both. The community has the right to express their opinion in mentioning what program activities that they want in the revitalized green open space. These community opinions were collected and analyzed during the charrette process.

4.1. Green open space community program activities

The local community program activities are shown by the community wish list, either the element is considered to be dominant in human health, environmental sustainability benefit or both. In fact such differentiation were never been asked to the community explicitly. Through the pre-design phase communications, this study underlines the urban green space awareness for the community (Figure 2). This study found that the wish list shows the dominance more on the human health benefit sides. From 13 types of program activities, 11 were considered benefiting human health, 6 were considered benefiting environmental sustainability, 5 were considered benefiting both, and 1 is not relevant to both (Table 1).

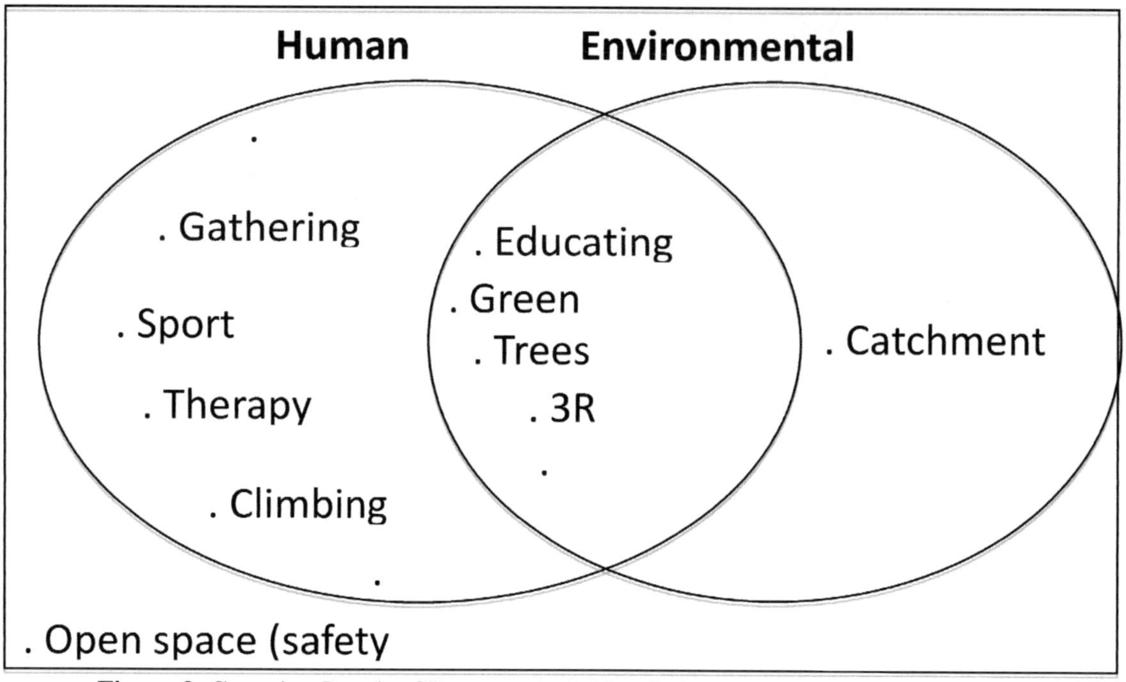

Figure 2. Grouping Result of Program Activities during Pre-Design Phase (Table 1)

4.2. *Community priorities on human health program activities*

On the human health context (Table 2), it is found that sport area, play ground, gathering place and green environment are the 3 highest priorities for the community. All the 7 locations mentioned Sport Area in their wish list, whereas 6 locations mentioned playground and 5 locations mentioned gathering place and green environment. The least mentioned is toilet, climbing wall, 3R facility and drainage. The climbing wall, toilet and drainage were only come out once, and all in location 4, in Waringin Jaya 2016, whereas the 3R facility was mentioned by location 2.

4.3. *Community priorities on environmental sustainability program activities*

On the environmental sustainability context (Table 3), the 3 highest priorities are green environment, water catchment, educating area, and tree plantation. Five locations mentioned green environment, 4 locations mentioned tree plantation, and 3 locations mentioned water catchment and educating area. The least mentioned was also 3R facility and Drainage.

5. Conclusion and suggestion

5.1. *Conclusion*

One of the main ideas of green open space revitalization is to create the microclimate and to reduce the air pollution potential from surrounding activities, to conserve the natural resources and to serve as evacuation areas during disasters. The Indonesian Law no.26/2007 is also emphasizing the aim of green open space planning to empower the local communities, preserve the quality of environmental sustainability, conserve heritage and maintain the balance of cities-regencies development. In this way, Indonesian Government is also supporting community to contribute to sustainability.

This study found that the word "green open space" gives more awareness on human health for the community and it is proven by the dominance of human health related program activities. However environmental awareness also showed up, but not explicitly as shown in Figure 3.

In terms of program activities related to human health, program activities like sport area, playground, gathering place and green environment are among the top priorities according to the community. In terms of program activities related to environmental sustainability, the green environment, tree plantation, water catchment area and the education area are among the top priorities. On both, the 3R facilities and drainage showed up as the least priorities according to the community's wish list.

Figure 3. The green open space with banks and water catchment (right) and sport (left) area

Table 2. Community priorities for human health

Elements	**Location**							**Sum**
	1	2	3	4	5	6	7	
Playground	√	√	√	√	√	√		6
Gathering Place	√	√	√			√	√	5
Sport Area	√	√	√	√	√	√	√	7
Therapy Area			√	√		√		3
Climbing Wall				√				1
Toilet				√				1
Educating Area			√			√	√	3
Green Environment	√	√	√	√			√	5
Tree Plantation	√	√			√	√		4
3R Facility		√						1
Drainage				√				1

Table 3. Community choices for environmental sustainability

Elements	**Location**							**Sum**
	1	2	3	4	5	6	7	
Water Catchment Area	√		√		√			3
Educating Area			√			√	√	3
Green Environment	√	√	√	√			√	5
Tree Plantation	√	√			√	√		4
3R Facility		√						1
Drainage				√				1

5.2. Suggestion
For the community, explicitly the green open space is a mean of awareness for health whereas includes awareness of environmental sustainability behind it (implicitly). This study found that in the context of urban green open space that is located in the urban area, the community is still having their mind set on a green open space area as playground, sports and gathering points under the green environment. It means currently the green open space represents more on the human physical activities. When an initiative program about green open space is about to launch, it is therefore necessary to inform the limitation for human activities and give portion of land for tree plantation (minimum 30%). For effective result, attention must be given to educational aspect, such as public hearing, community based action plan and trainings [10]. Indeed, the green open space exists to serve the community needs of interaction and gathering activities, but the proposal behind is also to reduce the air pollution potential from surrounding activities and to protect the environment towards sustainability.

6. Reference
[1] Gies, F, 2006. *The Health Benefits of Park*, P.7. Available from: http://conservationtools-production.s3.amazonaws.com/library_item_files/196/235/HealthBenefitsReport_FINAL_010307.pdf?AWSAccessKeyId=AKIAIQFJLILYGVDR4AMQ&Expires=1501562039&Signature=F7fEjICHs1Q%2F4I11B8TRZen4plM%3D. [Accessed: August 1st, 2017]

[2] Department of Public Health Los Angeles County, 2016. *Parks and Public Health in Los Angeles County*, P.3. Available from: http://publichealth.lacounty.gov/chronic/docs/Parks%20Report%202016-rev_051816.pdf. [accessed: August 1st, 2017]

[3] Undang-Undang Penataan Ruang Republik Indonesia No.26 Tahun 2007 Tentang Penataan Ruang or Indonesian Law No. 26 Year 2007 About Spatial Planning

[4] Dewi, O C., Shanti, LP., Atmoko, TU., Napitupulu, A., Anggraeni, M. 2017. *Sustainability Assessment of Community Park Revitalization, ASEAN Journal of Community Engagement*, Volume I. Number 1, 2017, p.83-95.

[5] Penatan Ruang (2017). *Ruang Terbuka Hijau*. Available from http://www.penataanruang.com/ruang-terbuka-hijau.html accessed: [May 27th, 2017]

[6] Haq, S M A, 2011. *Urban Green Spaces an Integrative Approach to Sustainable Environment*. Journal of Environmental Protection (2), 601-608.

[7] The Free Dictionary, 2017. *Human Health*. Available from: http://medical-dictionary.thefreedictionary.com/Human+health accessed: [August 1st, 2017]

[8] The Free Dictionary, 2017. *Environmental Sustainability*. Available from: http://encyclopedia2.thefreedictionary.com/environmental+sustainability accessed: [August 1st, 2017]

[9] The Town Paper, 2017. What is Charrette? Link: http://www.tndtownpaper.com/what_is_charrette.htm

[10] Dewi O.C. (2013). Decision Support Model for Waste Management in Support of Developing Low Carbon, Eco Region: Case Studies of Densely Populated Kampung Settlements in Urban Areas in Jakarta, Dissertaion, Hamburg University of Technology. Hamburger Berichte zur Sidelungswasserwirtsschaft.

Acknowledgments

Authors would like to express their greatest gratitude to American Red Cross, Indonesian Red Cross, Direktorat Riset dan Pengembangan Masyarakat Universias Indonesia (DRPM UI), and Social and Political Science Faculty, Department of Anthropology Universitas Indonesia (FISIP UI) especially Dr. Aji, Wahyu Putra, S. Kom., and Hafiz Awlia, Faculty of Psychology, Department of Psychology Universitas Indonesia especially Dr. Wustari Gunadi Mangundjaya., for their initiative, contribution and commitment in the program. Our sincere gratitude also goes to Architecture Department Faculty of Engineering Colleagues and Students for the unconditional and extensive support for the program, especially to Teguh Utomo, MURP, Evawani Ellisa, PhD., Lathiyfah P.S., Istiqamah, M.Ars., Muhammad Ichsan, S.Ars., Luthfi Eryando, S.Ars. Also our biggest thanks goes to the SIBAT community in Pondok Rajeg, Karadenan, Waringin Jaya, Sukahati, and Kedung Waringin including the Kabupaten Bogor Local Government for their worthy contribution and support during the program.

IOP Conf. Series: Earth and Environmental Science **120** (2018) 012015 doi:10.1088/1755-1315/120/1/012015

The Costs of Climate Change

Jason Guo[1]

1 High School Student, The Paideia School, Atlanta, USA
E-mail: guo.jason@paideiaschool.org

Abstract. This research paper talks about the economic costs of climate change, as well as the costs involved in responding to climate change with alternative fuels. This paper seeks to show that climate change, although seemingly costly in the short run, will both save future generations trillions of dollars and serve as a good economic opportunity. Scientists have long argued that the fate of humanity depends on a shift towards renewable energy. However, this paper will make clear that there is also an economic struggle. By embracing alternative fuels, we will not only lessen the danger and the frequency of these natural disasters but also strengthen the world's financial state. Although a common argument against responding to climate change is that it is too expensive to make the switch, this research shows that in the future, it will save millions of lives and trillions of dollars. The only question left for policymakers is whether they will grasp this energy source shift.

1. Introduction

Today, corporations thrive on the fossil-fuel industry. While doing so may help these corporate CEOs reap great profits, this reliance may also hurt the planet. Corporations gain by harming the Earth to produce the energy we rely on, and because we rely on this generated non-renewable energy so heavily, corporations in turn earn substantial revenue from this reliance. Today, we know that there are other energy alternatives that could save Earth, like solar and wind power. However, few businesses have embraced the idea of clean energy; this reluctance to switch suggests that switching from a coal driven industry to a clean one is perhaps not as simple as one might think. It is true that switching to clean energy would be a crucial and substantial step towards saving our earth from trajectories that would soon kill us. But at the same time, many political and private actors feel that jobs could be lost, and thus unemployment could rise. So how can we convince our leaders to step up and respond to climate change if they feel so entitled to defend their economy in accordance with these fears?
In order to do so, we need to demonstrate that great possible economic opportunities lie in store only if we respond to climate change.

2. Consequences of inactivity:

Firstly, we should recognize why climate change so urgently needs a response.
If our leaders decide not to respond to climate change, the climate will literally continue to drastically change. This changes will likely be palpably felt through "many heat waves, create more intense rainstorms and the disintegration of coral reefs" (Friedman, 2017) [1] in addition to the melting of huge ice sheets. As the ice sheets melt, the sea level will rise, which would thus increase the severity and occurrence of natural water-related disasters. According to NASA [2], the global average sea level has already risen by 86.4 millimeters [Figure 1] since 1993.

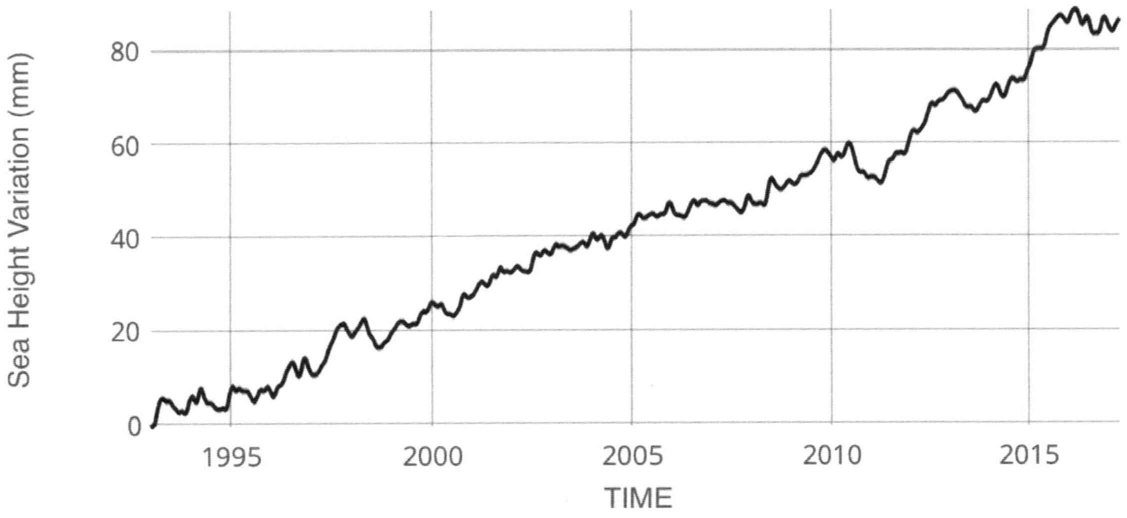

Figure 1: Global Sea Height Variation [2]

What does this mean for the economy? For starters, the infrastructure in coastal cities will be deeply damaged due to the increasing hurricanes – in both strength and size – caused by climate change. Governments in civilizations around the world will need to spend billions in order to simply repair the groundwork so many citizens rely on. What's worse, the many countries whose coastal cities are in peril due to rising sea levels will also be unable to trade with other countries because they will have more pressing economic and survival priorities. Even then, the consequences to climate change do not stop there. From a human standpoint, climate change and natural disasters often lead to many unfortunate and unnecessary deaths around the world.

If we do not respond to climate change by cutting our carbon emissions, there are no other simple and cheap alternatives. If we do not reduce our carbon footprint, the next generation will have to pay up to $535 trillion to clean the atmosphere, according to a new study by Anthropocene [3]. These costs would include the removal of carbon dioxide emissions from the air using negative emission technologies to avoid only the worst consequences of climate change.

Simply by not responding to climate change, mankind will negatively impact important aspects of society such as the economy; mankind will also experience negative ramifications through severe losses of human life. From an economic standpoint, we would deeply damage the economy by not responding to climate change. In other words, it would be economically beneficial if we responded to climate change, especially given the consequences of inaction.

However, even if 97 percent of scientists agree that climate change is real and requires an urgent response, such facts and near-universal agreement within the scientific community have clearly not been enough to convince politicians to feel the same way. These "climate deniers" disregard climate change simply because they believe – or fear – that responding to climate change is harmful to the economy. Such fear is primarily founded on the structure of our two-party election system, which forces elected officials to take both their party's platform and public opinion into account when seeking re-election. By providing evidence that responding to climate change will in fact strengthen the economy instead of draining it, politicians may find that there should be no reason to not respond. In fact, they should reframe their messaging to show that the economic boost from responding to climate change constitutes another incentive in addition to the environmental benefits.

3. Infinitesimal supply and high demand for renewable energy

The end goal of successfully responding to climate change is to reduce carbon dioxide emissions. And what better way is there to accomplish that reduction than by using renewable energy? By definition, renewable energy should be "infinite." On the other hand, non-renewable energy is limited in supply and thus does not replenish in a short amount of time. For the most part, non-renewable fuel is the main factor in causing climate change, as it emits carbon dioxide when this fuel is burned. On the other hand, most renewable energy sources do not emit carbon dioxide, so as a result, alternative energy, or renewable energy that does not emit carbon dioxide, can act as a desirable substitute to fossil fuels.

There are two reasons why the switch to renewable energy is absolutely necessary. Firstly, the switch has an immediate benefit simply because by using alternative fuels, carbon dioxide emissions will fall and the effects of climate change may slow to a halt. Another reason involves the magnitude of the growth of the human population. For perspective and context, the world population has grown seven times since the 1800s [Figure 2], according to the United Nations Population Division. [4]

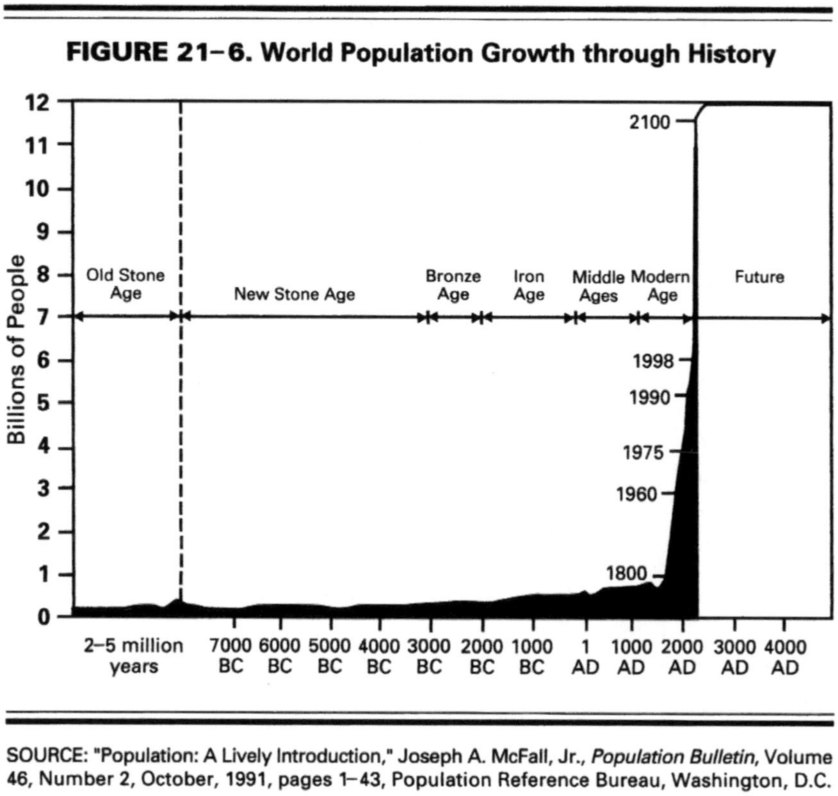

Figure 2: World Population Growth Through History [4]

With so many people, each household today requires massive expenditures of energy to conduct daily operations. Currently, Earth's population relies mostly on non-renewable energy, which is limited and not quickly replenished. Given high demand and the limitations of non-renewable energy, renewable energy sources need to be increased and contribute more to total energy generation, especially in the United States. Currently, renewables account for 10% [Figure 3] of the total energy generation for the United States. [5] If we do not make the switch to cleaner energy, the high demand for energy will mean continued emissions of large amounts of carbon dioxide, which will then contribute to climate change.

On the other hand, renewable energy is an infinite energy source that cannot be depleted. If we were to implement appropriate technology and infrastructure to capture and generate this renewable energy on a large scale, the economic and environmental benefits could be substantial. We have only scratched at the surface of the potential of renewable energy. Clearly, it is a new lucrative business simply because the source (renewable fuels) and the product (energy) are both infinite. Given that the high demand in mind for such energy sources, it would make sense for us to capture the infinite supply that exists to feed that large population demand. As it stands, we currently have a high demand with an increasingly diminishing supply, while at the same time the effects of climate change increase.

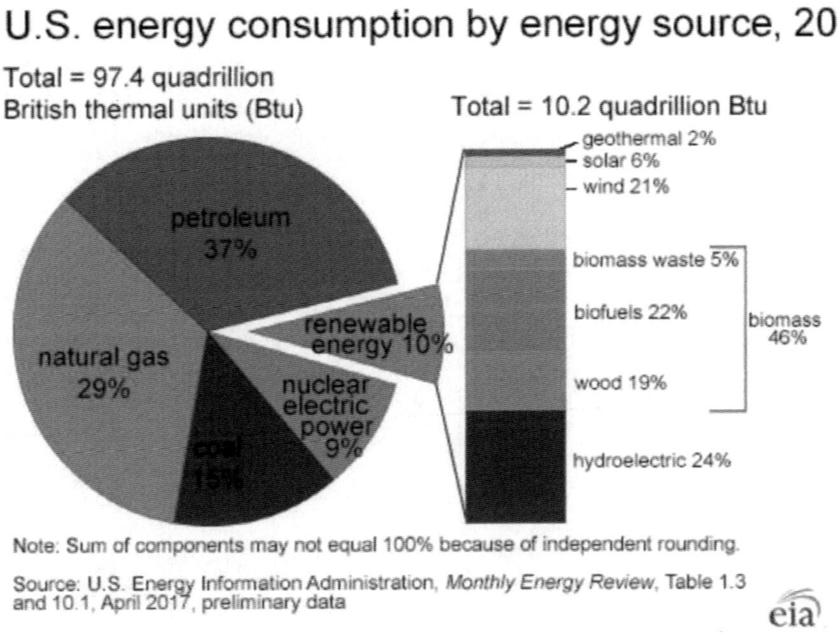

Figure 3: U.S. energy consumption by energy source [5]

4. Steady job growth for renewable energy

An economy cannot run without its respective employees. These jobs fuel businesses, which in turn drive the economy. Jobs have also been the main assumed pretext for not responding to climate change; many have pointed out that coal jobs will be lost and thus the unemployment rate will soar. Because of renewable energy's demonstrated potential, politicians are quite worried that renewable energy will become a replacement for non-renewable energy. Coal mining companies are especially worried that their own jobs will be replaced by solar and wind power. This dilemma could not have been more clearly demonstrated than in our current U.S. president Donald Trump's campaign for office.

Donald Trump's administration has promoted active falsehoods about the state of climate change. Specifically, his EPA administrator Scott Pruitt has lied about the jobs of the coal industry. According to Politifact [6], Pruitt claimed that 50,000 more coal mining jobs have been added since last year. Politifact rates this claim as "mostly false" [7]. In fact, the actual increase in coal jobs was 1,300 jobs [8]. However, the lies stem from political motivation; specifically, Donald Trump wants to revive the coal industry with coal jobs for people.

However, this is a near impossible feat. U.S.'s Coal Mining Employment, [9] has been at a steady decrease from 900,000 jobs in 1923 to 100,000 in 2016 [Figure 4].

There are numerous reasons for this, the drop in natural gas prices being the most paramount. The rise of renewable energy, as well as the advent of automated machines in place of human miners has contributed to the decrease. Given these and other reasons, there does not seem to be a need for coal miners anymore.

At the same time, renewable energy is on the rise. This kind of energy has become much cheaper in the last eight years. Solar energy's costs have fallen 85 percent between 2008 and 2016 [Figure 5] and

IOP Conf. Series: Earth and Environmental Science **120** (2018) 012015 doi:10.1088/1755-1315/120/1/012015

wind costs have fell 36 percent [Figure 6] (Columbia, Houser) [10]. As a result, renewable energy has clearly emerged as a new potential market ready for international trade.

Source: Mine Safety and Health Administration, includes office workers starting in 1973.

Figure 4: US Coal Mining Employment [9]

Dollars per Watt (Left Axis) and Billion kWh (Right Axis)

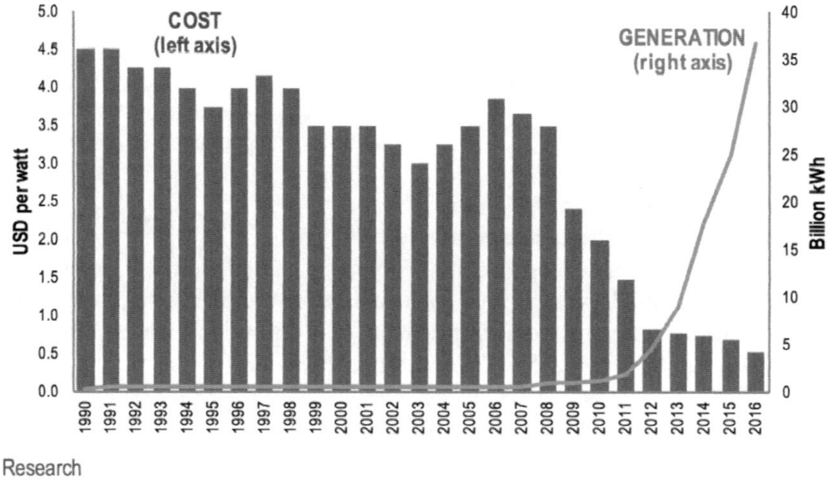

Source: EIA, GTM Research

Figure 5: Solar PV Module Costs and Generation [9]

With such a profitable market, there must be jobs to support these developing businesses. There are currently 260,000 solar energy jobs across the country, twice as many jobs as the total coal jobs in the U.S. (Plumer, Vox article) [11]. As coal plants have closed in recent years, renewable energy could be a good energy replacement, especially because renewable sources are well-suited to responding to climate change by lowering carbon emissions. However, renewable energy jobs are not in the "right" locations. According to the Solar Foundation [12], West Virginia and Kentucky, states that still contain coal jobs, are ranked at the bottom ten states in solar jobs per capita. These coal jobs have not been replaced by renewable energy, simply because renewable energy jobs are not where coal once was.

However, renewable energy still possesses massive potential. With the costs decreasing and the jobs increasing, people should embrace renewable energy as a solution to climate change and a solution to replacing lost closed coal plant jobs.

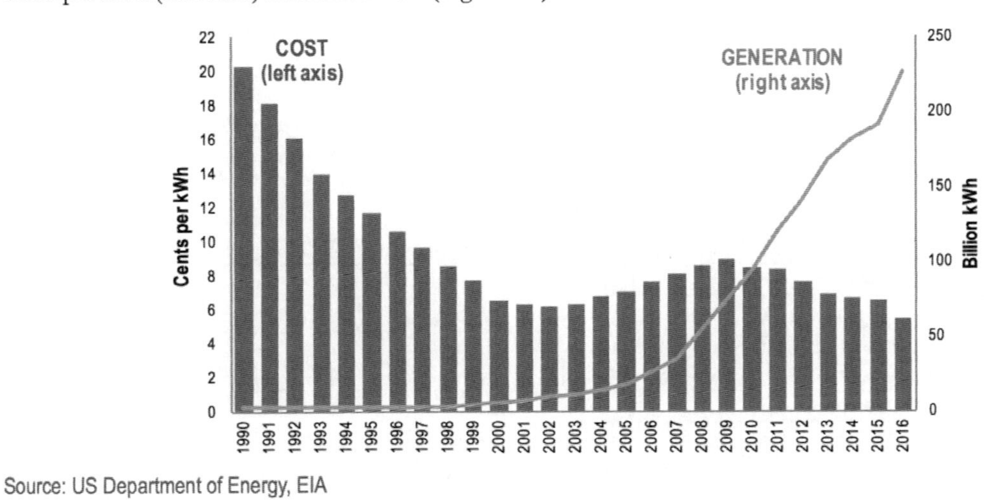

Figure 6: Wind Energy Costs and Generation [9]

5. Conclusion

As a result, there is no other alternative than to respond to climate change. Disregarding the proven facts will only exacerbate global warming, and as described above, likely will lead to a chain of natural disasters. Lowering carbon emissions is the only way to truly stop climate change and its impending disasters. Other solutions like negative emission technologies are very expensive and at the same time inefficient because they only prevent a small number of the many consequences climate change could entail.

On the bright side, responding to climate change leads to substantial benefits. By using renewable energy to power the world, it is possible to reduce carbon dioxide emissions and thus gradually stop climate change. It will not be a quick process to switch. But substantial efforts to transfer from one energy source to another are needed. Climate change requires an immediate call to action. Without a response, the human population will suffer under preventable yet terrible disasters. An energy-craven population relying on renewable energy is the best solution to lower carbon dioxide emissions and thus stop climate change. Given its economic benefits and effectiveness in combating climate change, it is important for the world to wholeheartedly embrace these renewable solutions. If the world can agree and cooperate with each other, the switch to renewable energy will be easier. Otherwise, without a switch, the world is set up for failure.

6. References

[1] Friedman, Lisa. "Government Report Finds Drastic Impact of Climate Change on U.S." *The New York Times*. The New York Times, 07 Aug. 2017. Web. 13 Aug. 2017. https://www.nytimes.com/2017/08/07/climate/climate-change-drastic-warming-trump.html?mcubz=3

[2] "Key Indicators: Global Mean Sea Level." *NASA*. NASA, 31 July 2017. Web. 13 Aug. 2017. https://sealevel.nasa.gov/understanding-sea-level/key-indicators/global-mean-sea-level

[3] Patel, Prachi. "Climate inaction will leave our kids a trillion dollar debt." *Anthropocene*. Future Earth, 3 Aug. 2017. Web. 13 Aug. 2017. http://www.anthropocenemagazine.org/2017/08/climate-inaction-will-make-our-kids-pay-hundreds-of-trillions-of-dollars/

[4] "United Nations Population Division" | Department of Economic and Social Affairs." *United Nations*. United Nations, n.d. Web. 13 Aug. 2017. http://www.un.org/en/development/desa/population/

[5] "U.S. Energy Facts Explained." *U.S. Energy Facts - Energy Explained, Your Guide To Understanding Energy - Energy Information Administration*, U.S. Energy Information Administration, 19 May 2017, Accessed 30 Aug. 2017. **https://www.eia.gov/energyexplained/?page=us_energy_home**

[6] Jacobson, Louis. "Are coal mining jobs up by 50,000 since last year?" *Politifact.* Tampa Bay Times, 05 June 2017. Web. 13 Aug. 2017. **http://www.politifact.com/truth-o-meter/statements/2017/jun/05/scott-pruitt/are-coal-mining-jobs-50000-last-year-not-exactly/**

[7] Jacobson, Louis. "Are coal mining jobs up by 50,000 since last year?" *Politifact.* Tampa Bay Times, 05 June 2017. Web. 13 Aug. 2017. **http://www.politifact.com/truth-o-meter/statements/2017/jun/05/scott-pruitt/are-coal-mining-jobs-50000-last-year-not-exactly/**

[8] Jacobson, Louis. "Are coal mining jobs up by 50,000 since last year?" *Politifact.* Tampa Bay Times, 05 June 2017. Web. 13 Aug. 2017. **http://www.politifact.com/truth-o-meter/statements/2017/jun/05/scott-pruitt/are-coal-mining-jobs-50000-last-year-not-exactly/**

[9] Houser, Trevor. "Can Coal Make a Comeback?" *Columbia | SIPA Center on Global Energy Policy | Can Coal Make a Comeback?* Columbia University, 25 Apr. 2017. Web. 13 Aug. 2017. **http://energypolicy.columbia.edu/publications/report/can-coal-make-comeback**

[10] Houser, Trevor. "Can Coal Make a Comeback?" *Columbia | SIPA Center on Global Energy Policy | Can Coal Make a Comeback?* Columbia University, 25 Apr. 2017. Web. 13 Aug. 2017. **http://energypolicy.columbia.edu/publications/report/can-coal-make-comeback**

[11] Plumer, Brad. "There are now twice as many solar jobs as coal jobs in the US." *Vox.* Vox Media, 07 Feb. 2017. Web. 13 Aug. 2017. **https://www.vox.com/energy-and-environment/2017/2/7/14533618/solar-jobs-coal**

[12] "National Solar Jobs Census." *National Solar Jobs Census.* Solar Foundation, 2016. Web. 13 Aug. 2017 **http://www.thesolarfoundation.org/national/?mc_cid=eaf6332238&mc_eid=%5BUNIQID%5D**

IOP Conf. Series: Earth and Environmental Science **120** (2018) 012016 doi:10.1088/1755-1315/120/1/012016

Impact of Industrialization on Environment and Sustainable Solutions – Reflections from a South Indian Region

Rasmi Patnaik

Center of Excellence for Sustainable Development, Indian Institute of Corporate Affairs, Manesar, Gurugram, India
E-mail: rasenvs@gmail.com

Abstract. Industrialization has brought economic prosperity; additionally it has resulted in more population, urbanization, obvious stress on the basic life supporting systems while pushing the environmental impacts closer to the threshold limits of tolerance. With booming industrial growth and relatively low land mass, environmental sustainability is now becoming a significant deciding factor in industrial development process. Accumulating evidences constantly indicate that the transition of the existing industries into eco-industrial network through successful implementation of green approaches provides a viable solution to preserve the natural resources of the region while concurrently enhances the regional economy on a sustainable basis. It calls for an appropriate planning and integrated framework in harmony with the environment, after careful assessment of past and prevailing conditions. The empirical knowledge on affected area helps understanding the local context and developing further course of action based on ground realities. With this aim, a study was conducted on the current industrial pollution and environmental setting of Puducherry. A causal chain analysis indicated severe impacts of industrialization on local environment while highlighting its immediate and root causes. The findings form a base for suggesting sustainable solutions to curb rampant pollution in Puducherry region and similar scenarios found across the world.

1. Introduction

Since ages, industrial growth has started to affect environment with severe downside problems. It causes tremendous stress on the entire bionetwork and natural system components like water, air, soil, bio-diversity including surrounding eco-system. Realizing the severity of the problem, impacts of industrialization on the environment need to be analyzed with more intensity and feeling [1]. Industrial effluents contain numerous essential nutrients or possess properties which can easily be utilized for many value-added purposes with commendable benefits to society and environment. Application of green approaches based on 6Rs technologies (reduce, reuse, recycle, recovery, redesign, rethink) and closed-loop systems within the integrated framework of industrial ecology (IE) provides an excellent opportunity to preserve the natural resources of the region while concurrently enhances the regional economy on a sustainable basis. It calls for an appropriate strategic planning encompassing technical, ecological, socio-cultural and economic driving factors that can affect the industrialization process.

Prior to evolving a practical approach towards strategic development, it will always be indispensable to derive lessons from other similar situations. With a systematic way of looking at events occurred with identical consequences, critically analyzing the situations, policies, activities and overall scenarios related to the problem, it may help to gain a sharpened understanding of the ways the instance has happened [2]. In this backdrop, industrial profile and impacts of industrialization on environment were examined in Puducherry, which has emerged as an industrial nerve centre of South

IOP Conf. Series: Earth and Environmental Science **120** (2018) 012016 doi:10.1088/1755-1315/120/1/012016

India over past few years. Based on the lessons learned, this study indicates how a strategic green approach may be generated and applied in Puducherry region and matching context in other parts of the world.

1.1. Industrial profile and local context of Puducherry

Puducherry is a peaceful place located in a scenic location in South India with a cosmopolitan culture. The Union Territory (UT) of Puducherry covers an area of 492 km^2 and comprises four interspersed geographical entities, namely Puducherry, Karaikal, Mahe and Yanam. The Puducherry region is the largest of the four, which covers 293 km^2 (60.66% of the UT).

During the last couple of years, Puducherry has attracted a large number of industries due to the easy availability of land, water, labour, power and various incentives and concessions given by the government in the form of tax concessions, power subsidies, tax holidays and soft loans that were not available in most of the neighbouring states. With annual industrial growth varied from 1.86–12.28%, by mid-2016 the UT of Puducherry had an impressive establishment of 8562 industries representing a cross-section of diverse industries. Almost 80% of the total industries are located in the Puducherry region followed by 13% in Karaikal, 4% in Yanam and 3% in Mahe. Currently, seven well-established industrial estates are propelling the industrial growth of the UT of Puducherry. Out of these, six industrial estates, namely Thatanchavadi, Mettupalayam, Sedarapet, Kirumambakkam,Kattukupam and Thiruvuvanai are located inside the Puducherry region. Due to the industrial friendly climate, new industrial estates are also being planned while many large industrial companies of national repute and several renowned multinational corporations are showing keen interest to establish their factories in Puducherry in the near future [3], [4].

This rampant industrialization, alongside economic prosperity, has resulted in sudden increase of population, urbanization, obvious stress on the basic life supporting systems, pushing the environmental impacts closer to the threshold limits of tolerance. Based on the nature and severity of the pollution, industries in Puducherry have been categorized as Red, Orange and Green, referring to high, medium and low pollution potential, respectively. At present, 19% of the total industries are grouped in the red category, 21% orange and 60% green [5].By mid-2016, the estimated generation of hazardous waste in the UT of Puducherry was approximately 36,768 tonnes per annum (TPA), of which Puducherry alone generated 33,305.63 TPA[4]. Except only one hazardous waste recycling unit and no secured landfill site to take adequate care of industrial waste, some industries are selling their waste to other regions, while in most of the units waste disposal is being done in an improper manner, exerting a significant level of environmental burden. Uncontrolled, inadequate and unsystematic handling of this waste poses a great risk towards the environmental health of this region [4-8].

During the years 1890 -1900, the Puducherry region received only 98.4 tons of total pollution load, while during the period 1960-1970, the total pollution load reached 1794.2 tons, an eighteen fold increase which continuously escalated in subsequent years. According to a preliminary study by Ramesh (2005), this may be directly attributed to the rapid pace of industrialization during the same period [9].

2. Methodology

The nature of the study needs the collection and analysis of primary and secondary data on environmental, technological, economic, and policy aspects of industrial pollution. Hence, to obtain first hand information, it requires direct interactions with the Government as well as local non-governmental organisations, closely working and actively involved in the field of environmental pollution.

Causal chain analysis is an ideal apparatus to study the impacts of industrialization on a region and its causes. It is a step-by-step process that identifies the most important causal links between the environmental and socio-economic impacts resulting from the issues and concerns related to the field, their immediate causes, the factors responsible and finally the root causes that determine the behaviour of the concerned sector [10].In a nutshell, it looks at the overall scenario related to the concerned problem with a view to identifying underlying roots and driving factors that can stimulate sustainable growth in line with green approaches.

Application of this analytical approach (adapted from GIWA methodology, 2002) in Puducherry region not only provided a holistic overview of impacts of industrialization on environment, but also it helped immensely to identify the underlying causes of industrial pollution prevalent in this area.

In the first phase work, after obtaining the necessary permission from Puducherry Pollution Control Committee and Department of Science, Technology & Environment, the study initiated a situational analysis with the collection of required data from primary as well as secondary sources, such as libraries, journals, feature articles, magazines and publications from relevant government departments, NGOs, academic & research institutes and business associations. Impacts of industrialization on Puducherry ecosystem, water, air, biodiversity and its causes were analyzed comprehensively and the results were documented through a causal-chain diagram (figure 1). Information on the driving forces, such as actors and institutions, physical environment, policy support, management options, market forces, trade and legal requirements, challenges and infrastructure facility available for the formation of an eco-industrial network were also collected for a better understanding to formulate a strategic planning and to develop sustainable solutions for the pollution problems.

3. Results and discussions

Causal chain analysis in Puducherry portrayed a detailed picture of environmental and allied socio-economic impacts arising out of rapid and unplanned industrialization and tried to find out its major immediate causes as well as route causes. Attempts have been made to find out key drivers which have an effect on the industrialization process starting from industrial evolution era to present period.

3.1. Impacts of industrialization on Puducherry environment

3.1.1. Impacts on water - A detailed analysis of the environmental impacts of industrialization revealed that industries set up before the 1990s included mostly textiles, sugars and distilleries that were water intensive and had a higher pollution potential, exerting enormous pressure on the environment. The indiscriminate discharge of industrial effluent alongwith municipal solid waste disposal is the principal source for surface water contamination [11], [12]. The heavy metals, salts and fluoride effluents from the industries of chemicals production, metal processing and paper manufacturing from the Mettupalayam Industrial Estate, established during 1979 on the fringe of Puducherry, were found to have contaminated the surrounding groundwater[4]. This polluted groundwater is moving at an average velocity of approximately 30m a year, migrating to the Mettupalayam well field which houses the main wells for Puducherry's drinking water supply. By comparing the groundwater quality over the past 20 years, i.e. from 1981 to 2001, Ramesh (2005) noted that there was an alarming drop in the water level from 6 m to 26 m. The study by Sivasankaran (1997) on the status of pollutants in groundwater of the Puducherry region highlighted the presence of pesticides and heavy metals. The study on heavy metal concentrations in the industrial area of Puducherry by Aruna Devy (2002) indicated that cadmium and lead were present at higher levels beyond the prescribed standards in various industrial areas. Several studies have found that the accumulation of these metals in the food chain had become an enormous environmental hazard [13-15].

Various studies carried out to assess the physico-chemical characteristics of water indicated that the values of conductivity, total dissolved solids, calcium, magnesium, hardness and sulphate in groundwater have increased twenty fold since past few years, the values of chloride and sodium have increased fifty fold and potassium by six fold [16].Based on the Linear Regression Forecasting by Ramesh(2005), it is estimated that by the year 2020 the discharge of the total suspended solid (TSS) load will reach 592.18 TPA, the Biological Oxygen Demand (BOD) load will be 1958.93 TPA, and the Chemical Oxygen Demand (COD) load will be 4667.27 TPA. Discharge of oil and grease load will reach 66.65 TPA during this period. The projected annual groundwater utilisation by the industrial sector will reach 5727715.75 KL (5.727 mcm). However, the annual replenishment would remain at 155 million cubic metres (mcm). Thus the deficiency will increase to 3.6 mcm by 2020. It will lead to

IOP Conf. Series: Earth and Environmental Science **120** (2018) 012016 doi:10.1088/1755-1315/120/1/012016

further depletion of the groundwater table in the north western part and salt-water intrusion in the south eastern part of Puducherry [9]. In view of this situation, stringent action has to be taken by the government for the sustainable use of water resources for future industrialization.

3.1.2. Impacts on air - The air quality is equally affected by industrialization. By the end of the eighteenth century, Puducherry received 499 tons of suspended particulate matter (SPM), 2.88 tons of sulphur dioxide (SO_2) and 1.99 tons of nitrogen dioxides (NO_x) per year. At the end of the nineteenth century, several air pollution potential facilities such as M.S. ingots, ferro alloys, and calcium carbide were established and the pollutant level increased nearly ten fold for SPM and NO_x and fifteen fold for SO_2. The total emissions from these industries are still vast, even after applying the advanced air pollution control devices such as bag filter, double way scrubber and multiple cyclone. By 2005, 65 industries were identified as 'point source' (Stack emission) of air pollutant releasing units in Puducherry. By 2012, pollutants such as SPM, CO, SO_2 and NO_2 were found to have either exceeded or nearly reached the limits[17], necessitating the immediate installation of a continuous monitoring and control mechanism. According to the predictions by Ramesh (2005), the emissions of SPM released by the industries in the Puducherry region will be approximately 7834 tons in 2020, while the SO_2 released will reach 54.7 tons and the NO_x released will be 32.4 tons. Thus, air quality may deteriorate further in due course.

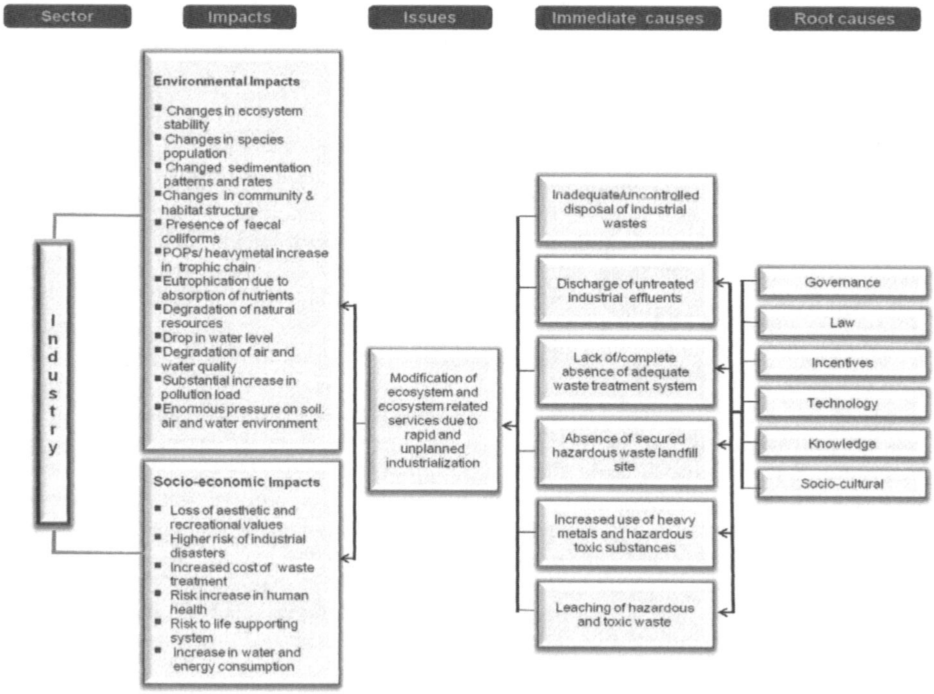

Figure 1. Causal chain diagram illustrating the causal links for Industrial Pollution in the Puducherry region (based on Giangiobbe *et al.*, 2002)

3.1.3. Impacts on ecosystem - This uncontrolled industrialization brought changes in community and habitat structure [18]. Forest cover of Puducherry has become very less, posing a threat to current ecosystem [8]. With this sudden boom in industries, the pattern of resource utilization in the form of energy, water usage etc increased drastically leading to the inevitable resource disturbance and imbalance. Today, the land bears no resemblance to its past, except in a few pockets of the region. It

IOP Conf. Series: Earth and Environmental Science **120** (2018) 012016 doi:10.1088/1755-1315/120/1/012016

already affected the sedimentation patterns, the distribution of major and trace elements and rate of soil formation, which in turn brought significant changes in ecosystem stability [9].
Regular inflow of industrial effluents into the ponds disturbed their ecological equilibrium. Consequently, the water which previously could be used for domestic and agricultural purposes became exceedingly unusable. This in turn affected the cropping pattern. Though not conspicuously alarming, the slow and steady pollution by heavy metals in the environment is quite hazardous. They reach the aquatic environment and being non-degradable, remains suspended or partially dissolved in water and subsequently accumulated in organisms [13].

3.1.4. Impacts on flora and fauna - Due to the changing climate of industrialization and in the absence of forest cover in most of the areas of Puducherry, several endangered and threatened plant species such as *Derris ovalifolia, Mallotus philippensis, Atlantia monophylla, Pamburus missicnsis Glyccsmis pentaphylla, Lepisanthus tetraphylla, Diosypyros ebnum,Gloriosa superba Gymnema sylvestre,Combretum ovalifolium,Derris scanden* and varieties of mangroves found in this region are under great threat [18]. The peri-urban estuaries where mangroves are located, receive a large quantity of untreated sewage and solid wastes that degrade the habitat and threaten the flora and fauna [19-20].
It also poses great risk to the very survival of rare and endangered species of wild animal population. Among the five species of sea turtles reported in Indian coast, leatherback (Dermochelys coriacea), loggerhead (Caretta caretta), olive ridley (Lepidochelys olivacea), hawksbill (Eretmochelys imbricata), and green olive ridley turtles (Chelonia mydas) are the common species nestling along Puducherry coastal belt. Possible adverse effects of noise on these turtles can range from behavioural modification including mild disturbances, disruption, displacement from key habitat, disorientation, capillary damage, loss of motor control, even death at some time [18].

3.2. Immediate causes
The current study identified inadequate/uncontrolled disposal of industrial wastes, discharge of untreated industrial effluent, lack of/complete absence of proper facilities for the waste treatment system, absence of secured hazardous waste landfill site, increased use of heavy metals and hazardous toxic substances, leaching of hazardous and toxic wastes as immediate causes of industrial pollution in Puducherry region (figure 1). With rapid industrialization, there has been a substantial increase in generation of various hazardous wastes, discharge of industrial effluents and emission contaminating groundwater resources [7].
The crisis of aquatic pollution and its impact on the health of organisms were an important current problem mainly because of the indiscriminate discharge of industrial effluents containing heavy metals like Iron, Chromium, Zinc and Lead etc. [11]. As the region's industrial structure shifts into highly polluting sectors, industrial effluents contain more and more heavy metals and non-biodegradable toxic and hazardous waste, leaching of which increasingly contaminate ground water resources both chemically and microbially [12].
Currently industries such as electroplating, pharmaceuticals, chemicals, steel, textiles, electronics, distilleries are major polluting industries, mainly responsible for generating hazardous waste of diverse nature. ETP sludge, waste oil, zinc oil, dichromate waste, spent pickling acid are major wastes or pollutants of concern. Uncontrolled, inadequate and unsystematic handlings of these wastes pose a greater risk and significant environmental burden in this region.

3.3. Root causes
Various issues related to governance, law, knowledge, technology and socio-cultural aspects have been identified as major root causes of industrial pollution in this region. Such causes include lack of long-range commitment from businesses to the environmental-friendly approaches, insufficient management systems, ineffective institutional mechanism for strategic planning, development, implementation and monitoring of pollution management, lack of access to appropriate technology and know-how, a failure to understand the specific potential of green approaches, lack of time and resources, corporate conservatism and a reluctance to drastically change business practices regardless of the economic and environmental benefits. Further challenges include the corporate belief that

pollution reduction is always a cost and never a saving and a business orientation toward sales maximization rather than cost minimization.

Hence, in this alarming situation if we fail to 'green' the industrial activities, then industrial pollution will be much more serious in this ecologically sensitive coastal region. Puducherry now needs to refurbish its infrastructure base to meet increasing sociological, economic and environmental demands simultaneously, without any further delay.

4. Recommendations for sustainable solutions

Strategic planning is necessary to address the environmental impacts of economic development to an acceptable level. In this scenario of degrading environment and pressure for constant industrialization, ecoindustrial network development provides viable solutions to move towards sustainable industrialization. However, it will require a supportive infrastructure base to make it happen. Industry will need a perfect combination of incentives, regulations, management mechanisms, information and other infrastructure facilities to provide the conditions in which industrial symbiosis (IS) can thrive [21].

A subsequent SWOT analysis by this author and her doctoral research broadly highlighted that significant potential exists for development of industrial clusters in different sectors in Puducherry. For businesses looking into inter-organisational interfaces for by-products and resources, potentials appear attractive. 5 types (one-one, one-many, many-one, many-many and one-all) and 28 total numbers of potential synergies were identified for by-product and resource exchanges during the same research. This can be supported by policy elements that emphasise and encourage the companies to move in directions aligned with IS principles [22], [23]. Regional development agencies and local Govt, which have increasingly accepted responsibility for balancing economic development with other pillars of sustainability, have a welcoming approach to the concept[24].Industrial diversity,continuous waste production, existing motivation in work environment, willingness of the govt., industry friendly incentives/subsidies, huge investment on industrial sector, good transport access, close proximity of industrial participants, local academic skill/expertise provide promising opportunities for successful implementation of IE principles[25-28]. All these create a supportive atmosphere for creation of eco-industrial park in this region based on material and by-product exchanges.

5. Conclusion

Rampant industrialization and urbanization have been very important causes for putting pressure on natural resources and also causing various degrees of environmental degradation. Puducherry, the south Indian industrial base has been experiencing similar kind of situation since many decades. There is urgent need of initiatives that ensure that industrialization is sustainable both in terms of taking measures to prevent damage to the environment and also promoting more environmentally friendly industries. A transition of the industries into eco-industrial network has emerged as a dynamic approach to preserve the natural resources of the region. Forging strong linkages between industry and related services can accelerate this transition to a diversified, high income economy while providing a safer environment. Situational analysis of Puducherry region pointed out existence of significant potential for industrial clusters development and inter-organisational interfaces for by-product and local resources. Key lessons learnt from Puducherry may act as a good basis for the development of a strategic action for the areas with similar environmental problems. From a comprehensive analysis of local environmental settings, strategies can be evolved to overcome the problems, by modifying or introducing appropriate approaches and filling the missing aspects of the related policies for strengthening sustainable industrialization.

Nevertheless, there is no one system that will work for every case. The same framework by itself will not bring a successful eco-industrial model. In light of this, the case of the each industry needs to be analyzed separately; however, other examples of successful strategies can be used as a basis to draw inspiration. The achievement of eco-industrial network depends on the in-depth study of each industrial characteristics and driving factors. It demands a strong determination and complete

dedication of major stakeholders, especially participating companies. It is important to make the participating companies understand that eco-industrial network will undoubtedly provide them with opportunities to gain environmental and socio-economic benefits in a long term and will greatly increase the sustainability of the system.
However, we need to address a number of concerns about implementation, dissemination and uptake of new management approaches to move industries in the direction of more sustainable practices and make them ready for the acceptance of recycled products and materials. Such solutions cannot be imposed from outside and need a holistic approach that is widely understood, or at least accommodated by local industrial systems. This will further allow to be readily absorbed into the local business culture while simultaneously helping to improve overall effectiveness, and thereby possibly assisting in a benign transformation of an existing ‘business-as-usual culture’ to a ‘sustainability-based culture’[29]. All that is needed to achieve this is a logical way to approach the strategy, through binding all the essentials rooted in social, economic, environmental, technical, scientific, cultural and intellectual green components in an integrated fashion. Implementing a feasible green approach along these lines will create the much desirable integrated structure, needed for the development of sustainable industrial base in the regions with similar features.

6. References

[1] Webnote 2017 Impact of industrialization on the environment *Environment Insider.* http://environmentinsider.com/impact-industrialization-environment/

[2] Zucker D M 2009 How to do case study research *School of Nursing Faculty Publication Series Paper 2.*

[3] Directorate of Industries and Commerce 2010 Government of Puducherry

[4] Raj S 2016 Environmental issues of Puducherry UT an environmentalist’s outlook *Best:* [5] *International Journal of Humanities, Arts,Medicine and Sciences* 44 171-196

[5] Annual Environment Survey Report 2010 –2011 Govt. of Puducherry

[6] Puducherry government inventory 2008. Inventory of Hazardous Waste Generation and the *Hazardous Waste Generating Units in the UT of Puducherry* PPCC Government of Puducherry

[7] PIPDIC handnote 2009 *Puducherry – The gate way to success*

[8] PondyCAN final report 2012 Sustainable regional planning framework for puducherry,viluppuram, auroville & cuddalore *Sustainable Regional Plan* Project Number 1009C0056

[9] Ramesh N 2005 Evolving an Integrated Environmental Management Plan for Sustainable Industrialization of Pondicherry Region PhD dissertation Pondicherry University

[10] Giangiobbe S, Gonzalez S and Pacheo F 2002 Causal chain analysis *GIWA Regional Assessment* 38 Patagonian Shelf Kalmar Sweden: University of Kalmar

[11] Murugesan A G, Baladhandayuthapani M and Sukumaran N1999 Immunotoxicity of lead and modulatory effect of extract of aegel marmelos to fish, cyprinus Proc of Int Wrkshp on Environmental Impacts of Metals (Coimbatore India: TNAU) 61–67

[12] Abbasi SA, Ramesh N and Chari KB 2002 Studies on environmental management of Pondicherry: The water and air resources *Journal of the Institution of Public Health Engineers* 3 20–24

[13] Rao U and Govindarajan S 1992 Transfer of copper and zinc through a marine food chain *Acta Botanica India* 2 71–75

[14] Senthil Nathan D Kumar RM Reddy SS Sivasankaran MA and Ramesh R 2012 Trace elements in groundwater of coastal aquifers of Pondicherry region India *Journal of Environment* 1 4 111–18

[15] Jameel AA Sirajudeen J and Vahith RA 2012 Studies on heavy metal pollution of groundwater sources between Tamilnadu and Pondicherry India *Advances in Applied Science Research* 3 1 424–29

[16] Ramesh NE, Ramasamy V and Kothandraman R1994 Salinity intrusion in the coastal aquifer of Pondicherry Nat Semnr Clean Water Great Needs Greater Challenges(Puducherry India: Pondicherry University) 23-24

[17] Balashanmugam P, Ramanathan AR, and Kumar VN 2012 Ambient air quality monitoring in Puducherry *International Journal of Engineering Research and Applications* 2 2 300–07
[18] EIA report 2010 Design and construction of work shelters in fishing villages along the Puducherry and Karaikal coasts *Emergency Tsunami Reconstruction Project _ Environment Impact Assessment Report* Government of Puducherry
[19] Satheeshkumar P and Senthilkumar D 2011 Identification of heavy metals contamination by multivariate statistical analysis methods in Pondicherry mangroves *India Journal of Environment and Earth Science* 1 1 30–48
[20] Palanisamy S 2012 Puducherry mangroves under sewage pollution threat need conservation *Current Science* 102 1 13–14
[21] Cote RP, Grant J, Weller A, Zhu Y and Toews C 2006 Industrial ecology and the sustainability of Canadian cities *Eco-Efficiency Centre*
[22] Patnaik R and Poyyamoli G 2015 Developing an eco-industrial park in Puducherry region India – a SWOT analysis *Journal of Environmental Planning and Management* 58 6 976-96
[23] Patnaik R 2012 Studies on the Application of Industrial Ecology Principles in Puducherry Region PhD dissertation Pondicherry University
[24] Amurthalingam A 2016 Environmental protection through NGOs in Puducherry region International *Journal of Interdisciplinary Research in Arts and Humanities* 1 1 67-69
[25] Mirata Mand Pearce R2006 Industrial symbiosis in the UK. In *Industrial Ecology and Spaces of Innovation*, ed K Green and S Randles (Cheltenham: Edward Elgar) pp 77–105
[26] Chiu ASF and Yong G 2004 On the industrial ecology potential in Asian developing countries *Journal of Cleaner Production* 12 1037–45
[27] Banerjee SB 1998 Corporate environmentalism: perspectives from organizational learning *Management Learning* 29 147-64
[28] Chertow M 2000 Industrial symbiosis: literature and taxonomy *Annual Review of Energy and Environment* 25 313–37
[29] Clayton A, Muirhead J and Reichgelt H 2002 Enabling industrial iymbiosis through a webbased waste exchange *Greener Management International* 40 93–106

Acknowledgments

The author is grateful to the Director, DSTE, Government of Puducherry, Dr N.Ramesh, Environmental Engineer, Puducherry Pollution Control Committee, Assistant Director, Directorate of Industries and Commerce, PIPDIC for facilitating data collection from industries and for providing secondary data. The author gratefully acknowledges valuable suggestions and inputs given by Prof. G. Poyyamoli from Pondicherry University for improving the content of this background research. The author also thanks Pondicherry University for providing the necessary facilities to carry out this research.

International Trade, Pollution Accumulation and Sustainable Growth: A VAR Estimation from the Pearl River Delta Region

Hui Zuo and Lu Tian

School of Economics and Commerce, Guangdong University of Technology, Guangzhou City, 510520, China
E-mail: zuohui@gdut.edu.cn

Abstract. In order to investigate international trade influence in the regional environment. This paper constructs a vector auto-regression (VAR) model and estimates the equations with the environment and trade data of the Pearl River Delta Region. The major mechanisms to the lag are discussed and the fit simulation of the environmental change by the international impulse is given. The result shows that impulse of pollution-intensive export deteriorates the environment continuously and impulse of such import improves it. These effects on the environment are insignificantly correlated with contemporary regional income but significantly correlative to early-stage trade feature. To a typical trade-dependent economy, both export and import have hysteresis influence in the regional environment. The lagged impulse will change environmental development in the turning point, maximal pollution level and convergence.

1. Introduction

The Pearl River Delta, known as the "world factory", contributes about 30% of aggregate value international trade in China annually in the past ten years, which is considered as world-shaking for over 30 years for its amazing reform and opening. However, the Pearl River Delta's international trade is based on the high input, high consumption, high pollution and developing mode of pollution first and treatment last. Pollution and resources shortage has become a double-edged sword to the sustainable development of economy in the Pearl River Delta region.

Environment Kuznets curve (EKC) reflect the interrelationship between economic growth and environment change. EKC is often shaped as inverted-U because environment deteriorates as economic growth at early stage and improves later. Most of researchers regard that inverted-U is the result of sequential development of three effects, i.e. scale effect, technology effect and structure effect (Grossman G M and Krueger A B, 1993) [1]. In the very beginning of industrialization, scale effect is overwhelming and dominates the change of environment. The other two effect is so weak that the environment changes mainly by the economic scale. It is the deterioration stage in the inverted-U EKC. In this stage economic growth exploit environment and resources at little residential consciousness and very low cost. Afterwards, structure effect and technology effect comes to stronger and replace the leading position of scale effect, which will guide the interrelationship between economic growth and environment to virtuous circle (Lu Y and Guo L, 2008) [2]. Usually, they play effectiveness simultaneously, because technology effect is interrelated to structure effect. The former one will help the latter one to perform better and advance faster. As to the reasons of the replacing, much research work has been done from different perspectives including market mechanism (Thampapillai et al., 2003) [3], consumer preference (Lopez R and Mitra S, 2000; Lopez R et al., 2007) [4], [5], environmental policy (Torras M J, 1998) [6] and international impacts (Copeland B R and Taylor M S, 1995, 2004) [7], [8]. The explanation varies greatly but the replace order and consequence is quite similar in most of empirical research.

A great deal of empirical study testified the mechanism of inverted-U EKC. They have selected different kinds of data in cross-section, time series and panel structure from different countries and territories in different periods. However, there are many differences among not only methods but also results. The discrepancies include shape, slope and turning point. As to the shape of EKC, some research work disagrees that all of EKC is shaped as inverted Egli H and Steger T M (2007) [9] regarded that EKC is probably multi-shaped. That is environment may deteriorate continuously, or, in some circumstance, improve after the first turning point and deteriorate at the turning point. Therefore, EKC may also be a line with positive slope or N shape (Fu M, 2008) [10]. In the first case, there is no turning point and in the second case, there are two turning points. Some panel data regression results show that the second turning point does exist in some developing countries and that whether or not turning point will appear depend on the detailed circumstances of selected countries and periods as well as the pollution measurement (He J and Wang H, 2012) [11].

As to the position of the EKC turning point, there is no convincing result because the analysis from different paper shows that different regime or geographic location may result in different turning point (Zhu P H et al., 2010) [12]. Indicator variety or different econometric method also influence the turning point estimated, which makes the comparison of the turning point among different countries difficult. If take the stochastic factor into consideration, EKC will be more complex among countries (Prieur F, 2009) [13]. Suppose that pollution accumulation is some kind of Brownian motion, up to three threshold value may arise in environment changing process, which will lead to M or more complex shaped EKC.

Regarding interrelation between trade and environment, there are three kinds of viewpoints. The first one is that trade liberalization is beneficiary to the environment. Lucas R et al. (1992) [14] made the regression analysis on the relationship between the intensity of pollutants in measured in emission per unit of output and economic growth by multiple countries data from 1960 till 1988. The result shows that pollutants agglomeration declines in some countries after reaching a certain income level after rapid growth of pollution emissions and pollution intensive industry scale in the developing world. The second viewpoint regards that international trade accelerate the pollution of the environment. Chilchilniky G (1994) [15] argued that free trade to developing countries is more destructive than to developed countries. Trade liberalization will accelerate the deterioration of environment as well as resources exhaustion in developing countries, which threatens global environment and ecosystem if private property rights are not clearly defined, in another word, the polluter thinks emission as public and will surly average the cost and damage to every country. The third viewpoint is that the impact of international trade on the environment is uncertain. Copeland B R and Taylor M S (1995) [7] introduced a South-North model and disintegrate the influence to environment into three effects, which are scale effect, structure effect and technology effect respectively, so as to analyse the impact of international trade on the environment. The result suggests that free trade makes the South environment destructed and the North, improved.

Some domestic study of the relation between trade liberalization and environment is more precise in analytical regional-concerned result. Some scholars think that the interrelation between international trade and environment can be explained by some kind of game theory and apply to the law of the unity of oppositeness. Wu G M and Wu G S (2011) [16] introduced a game theory model of the prisoner's dilemma to study the environmental game process and trade game process among the partners of the international trade. He found that the relationship of environment and trade is a stable Nash equilibrium rather than an irreconcilable contradiction. Expanding the mechanism of trade liberalization influencing pollution, some domestic scholars regarded that four effects should be taken into consideration: scale effect, structure effect, technique effect and law effect. Li B et al. (2006) [17] constructed a general equilibrium model of international trade and environment so as to decompose the environmental effect by trade into scale effect, structure effect, technology effect and law effect. He found that scale effect aggravates the deterioration of environment in China, and the structure effect and technology effect improves environment.

The discrepancies in empirical study are partly due to ignorance of endogenous interrelationship between economic growth and environment. Most of research work put emphasis on one-way impact on environment by economic growth rather than reaction of environment (Andreoni J and Levinson A, 2001; Dinda S, 2005) [18], [19]. If economic growth results in scale effect, structure effect and technology effect, which have effects on environment, will environment change has feedback and influence economic growth? In fact, when regarded as factor of production, environment will affect economic growth by all of these effects. In our early research, it was found that economic growth is affected by the environment. Environment quality is also featured as a flow factor for output, which implies that more rapidly the economy does grow, more costly the environment is. Considering that high environment quality can provide abundant factor endowment for production, a country before turning point obviously has different structure effect from one after turning point. Such kind of upturn will happen to one country in different economic stage and change all of effects on environment by economic growth.

This paper will focus on the inner mechanism and its effects on relationship between the international trade and the regional environmental pollution by empirical analysis. Quite different from other research work, we introduce lagged factor so to provide more powerful explanation to auto-correlation in environment change in the Pearl River Delta. In the next section, we will introduce variables and data. Section three introduces the VAR model, test the stability and cointegration and estimate the coefficients. In the fourth section, we discuss the international trade hysteresis influence in environment and impulse to environment inverted U process. The last section is our conclusions.

2. Materials and methodology

To introduce endogenesis, we suppose that environment and economic growth will impact each, so each of them should be taken as endogenous variable decided by another endogenous variable and exogenous variables. We set the model structure in Figure 1.

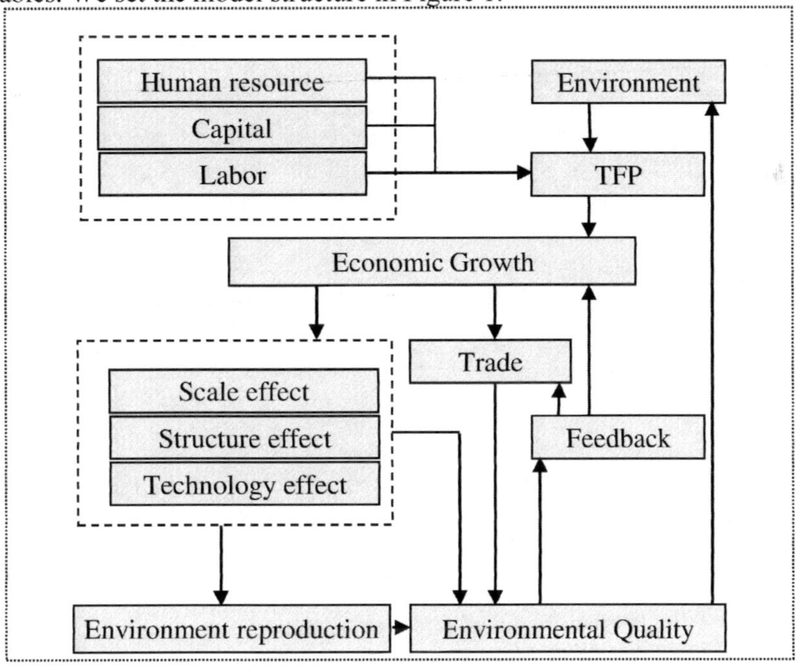

Figure 1. Model Framework

The model includes two endogenous variables, one is environment and the other is economic growth. Economic growth has four channels to influence environment, i.e. scale effect, structure effect and technology effect and trade. Scale effect is deleterious to environment and structure effect and technology effect is advantageous. Trade effect is variable. These effects not only influence environment but also change environment reproduction, which may speed up or slow down environment improvement. Economic growth relies on several factors, including capital, labour, human resources and especially, environment. Environment has effects on total factor productivity, which determines the output. Environment factor input quality and quantity is dependent to environment performance and the performance is influenced by economic growth through three effects. It is very important that environment has feedback to economic growth and trade. And, economic growth and trade influence the environment in turn. In our early research, we have construct a simultaneous equation group to simulate the interrelationship among them. We have estimated the coefficients in the equation group and transform the group to single equation. Considering the feedback will have hysteresis effect, we include the feedback in the lag item of economic growth. The model is set as

$$y = f(y, e, L.y, L.e, tr, ot) \tag{1}$$

$$e = g(y, e, L.y, L.e, tr, ot) \tag{2}$$

in which y, e, tr and ot is GDP, environment, trade and other exogenous factors respectively and L is lag operator. The effect of feedback is included in variables L.y and L.e. Other exogenous factors include foreign direct investment (FDI), energy and industry structure. Variables of trade include export ex and import im. Lag term is tested up to 4th order. Both y and square y are included in the equation group.

For the convenience of analysis and data availability, this paper selects exports and imports as indicators of international trade. In order to highlight the influence by trade structure, we use product of ratio and GDP. However, the ratio is not the simple proportion of aggregate import or export in GDP. Considering that technology progress will impact international trade, we weigh the ratio by industry and give technology intensive and less pollution intensive industry more weight. Pollution is measured by emission of wastewater, industrial waste gas and industrial waste solid. We collect the environment quality year report 2004 till 2014 from GIS in Guangdong Environmental Protection Department Data Centre. According to the emission in each city where monitor station lies in, we weigh environmental quality data to an index indicator from 0 to 1, smaller is better. The Pearl River Delta's pollution and trade data, covering the period from 2002 to 2014, are collected from statistical yearbook and the economy development statistics report in the Guangdong and the Pearl River Delta. Some data before 2005 are calibrated by other statistic material.

We will pay more attention to the interrelationship between the growth rate of international trade and environment. Primary regression also suggests that logarithmic indicators perform better in our model considering goodness of fit and logarithm will not change the interrelationship but can reduce heteroscedasticity significantly and make the sequence into stationary series easily. Considering the dependent variables of pollution may have correlation with dependent variables of trade and in order to prevent estimation from multi-collinearity, we also calculate 1st class difference of each variable for test comparison and use differentiated logarithmic variables in our VAR and panel model.

3. Results

3.1. Time series test

First, we inspect the time series stationarity by ADF test for samples of time series using STATA. Table 1 shows the logarithmic results as well as the differentiated logarithmic results for comparison.

Table 1. ADF test[a].

Variable	Test	Statistics	Stationarity	Test Δ	Statistics	Stationarity
Log(ex)	c, t, 1	-1.406	No	c, t, 3	-5.533	Yes
Log(im)	c, n, 0	-1.196	No	c, t, 0	-3.676	Yes
Log(water)	c, t, 0	-3.092	No	c, n, 3	-4.028	Yes
Log(gas)	c, t, 1	-2.548	No	c, n, 0	-4.203	Yes
Log(solid)	c, t, 0	-2.61	No	c, n, 0	-4.409	Yes

[a] Test form c, t, k parameters denote constant term, time trend, and lag order number of unit root test. Δ represents the first order difference, The optimal lag order number is determined by the AIC criterion, when the absolute value of ADF is greater than 5% critical level, the sequence is judged as stationary series.

According to the rule of AIC and SC, each group of variables are tested up to 3rd lags. Table 2 shows the results.

Table 2. Vector cointegration test[a].

VECTOR	H0: NONE		H0: 1 AT MOST		COINTEGRATION
	Trace	Max Eigenvalue	Trace	Max Eigenvalue	
ex, water	15.9578	12.6442	3.3136	3.2274	1
ex, gas	19.1293	16.0673	3.0620	3.0010	1
ex, solid	12.6220	8.8423	3.7797	3.8525	0
im, water	15.9933	12.3866	3.6067	3.5503	1
im, gas	21.9442	18.4832	3.4610	3.1217	1
im, solid	12.0990	7.6466	4.4525	4.2564	0

[a] Null hypothesis is none cointegration and 1st order cointegration at most respectively. The 5% critical value for trace is 15.41 and for maximum eigenvalue is 3.76. The last column is the judgement order. Ex and im is export and import respectively. All tests are applied in 1st order differentiated logarithmic variables.

Table 2 suggests that both export and import are 1st order cointegrated with water and air pollution while there is no significant cointegration relationship between solid pollution and any trade, either export or import. Therefore, international trade has long run influence on the environment of the Pearl River Delta, despite that the environmental change may has intrinsic random features and some emission are not significantly observed tightly related to trade, e.g. solid. For comparison, we also test the logarithmic variables without 1st order difference. The result shows that no cointegration relationship exist among any groups of vectors.

3.2. VAR model and impulse response

According to the result of time series test, we focus on relations of international trade and pollution of water and air, so we construct two groups of VAR model as follows:

$$VAR\ 1:\ Z_1 = (d\log \text{water}, d\log \text{ex}, d\log \text{im})' \tag{3}$$

$$VAR\ 2:\ Z_2 = (d\log \text{gas},\ d\log \text{ex},\ d\log \text{im})' \tag{4}$$

Considering that increasing lag order will lower the freedom degree of estimation and in order to eliminate autocorrelation of residual to provider stronger explanatory capability, we set the maximum lag order to two. All variable are 1st order differentiated and logarithmic. VAR model test result for order is given in table 3.

Table 3. Lag order to VAR[a].

	lag	LL	LR	FPE	AIC	SC	HQ
	0	28.9099	--	1.5e-07	-7.1744	-7.1446	-5.89963
VAR1	1	49.8923	22.64	1.2e-07	-7.7544	-7.6353	-8.55817
	2	133.566	56.43	0	-8.0023	-7.6440	-8.83254
	0	27.1214	-	4.8e-07	-7.03036	-6.0005	-6.23128
VAR2	1	41.3329	28.423	1.8e-07	-7.33322	-7.2140	-8.13692
	2	117.856	35.34	-3.6e-36	-7.6743	-8.1543	-7.34643

[a] LL gives the likelihood function value of VAR model. LR is likelihood ratio. FPE represents the final prediction error, AIC refers to Akaike information criterion. SC means Schwarz' s information criterion, HQ denotes Hannan-Quinn information criterion.

Taking all criterions into consideration, we regard that 2nd lags for VAR1 and VAR2 are the optimal result. Estimation result of the model is listed in Table 4.

Table 4. VAR estimation[a].

		Con	L.Water	L2.Water	L.Export	L2.Export	L.Import	L2.Import
	Water	0.001	0.464	0.301	0.487	0.355	-0.501	-0.361
VAR1	Export	0.055	0.243*	-0.124	0.748	-0.002*	0.011	0.008
	Import	0.045	0.177	-0.093*	0.098	-0.012	0.726**	-0.001

		Con	L.Air	L2.Air	L.Export	L2.Export	L.Import	L2.Import
	Air	0.013	0.811	-0.028	0.425	0.218	-0.536	-0.009
VAR2	Export	0.036	0.235	-0.115*	0.806	-0.089	0.088	-0.011
	Import	0.043	0.215	-0.111*	0.154	-0.083**	0.808	-0.023*

[a] L is the 1st order lag operator and L2 is the 2nd order lag operator. Con is constant term. Critical level is 5%. * and ** is significant at 10% critical level only and not significant at 10% respectively.

Figure 2 shows the scatter plot and average line of water and air pollution estimated by the VAR model in 50 periods. We can find that the shape of line is quite similar with each other, but the pollution level is not exactly the same and this is only a matter of measurement. It can be found that the maximal pollution change rate of water and gas is 4.6860% and 6.4169% respectively. However, it is not the traditional EKC tuning point because the pollution keep growing at this point at its highest marginal rate. In these curves, pollution of water or air increase at beginning and the change rate of the pollution deterioration speed rate keep increasing until the 6th (to water pollution) and the 5th (to air pollution) period. After these point, environmental quality keep degrading and the degrading speed is increasing until the 17th (to water pollution) and the 18th period (to air pollution). Then, the change rate of environmental degrading rate of both water and air turn to negative, i.e. environment is deteriorating but the deteriorate rate stops from increasing and begins to decrease.

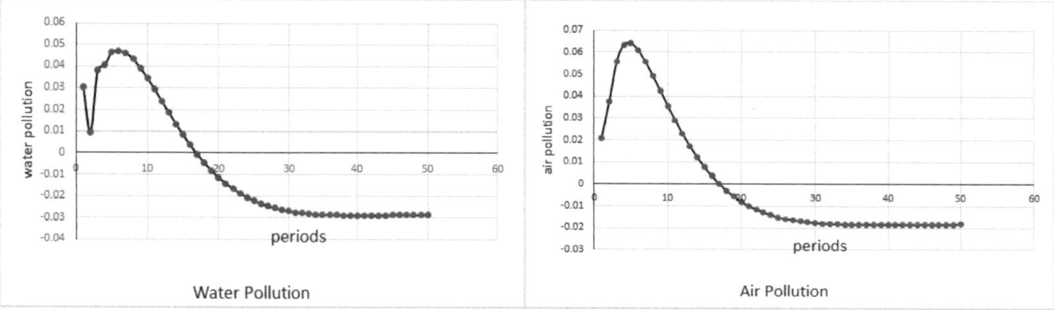

Figure 2. Rate of water and air pollution change percentage

Considering the initial point in the sample of this paper is in pollution deterioration stage, accumulated change of both water and air pollution change percentage needs to decline below zero, when environment improves. The accumulated change of water and air pollution change percentage is plotted in Figure 3. To water pollution, at the 39th period, the accumulated change is near zero and to air pollution, the 53th period. About 4-7 periods after those points, environment begin to improve, i.e. 43-60 periods of deterioration need to be experienced according to our estimation results.

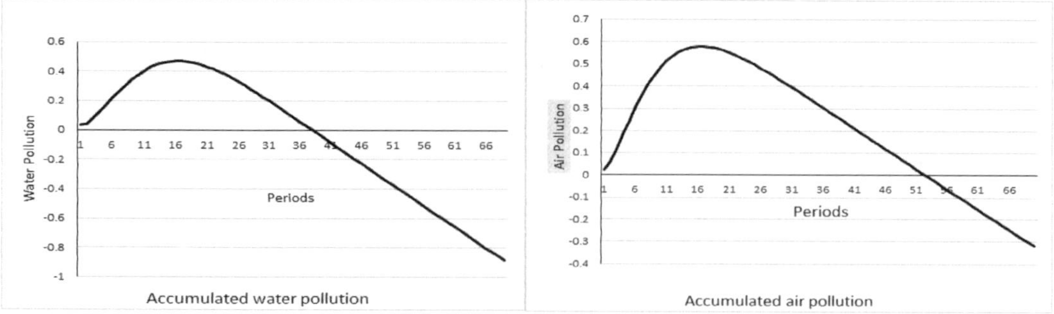

Figure 3. Accumulated sum rate of water and air pollution change percentage

3.3. Robust test

We have carried on 2 groups of robust test to the model estimation. The control variable is FDI and industry structure. The former one is for opening test, i.e. if another variable indicating opening can replace the agent position of international trade and also influence environment, will the model estimation result keeps robust. If FDI of similar type of products can change the fitness of the model or weaken the significance of the trade, the original estimation result is lack of robustness, which may imply that change of environment is not the specific result of trade.
Another control variable is industry structure. The structure effect of economic growth has already included this effect in the endogenous mechanism. If the model is robust, an additional structure variable will produce some collinearity and weaken the significance of trade. We also take the combined test for two control variables. The robust test result for water and air is listed in Table 5 and Table 6.

Table 5. Robust test for water[a]

variables	Water pollution equations			
	Trade	FDI	Structure	Combined
constant	0.001***	0.002**	0.001**	0.001*
L.water	0.464***	0.426***	0.501**	0.550***
L2.water	0.301***	0.315**	0.228**	0.013**
L.export	0.487***	0.409***	0.444***	0.334**
L2.export	0.355***	0.289**	0.295°	0.283*
L.import	-0.501***	-0.435***	-0.488**	-0.465***
L2.import	-0.361***	-0.187***	-0.273*	-0.201°
FDI	--	-0.173***	--	-0.102***
Structure	--	--	0.005**	0.004°

[a] ***, ** and * indicates 1%, 5% and 10% of critical level respectively. $^{\circ}$ is not significant at 10% critical level.

Table 6. Robust test for air[a]

variables	Air pollution equations			
	Trade	FDI	Structure	Combined
constant	0.013***	0.008*	0.010***	0.001*
L.water	0.811***	0.600**	0.655**	0.477**
L2.water	-0.028***	0.012**	-0.001**	-0.023***
L.export	0.425***	0.384**	0.419**	0.390**
L2.export	0.218**	0.180**	0.224°	0.190*
L.import	-0.536**	-0.334**	-0.451**	-0.440**
L2.import	-0.009***	-0.240*	-0.215***	-0.207°
FDI	--	-0.094*	--	-0.056**
Structure	--	--	0.010*	0.009°

[a] ***, ** and * indicates 1%, 5% and 10% of critical level respectively. $^{\circ}$ is not significant at 10% critical level.

The robust test result shows that both FDI and industry structure has slight influence on the model because overall significance lower down slightly in the test. However, the significance of lag term of environment has little change and it suggests that the hysteresis effect in this EKC VAR model is still robust. FDI effect on environment is quite similar to trade, which suggest opening effect may partly be resulted from investment.

4. Discussion

In this paper, we will focus two key points. One is international trade hysteresis influence in environment and the other is how impulse will change environment inverted U process.

4.1. International trade hysteresis influence in environment

Regression results in Table 4 show that both water and air pollution is highly related to international trade hysteresis effects. One of the important founding is that both the 1st order and the 2nd order

IOP Conf. Series: Earth and Environmental Science **120** (2018) 012017 doi:10.1088/1755-1315/120/1/012017

coefficients of pollution intensive export in either VAR1 or VAR2 are positive, which means the impulse of such export is deleterious to the environment continuously. To most export intensive developing countries or regions, resources exhaustion, extensive manufacturing and poor regulation in pollution will all accelerated the pace to worse environment. However, lots of research done can only find such relationship as between pollution and contemporary international trade rather than pollution and lagged international trade. EKC empirical research by us earlier also suggests that the correlation between international trade and water environment is so weak that contemporary economic growth and corresponding structure, technology and scale effects are much more decisive than international trade in environmental change and the turning point. This is true only when considering the synchronous effects. Export and import effect on environment need more time to be seen.

One of the key explanations to this point is comparative advantage and factor endowment. When a region finds some factor endowment can bring comparative advantage in their export department, e.g. Foshan city in the Pearl River Delta in our sample find they can produce and export many ceramic goods for its endowment in labour and kaolinite, it tends to build strong producing capacity and export in large scale. In the early stage, it can be found that rapid growth in export doesn't bring much environment destruction because the exhaustion of resource and its environmental damage is covered by natural recovery of eco-system. Thus we can only find insignificant influence in environment by export industry. And, this is stimulation to absence of environmental regulation as well as faster expansion of such export. With the development of the industry, environmental deterioration accelerates and cannot be restored by natural renewable process any longer. Therefore, the pollution deterioration is highly correlative to the influence of earlier stage export.

Another reason to export lag lies in the industry cluster led by export. Like the Pearl River Delta, most export-led regions in China or Asia attract a lot of enterprises for preferential policies, mature infrastructure, brand effect and technology spill over. When export booms, such kind of region has more rapid growth rate and attractiveness than other regions. More and more enterprises will invest here, whether they do international trade or not. Therefore, export growth in early stage may bring continuous prosperity as well as increasing pollution for the region. In some cases, environmental pollution will not decline even though export dwindles in several years because a large part of output for exporting will be redirected to regional or domestic market, or kept as stock. So export fluctuation is not the decisive factor to contemporary environment, but the important factor to future environment. The pollution is correlative to production in earlier stage, and part of which would have be arranged as the export, but now as regional or domestic sale.

Quite different from export, the influence of import in environment is advantageous rather than deleterious. But the lag does also exists. The negative coefficients of all lag terms of import suggest that environment will improve with the import growth although such improvement is partly offset by pollution in consumption process. Similar with export performance in environment, there is even inferior correlation between environment change and contemporary import growth. According to comparative advantage theory, if developed region imports a great amount of pollution intensive goods, which means the production process is already shifted abroad with regional industry upgrade, and environment improves with this process. Nevertheless, the improvement does not happen synchronously in significance due to the hysteresis effect. However, hysteresis in import is completely different from it in export. It is generated by industry shift lag and consumption lag.

Industry transfer lag is the time between pollution intensive transfer and environment improvement. When a region decides or is obliged to transfer its pollution intensive industry abroad, e.g. to some lower development region, the transfer will not recover the environment very soon. There are several reasons to this point. First, recovery of ecosystem needs a relative long period, so the water and air quality does not improve as soon as pollution emission decline. Second, the transfer of pollution intensive industry is always later than import increase. To some regions, loss of factor endowment as well as competitive power are the reasons of such industry transfer. The diminishing of export and growth of import happen along with the loss, surely ahead of transfer, and environmental improvement happens even later to it. Third, some regions will kept their damaged environment unrecovered even when they can restore environment after the pollution intensive industry transfer because of huge costs in restoration. The environment improvement will slow down with the inaction of government. Fourth, some consideration in regional employment makes the pollution intensive industry transfer halfway, so the increase in import is limited effective to pollution control.

Consumption lag is the lag of the environmental improvement due to consumption upgrade from more pollution intensive domestic products to more environmental domestic products. When a region needs to upgrade its growth mode to more technology intensive and environmental friendly type, contemporary environment improvement is not so significant. There are many reasons to this point and we discuss four of them.

First, the government always restricts the production with uncontrolled emission and these manufacturing must transfer to other less developing regions. However, such now imported goods are even less environmental than produced regionally earlier in consumption process, because the current producers in less developing regions usually has much less environmental regulation. That's also why many researchers thought that there are may be some Pollution Heaven in the industry transfer. In the relative early stage, a region has less control for such pollution intensive import in order to satisfying regional demand. So environment does not improve along with such import increase.

The second reason to that point is that the regional producers may cut their producing cost to compete with other region producers, and the cut parts may include emission clean and pollution control. The regional producer's response aggravates environmental destruction until it pulls out of the regional market. Before that, they produce more such goods with lower costs in order to recoup investment. So the environment will not improve with the import increase.

The third reason is something like the reason discussed in export, i.e. the import increase does not definitely decrease the regional production for the reasons of regional employment and tax. The adjustment usually needs three to ten years or even longer. The improvement of ecosystem is influenced by this process.

The last reason lies in the time between import and consumption, and this lag is not too long. Imported goods production lies in other region and will not pollute native region, but the consumption and waste recycling may pollute regional ecosystem. Therefore, the offset effect does not happen before the final consumption. Some imported goods are intermediate goods and it will spend some time before it changes into final products and be sold to consumers.

4.2. International trade impulse to environment inverted U process

Pollution rate dynamic fitted line in Figure 1 suggests that Environmental Kuznets Curve as well as its stage and turning point alteration may be explained as autoregressive process rather than cubic polynomial fitting of income. A great deal of literature regard that the economic growth makes the pollution deteriorate in early stage and turns to improve environment when income reaches a certain level. As stated in many surveys, the technology effect, scale effect, structure effect and regulation effect are adjusting tools to interrelationship between income and pollution. International trade influence in environment relies on its impact on economic growth and structural change in it.

However, the conclusion in this paper implies that environmental change in highly opening region may be less correlative to economic growth but more to international trade. It is a process of endogenous auto-development, and most attention-getting point is that trade has direct effects on environment, regardless of regional income level. International trade can change the environmental curve considerably. That also means the turning point in different countries or regions may vary considerably with their international trade mode and structure. Many research work has questioned that why the turning point in EKC is so different among countries that the maximal turning-point-corresponding income is twice or more of the minimal value. We regard the reason lies in this endogenous dynamic process. Estimation result suggests that either water or air pollution change rate (attention, change rate rather than absolute level) will rise in the early periods. After a certain amount of periods (depending on the initial point and actual situation, which is 17 to 18 in our research region), pollution rate comes to the turning point and then environment deterioration rate reach maximum. After about 25 periods, the rate altercation converges to a certain level (-0.0286 to water and -0.0182 to air in the Pearl River Delta region in our research). Therefore, the corresponding pollution growth rate has an accelerative rise in early stages and the rise rate will reach the maximal value after 17 or 18 periods. After that point, the pollution will not abate immediately, but keep increasing with decreasing deterioration rate until this rate lowers to zero. That is 39 and 53 periods to both water and air in our sample of the Pearl River Delta. Then, environment begins to improve continuously and the improving rate will also rise until converge to the certain rate.

This inverted U shape is completely an endogenous auto regressive progress, in which trade has great influences in environment directly. In this paper, we only discuss the change of the turning point, zero point and convergence of environment. In our estimation and fitting, it is found that an increase of the 1st or 2nd order lag positive impulse of export (or negative impulse of import) has little change on the turning point period of the environment change rate, but significantly increases the maximal pollution level. Such kind of impulse change of export will delay the period to zero pollution rate. It may decrease the final convergence environmental improvement rate, or even make environment converges to a steady deterioration rate and never improve. Regions as the Pearl River Delta is usually featured with scale pollution intensive export. Due to poor regulation, such export has tremendous and long-lasting damages to regional environment. That is why we find impulse to environment of export in the Pearl River Delta region is higher than data in other regions in China. Extensive scale development in such export department leads to higher maximal level of pollution growth rate. When the impulse is not so strong that offset influence of other factors, e.g. import and other endogenous various, it will only delay the coming of environment improvement and dwindle the final improving rate. However, if the impulse overwhelm other influences, (may it happened in some cities of this region in case that pollution emission is not controlled properly). The zero environmental deterioration rate cannot be reached and regional environment will keep deteriorating until environment cannot support the economic growth any longer.

We also find the impulse of export will abate with lag period in both water and air pollution, which suggests that the export influence in environment abate as time goes go. The coefficient of the 2nd order lag export is smaller than the 1st order lag. We have tested the 4th order lag of the VAR for comparison, using same sample and find even weaker and less significant influence by 3^{Rd} and 4th order lag. Part of the influence of higher order lag export has been included in the lower lag order term, but more important, the region is motivated to change the industrial layout, stimulate export upgrade and impose tough sanction and regulation against illegal emission. So the marginal influence in environment by early export growth decreases. This also happens to import.

5. Conclusions

In this paper, we investigate the international trade influence in environmental change, using the data of the Pearl River Delta. We find that pollution is highly related to international trade hysteresis effects. Impulse of pollution intensive export deteriorates the environment continuously and such impulse of import improves environment. Such effects as well as their influences in environment are insignificantly correlated with regional income, which implies that environment can be improved with trade structure and mode advancement in highly opening or trade-dependent economy, even in relative low-developed stage. This also suggests that rapid economic growth is not definitely helpful to an economy to cross the EKC mountaintop because it may has much higher turning point than other economies. International trade hysteresis effects can move auto-regressive EKC turning point, make stages shift and change the maximal value of environment as well as convergence feature. Long-lasting effect of impulse of trade can decide the way of environmental sustainable development or a vicious circle to some extent.

One of the most interesting conclusions in this paper is that environmental effect lag are quite different between export and import. The export lag is caused by comparative advantage, factor endowment and industry cluster. While the import lag is cause by industry transfer lag and consumption lag. We offer some major mechanisms. However, whether or not they have effect equivalency is still questioned.

We are not optimistic about the environment performance in the Pearl River Delta region and other such kind of export intensive economic. According to the estimation result in the paper, this region needs at least 30-40 years before its environment coming to virtuous circle. Currently, the government of this regions is planning to apply stringent specification and control to pollution. To this trade dependent economy, before structural change, these stringent rules are merely teeter-totter between environmental protection and economic growth.

IOP Conf. Series: Earth and Environmental Science **120** (2018) 012017 doi:10.1088/1755-1315/120/1/012017

6. References

[1] Grossman G M and Krueger A B 1993 Environmental impacts of a North American Free Trade Agreement *The Mexico-U.S. Free Trade Agreement* ed Garber P M (Cambridge: MIT Press) pp. 13-56

[2] Lu Y and Guo L 2008 A neo-classical growth model for pollution of inverted u-shaped curve and environmental expenditure of s-shaped curve *The Journal of World Economics* **31** pp. 82-92

[3] Thampapillai D J et al. *The Environmental Kuznets Curve Effect and the Scarcity of Natural Resources* (Sydney: Macquarie University NSW)

[4] Lopez R et al. 2007 Structural change and sustainable development *Journal of Environmental Economics and Management* **53** pp. 307-22

[5] Lopez R and Mitra S 2000 Corruption, pollution, and the Kuznets Environment Curve *Journal of Environmental Economics and* Management **40** pp. 137-50

[6] Torras M J 1998 Income, inequality and pollution: a reassessment of the Environment al Kuznets Curve *Ecological* Economics **25** pp. 147-60

[7] Copeland B R and Taylor M S 1995 Trade and environment: a partial synthesis *American Journal of Agricultural* Economics **77** pp. 765-71

[8] Copeland B R and Taylor M S 2004 Trade, Growth and the environment *Journal of Economic Literature* **42** pp. 7-71

[9] Egli H and Steger T M 2007 A dynamic model of the Environmental Kuznets Curve: turning point and public policy *Environmental and Resource Economics* **36** pp. 15-34

[10] Fu M 2008 Chinese Environmental Kuznets Curves: shapes, turning points and responsible factors *Journal of Quantitative & Technical Economics* **25** pp. 40-55

[11] He J and Wang H 2012 Economic structure, development policy and environmental quality: an empirical analysis of Environmental Kuznets Curves with Chinese municipal data *Ecological Economics* **76** pp. 49-59

[12] Zhu P H et al. 2010 Analysis of Chinese industry Environmental Kuznets Curve: empirical study based on spatial panel model *China Industrial Economics* **267** pp. 65-74

[13] Prieur F 2009 The Environmental Kuznets Curve in a world of irreversibility *Economic Theory* **40** pp. 57-90

[14] Lucas R et al. 1992 Economic development, environmental regulation and the international migration of toxic industrial pollution: 1960-88 *World Development Report* World Bank working papers 1062

[15] Chilchilniky G 1994 North-South trade and the global environment *American Economic Review* **84** pp. 851-74

[16] Wu G M and Wu G S 2011 Trade and environment: subject of international trade game and integration process *Commercial Times* **11** pp. 48-50

[17] Li B et al. 2006 Empirical Analysis of Trade Liberalization Influence on Environment *Business Research* **10** pp. 113-15

[18] Dinda S 2005 A theoretical basis for the Environmental Kuznets Curve *Ecological Economics* **53** pp. 403–13

[19] Andreoni J and Levinson A 2001 The simple analytics of the Environmental Kuznets Curve *Journal of Public Economics* **80** pp. 269-86

Acknowledgement

This work is financially supported by the National Natural Science Foundation of China (71203037), Ministry of Education in China Project of Humanities and Social Sciences (17YJC790216), Philosophy and Social Science Project of Guangzhou (2017GZYB23), Natural Science Foundation of Guangdong Province (2015A030313499) and the Foundation for Distinguished Young Talents in Higher Education of Guangdong (2012WYM-0049).

IOP Conf. Series: Earth and Environmental Science **120** (2018) 012018 doi:10.1088/1755-1315/120/1/012018

Structural Break, Stock Prices of Clean Energy Firms and Carbon Market

Yubao Wang and Junyu Cai[1]

1 School of Economics and Finance, Xi'an Jiaotong University, Xi'an, China
E-mail: jacky.mucklow@iop.org

Abstract. This paper uses EU ETS carbon future price and Germany/UK clean energy firms stock indices to study the relationship between carbon market and clean energy market. By structural break test, it is found that the 'non-stationary' variables judged by classical unit root test do own unit roots and need taking first difference. After analysis of VAR and Granger causality test, no causal relationships are found between the two markets. However, when Hsiao's version of causality test is employed, carbon market is found to have power in explaining the movement of stock prices of clean energy firms, and stock prices of clean energy firms also affect the carbon market.

1. Introduction
There have existed studies on the relationship between clean energy and typically traditional economic variables such as oil price, high technology stock and interest rate [1]-[8]. However, once taking the relationship between stock prices of clean energy firms and carbon market into account, few studies are applicable. For this paper, only two papers may be directly referred to in such a field: those of Kumar et al. (2012) [9] and Zhu and Kong (2016) [10]. However, their conclusions differ: Kumar failed to find a significant relationship between the two markets, while Zhu and Kong found that carbon price would affect the stock prices of clean energy firms.

2. Data
For studying relationship between stock prices of clean energy firms and carbon market, clean energy firms' stock prices, high technology firms' stock prices, oil price, carbon price and interest rates would be discussed.

2.1 Time frame and carbon price
Time ranged from 01/08/2014 to 12/14/2016, totally 154 weekly observations available. The Wednesday data for the variables discussed below were adopted and closing prices from the most recent trading session would make up for the missing data at that day. And for the continuity of data and reflection of the potential equilibrium of the markets, closing prices of carbon future were utilized. The EUA data were from WIND.

2.2 Index constructing
Index for certain purpose of research may not be available at hand but could be constructed effectively in some way (B.T. Katangodage and Wijeratne, 2016 [11]). And to avoid coming across survivor-ship bias discussed by Rohleder et al. (2011) [12], value-weighted approach proved to be more efficient than equally-weighted approach practically is adopted (Ibbotson et al., 2011[13]).

Four companies of DAX index were selected for Germany: companies of E.ON.SE, Infineon Technologies AG, RWE AG and Siemens Aktiengesellschaft. And UK is also checked. Seven companies from FTSE 350 were chosen: companies of Centrica plc, Drax Group plc, National Grid plc, Severn Trent, SSE plc, United Utilities Group plc and Vedanta Resources plc.
Four companies in the DAX were chosen for high technology firm stock index constructing: companies of Bayer AG, Deutsche Telekom AG, Merck KgaA and SAP SE., while UK all-share technology index was available.

2.3 Interest rates and oil price

EURIBOR three-month funds and Euro deposit rates (GBP-3 Month-rate value) were adopted to respectively stand for interest rates of Germany and UK. Oil price was obtained by the average of the closing price of West Texas Intermediate (WTI) and that of the BRENT crude oil futures contract.

3. Methodology and empirical results

All variables took logarithms with Germany and UK interest rates as exceptions. Interest rate for Germany turned into negative after April 15, 2015, which is inappropriate for logarithm. To keep consistency with Germany, UK interest rate would not be taken logarithm, either.

3.1 Statistics summary for investment choice

Table 1 shows the descriptive statistics of weekly continuously compounded returns for investments on different assets. Annual returns for each asset is obtained by multiplying 52. For investments in Germany and UK clean energy firms, the returns are 13.52% and 6.76%, respectively twice and four times their corresponding market index. Further, traditional Sharpe ratio shows that oil would not be appealing for investors in these years with negative Sharpe ratio, a difference from Kumar et al.(2012) [9]for whom oil ranked the first.

Table 1 Descriptive statistics of weekly returns.

Variable	Obs	Mean	S. Dev	Min	Max	Skewness	Kurtosis	Sharpe
Carbon price	153	5.7e-05	0.0554	-0.2289	0.2093	-0.5589	6.4588	0.005
Oil price	153	-0.0042	0.0506	-0.1440	0.1836	0.5102	4.5843	-0.079
Dax index	153	0.0011	0.0287	-0.0662	0.0711	-0.1693	2.7125	0.046
Germany high tech	153	-0.0010	0.2627	-0.8565	0.6697	-0.3108	3.9403	-0.003
Germany clean energy	153	0.0026	0.1152	-0.3280	0.3374	0.1702	4.1886	0.024
FTSE 350 index	153	0.0003	0.0209	-0.0647	0.0577	-0.0287	3.6310	-0.278
UK high tech	153	0.0027	0.0301	-0.0674	0.2159	2.1874	18.141	-0.114
UK clean energy	153	0.0013	0.0840	-0.3853	0.3427	-0.0778	6.6610	-0.057
Germany interest rate	154	-0.022%	0.0020	-0.316%	0.339%	0.1422	1.9176	/
UK interest rate	154	0.612%	0.0007	0.420%	0.790%	-0.4600	2.7854	/

3.2 Traditional unit root test

The results of ADF tests are displayed in Table 2, showing that only half of variables are stationary, i.e., variables of Germany high technology firms stock, Germany clean energy firms stock, FTSE 350 index, UK clean energy firms stock and UK interest rate.
Simply considering the traditional ADF test would urge the remaining five variables, i.e. carbon price, oil price, DAX index, UK high technology firms stock and Germany interest rate, to take first difference. However, Perron (1989) [14]emphasized that de-trended data in essence may be labeled with "non-stationary" when structural breaks exist. Further, Bai and Perron (2003) [15] proposed a model that finds the global structural breaking date with appropriate variables $\{Z_t\}$. Therefore, it is necessary to check whether structural break date exists in the five "non-stationary" time series before simply taking the first difference.

Table 2 Traditional ADF test.

	Level			First difference		
	Intercept and trend	**Intercept**	**None**	**Intercept and trend**	**Intercept**	**None**
Carbon price	-2.3811	-2.1324	-0.2603	-10.042***	-9.97***	-10.0***
Oil price	-0.8217	-1.3762	-1.1494	-11.617***	-11.5***	-11.52***
Dax index	-2.5044	-2.3121	0.4564	-12.800***	-12.8***	-12.9***
Germany high tech	-9.7411***	-7.784***	0.1996	-10.511***	-10.5***	-10.6***
Germany clean energy	-3.9345**	-1.6653	0.3471	-20.613***	-20.6***	-20.7***
FTSE 350 index	-2.8304	-2.8781*	0.1268	-14.600***	-14.6***	-14.7***
UK high tech	-2.3579	-0.4517	1.0794	-11.416***	-11.4***	-11.4***
UK clean energy	-5.6740***	-3.3866**	0.1116	-14.515***	-14.6***	-14.6***
Germany interest rate	-2.3378	-0.7388	-0.4201	-7.5904***	-7.6***	-6.69***
UK interest rate	-2.6516	-2.742*	-0.1494	-16.570***	-16.6***	-16.6***

***1 percent of significance
**5 percent of significance
* 10 percent of significance

3.3 Structural breaks

$\{Z_t\}$ is important for Bai and Perron (2003) [15] to find the global structural break dates. In their paper, $\{Z_t\}$ refers to the lag value of time series themselves, the time trend or just a constant term; and data generating processes for time series normally could capture these information. However, data generating process is hard to get. In order to get the data generating process, four models, used by Wang (2001) [16] and Wang and Chen (2007) [17], in Table 3 are assumed.

Table 3 Four models of data generating process.

	Model I $y_t = y_{t-1} + u_t$	**Model II** $y_t = c + y_{t-1} + u_t$	**Model III** $y_t = c + \beta t + u_t$	**Model IV** $y_t = c + \beta t + y_{t-1} + u_t$
carbon	0.9194	0.9206	0.0284	0.9219
	(-2.922)	(-2.910)	(-0.396)	(-2.901)
oil	0.9818	0.9819[a]	0.6941[b]	0.9817[c]
	(-3.107)	(-3.085)	(-0.253)	(-3.052)
dax	0.8535	0.8576	0.1151	0.8575[d]
	(-4.234)	(-4.236)	(-2.410)	(-4.210)
UK high tech	0.9691	0.9689	0.7205[b]	0.9699
	(-4.143)	(-4.111)	(-1.920)	(-4.119)
Germany interest rate	0.9978[b]	0.9982	0.9723[b]	0.9982[c]
	(-15.658)	(-15.811)	(-13.076)	(-15.792)

a: parameter of constant C is not statistically significant;
b: R^2 larger than DW statistics;
c: parameter time and constant C are not statistically significant;
d: parameter of time is not statistically significant.

Adjusted R square and Schwarz criteria are showed in the upper row and in brackets, respectively. And optimal model selection should satisfy the following conditions simultaneously: (1) highest adjusted R square; (2) lowest Schwarz criterion; (3) R squared value less than DW statistics in order to avoid problem of spurious regression (Granger and Newbold, 1974 [18]); and (4) significant coefficients. Overall, Model I suits carbon price, oil price and UK high technology index, and Model II suits Dax index and Germany interest rate. Therefore, for the former three variables, $\{Z_t\}$ employs {y(-1)}, while for the latter two, {1, y(-1)} is chosen, where "y(-1)" and "1" stand for lag term and constant term, respectively.

BP test is provided by E-views 8.0 for detecting structural breaks. From Table 4, no structural break dates exist in the five "non-stationary" time series. Such a finding justifies the conclusion in Table 2 that carbon price, oil price, Dax index, UK high technology firms stock and Germany interest rate do own unit roots and need taking first difference.

Table 4 BP test for seeking global structural break dates.

Variables	Sup F(0|1)	Z_t
Carbon price	2.377	{y(-1)}
Oil price	7.269	{ y(-1)}
Dax index	3.882	{1, y(-1)}
UK high tech	2.674	{ y(-1)}
Germany interest rate	0.0786	{1, y(-1)}

3.4 VAR analysis and Granger causality test

To select an appropriate lag length for VAR model, this paper resorts to Kumar et al (2012)[9]: if criteria of FPE, AIC, SC and HQ showed different lags, the Likelihood Ratio test would be adopted. In Table 5, the chosen lags by FPE, AIC, SC and HQ are the same. Therefore, the VAR is estimated with one lag for both Germany and UK.

Tabel 5 Lag for VAR.

	Germany					UK				
lag	LR	FPE	AIC	SC	HQ	LR	FPE	AIC	SC	HQ
0	NA	8.6e-20	-26.9	-26.8	-26.8	NA	3.6e-20	-27.7	-27.6	-27.7
1	264.6	2e-20*	-28.3*	-27.4*	-27.9*	523	1.4e-21*	-31.0*	-30.2*	-30.7*
2	56.61	2.2e-20	-28.3	-26.6	-27.6	49	1.5e-21	-30.9	-29.3	-30.3

Table 6 VAR model fit and Lagrange-multiplier test.

VAR model fit			
Equation	parameters	R^2	F-statistics
Germany			
Germany clean energy index	7	0.719427	61.96670
Germany high tech index	7	0.235464	7.442937
Oil prices	7	0.039411	0.991517
Carbon prices	7	0.055775	1.427509
Germany interest rate	7	0.209035	6.386747
DAX index	7	0.019190	0.472829
UK			
UK clean energy index	7	0.289163	9.830831
UK high tech index	7	0.050887	1.295690
Oil prices	7	0.032905	0.822270
Carbon prices	7	0.076428	1.999851
UK interest rate	7	0.753389	73.82857
FTSE 350 index	7	0.788946	90.33797

Lagrange-multiplier test				
lag	Germany		UK	
	LM statistics	Probability	LM statistics	Probability
1	50.05374	0.0599	50.83750	0.0516
2	38.59113	0.3533	35.22882	0.5021

H_0: no serial correlation at lag order.

In Table 6. As for the VAR model for Germany, only clean energy firms stock equation owns a high R squared value as well as high F statistics. For UK, interest rate equation and FTSE 350 index equation exhibit a high R squared values as well as F statistics followed by clean energy index equation with lower R squared value but higher F statistics. The Lagrange-multiplier test statistics in Table 6 reveal that VAR models are free from serial correlation at the significance of 1%. All of the eigenvalues are found to lie inside the unit circle; therefore, VAR model satisfies the stability condition and Granger causality test could proceed.
Even though this paper endeavors to capture the relationship between carbon price and corresponding European clean energy firms stock, stock prices of both Germany and UK clean energy firms do not "Granger" cause the carbon price. And carbon price does not "Granger" cause Germany or UK clean energy firms stock prices, either (see Table 7).

Table 7 Granger causality Wald tests.

Granger causality for UK					
Dependent variable: clean energy			**Dependent variable: carbon price**		
excluded	Chi-sq	prob.	excluded	Chi-sq	prob.
Carbon	0.000115	0.9915	Oil	0.008554	0.9263
Oil	1.339988	0.2470	FTSE 350 index	0.864378	0.3525
FTSE 350 index	0.759843	0.3834	clean energy	0.785536	0.3755
interest rate	0.615504	0.4327	interest rate	0.259615	0.6104
high tech index	4.730982	0.0296	high tech	2.783351	0.0952
All	5.674341	0.3392	All	5.759074	0.3304
Granger causality for Germany					
Dependent variable: clean energy			**Dependent variable: carbon price**		
excluded	Chi-sq	prob.	Excluded	Chi-sq	prob.
Carbon	0.026357	0.8710	DAX	0.234231	0.6284
DAX	3.728012	0.0535	Oil	0.018670	0.8913
Oil	1.103068	0.2936	clean energy	0.893183	0.3446
interest rate	0.280802	0.5962	interest rate	0.008955	0.9246
high tech	1.739218	0.1872	high tech	0.355254	0.5512
All	6.130239	0.2938	All	2.461510	0.7823

3.5 Hsiao's version of Granger causality test

It seems that there is no Granger causality between the two countries' clean energy firms stock and carbon price. However, VAR model is subject to chosen lag. To overcome such a problem, Hsiao's version of Granger causality test would be used. The formulas (1) and (2), (3) and (4) come from Hsiao's (1981) [19] and Altinay and Karagol (2004) [20], respectively.

$$y_t = \sum_{i=1}^{M} a_i y_{t-i} + \sum_{j=1}^{N} b_j x_{t-j} + u_t \tag{1}$$

$$x_t = \sum_{i=1}^{M} c_i x_{t-i} + \sum_{j=1}^{N} d_j y_{t-j} + v_t \tag{2}$$

$$FPE(m,0) = \left(\frac{T+m+1}{T-m-1}\right)\left(\frac{Q(m,0)}{T}\right) \tag{3}$$

$$FPE(m,n) = \left(\frac{T+m+n+1}{T-m-n-1}\right)\left(\frac{Q(m,n)}{T}\right) \tag{4}$$

where M and N donate the maximum lag orders, X_t and Y_t donate the two stationary time series, u_t and v_t stand for white noise.

Hsiao employs Akaike's final prediction error criterion to determine the lag order of auto-regressive models (1) and (2). Detailed conduct of running model (1) and (2), and FPE calculations of model (3) and (4) could be referred to Altinay and Karagol (2004). Only the judging rule is put here: when the minimum FPE from equation (4) is smaller than that from equation (3), then X_t is said to cause Y_t in the equation (1). Same rule applies for equation (2).

Here, X_t donates stock prices of clean energy firms of each country and Y_t donates the carbon price. 52 weeks are there in a year; therefore, the maximum lag order for Germany and UK is set as 52. In Table 8, for Germany, the minimum FPE of the controlled variable (carbon price) is found at lag 50. After adding the manipulated variable (stock price of clean energy firms) into the model with the lag order of the controlled variable as 50, the minimum FPE (0.000091) is found at lag 51, smaller than the initial FPE (0.001917); therefore, Germany clean energy stock index does "Granger" cause carbon price. And likewise, it is found that carbon market also affects stock prices of clean energy firms. Same results hold for UK.

Table 8 Hsiao's version of Granger causality test.

Direction	Controlled variables	Minimum FPE	Manipulated variables	Minimum FPE
Germany				
Equation (1): from X_t to Y_t	Carbon(50,0)	0.001917	Clean energy (50,51)	0.000091
Equation (2): from Y_t to X_t	Clean energy (50,0)	0.008006	Carbon(50,51)	0.000003
UK				
Direction	Controlled variables	Minimum FPE	Manipulated variables	Minimum FPE
Equation (1): from X_t to Y_t	Carbon(50,0)	0.001917	Clean energy (50,51)	0.000315
Equation (2): from Y_t to X_t	Clean energy (48,0)	0.003827	Carbon(48,52)	0.000093

4. Implication

Findings of this paper have important implications for China. As carbon market would be affected by the clean energy industry, the Chinese government should make sure the effective operating and reasonable pricing of clean energy industry, which allows the forthcoming carbon market to operate under an overall effective environment. Equally, carbon market should also be cultivated to be mature for the development of clean energy industry, due to the reason that clean energy firms are affected by the carbon price.

5. References

[1] Henriques I and Sadorsky P 2008 *J. Energy. Economics* **30** 998-1010
[2] Sadorsky P 2012 *J. Energy. Economics* **34** 248-55
[3] Broadstock D C, Cao H and Zhang D 2012 *J. Energy. Economics* **34** 1888-95
[4] Managi S and Okimoto T 2013 *J. Japan. And. The. World. Economy* **7** 1-9
[5] Wen X, Guo Y and Wei Y 2014 *J. Energy. Economics* **41** 63-75
[6] Reboredo J C 2015 *J. Energy. Economics* **48** 32-45
[7] Reboredo J C and Wen X 2015 *J. Renewable. And. Sustainable. Energy. Reviews* **45** 624-36
[8] Bondia R, Ghosh S and Kanjilal K 2016 *J. Energy* **101** 558-65
[9] Kumar S, Managi S and Matsuda A 2012 *J. Energy. Economics* **34** 215-26
[10] Zhu D, Kong Y 2016 *J. Ecological. Economy* **1** 52-7
[11] Katangodage B T and Wijeratne A W 2016 *J. International. Journal. Of. Agricultural Resources, Governance and Ecology* **12** 27-52
[12] Rohleder M, Scholz H and Wikens M 2011 *J. Review. Of. Finance* **15** 441-74
[13] Ibbotson R G, Peng C and Zhu K X 2011 *J. Financial. Analysts. Journal* **67** 15-25
[14] Perron P 1989 *J. Econometrica* **57** 1361-1401
[15] Bai J and Perron P 2003 *J. Applied. Econometrics* **18** 1-22
[16] Wang S 2001 *J. Statistics. And. Decision* **6** 4-5

[17] Wang Y, Chen N 2007 *J. Statistics. And. Information. Forum* **22** 69-72
[18] Granger C W J and Newbold P 1974 *J. Econometrics* **2** 111-20
[19] Hsiao C 1981 *J. Monetary. Economics* **7** 85-106
[20] Altinay G and Karagol E 2004 *J. Energy. Economics* **26** 985-94

Generation of OH Radical by Ultrasonic Irradiation in Batch and Circulatory Reactor

Yu Fang[1], Sayaka Shimizu[2] , Takuya Yamamoto[1] and Sergey Komarov[1]

1 Graduate school of environmental studies, Tohoku University, Sendai, Miyagi Prefecture 980-8579, Japan
2 Komatsu Ltd., Komatsu, Ishikawa Prefecture 923-0392, Japan
E-mail: fang.yu.p7@dc.tohoku.ac.jp

Abstract. Ultrasonic technology has been widely investigated in the past as one of the advance oxidation processes to treat wastewater, in this process acoustic cavitation causes generation of OH radical, which play a vital role in improving the treatment efficiency. In this study, OH radical formation rate was measured in batch and circulatory reactor by using Weissler reaction at various ultrasound output power. It is found that the generation rate in batch reactor is higher than that in circulatory reactor at the same output power. The generation rate tended to be slower when output power exceeds 137W. The optimum condition for circulatory reactor was found to be 137W output and 4L/min flow rate. Results of aluminum foil erosion test revealed a strong dependence of cavitation zone length on the ultrasound output power. This is assumed to be one of the reasons why the generation rate of HO radicals becomes slower at higher output power in circulatory reactor.

1. Introduction

1.1. Advanced oxidation processes in water treatment

With the expansion of urban city and modern industry, demand of water is phenomenally growing. Meanwhile, wastewater treatment and recycling has becoming an important issue, especially for recalcitrant organic wastewater, which mainly contains polycyclic aromatic hydrocarbon, halogenated hydrocarbon, heterocyclic compounds, organic pesticide and so on [1].

Advanced Oxidation Processes (AOPs), as a promising water treatment technology, has been investigated widely in recent years. By supplying energy and adding matters, such as ozone and H_2O_2, OH radicals can be produced in water, which attack the organic pollutants to promote their degradation into CO_2, H_2O, and inorganic salts [2], [3].

According to the mechanism of OH radical generation, various advanced oxidation processes have been widely investigated. Examples are Fenton process, ozone, ultrasound and plasma. Many hybrid processes have been also developed [4]-[15]. Some of them have already been industrialized for treating small amount of wastewater [16], [17]. Among these techniques, ultrasound water treatment has shown great potential. The main effect of ultrasound is related to the phenomenon called cavitation.

1.2. Cavitation

Cavitation is produced as a result of rupturing a liquid during the negative pressure half cycle of ultrasound wave [18]. Cavitation also can be defined as a process of generation, subsequent growth and collapse of the cavities (or cavitation bubbles) releasing large amounts of energy over a very small location providing very high-energy densities [19]. In this process, hot spots are generated in the

IOP Conf. Series: Earth and Environmental Science **120** (2018) 012019 doi:10.1088/1755-1315/120/1/012019

vicinity of cavities, which creates conditions of high temperature and pressures, usually up to few thousand Kelvins and few thousand atmospheres [20]. As a result, highly reactive free radicals, mainly OH radicals, are generated. There are four types of cavitation. Acoustic cavitation, hydrodynamic cavitation, optic cavitation, and particle cavitation. Among these, only acoustic and hydrodynamic cavitation can generate cavitation field of high enough intensity for water treatment [19].

1.3. Acoustic cavitation

Power ultrasound is capable of producing both chemical and physical effects in liquid. A typical frequency range for ultrasound is 20 ~100 kHz. In the case of acoustic cavitation, tiny bubbles are generated by the ultrasound wave, which is transmitted in liquids following compression and rarefaction cycles. When the negative pressure of rarefaction cycle exceeds attractive forces between the molecules of liquid, a void is formed. This void, or so-called cavity, is filled with s solvent vapor or dissolved gas. In the compression cycle, the cavity does not totally collapse but continues to grow to form larger acoustic cavitation bubbles. When some bubbles expand to an unstable size, they violently collapse and generate high temperature zone (hot spots) and shock waves.

Volatile and hydrophobic pollutants are degraded by thermal reaction in the hot spots of cavitation bubbles. Compounds which are more hydrophilic are decomposed in the bulk liquid by OH radicals generated in the cavitation bubble [21].

1.4. Acoustic streaming

Apart from generation of cavitation bubbles and hot spots, ultrasonic waves may also cause a steady flow in fluids, which is known as acoustic streaming [22]. Acoustic streaming is a kind of macroscopic flow caused by attenuation of ultrasound waves due to viscosity, thermal conductivity and interaction with cavitation bubbles [23]-[26].

Acoustic streaming is very important for wastewater ultrasonic treatment. The OH radicals, which have quite short life, are mainly generated in a small volume near the ultrasonic sonotrode tip, usually called cavitation zone. Therefore, targeted compounds should pass through the cavitation zone to be treated with OH radicals. Hence, to design an efficient ultrasonic wastewater treatment reactor, an optimized flow pattern inside the reactor is needed, especially for continuous reactors, adapted to application in industry.

Thus, in this study, generation of OH radicals in batch and circulatory reactors was investigated under various acoustic powers using a Weissler model reaction. Besides an aluminum foil erosion test was conducted to investigate intensity and dimension of cavitation zone under the sonotrode tip.

2. Experimental setup and instrumentation

2.1. Ultrasonic equipment and sonotrode characteristics

In this study, In this study, two kinds of ultrasonic system were applied. The first one was made by NIPPON SEIKI, for the sake of convenience, referred as to device A, and the other was made by TELSONIC, referred as to device B. Output frequency ranges for device A and B were 17~23kHz and 18.9~21.2kHz respectively. In these experiments, there was a need to know acoustic power transmitted into water. For this purpose, calorimetric measurements were performed using both devices.

The main dimension of sonotrodes is showed in Table 1. Both sonotrodes were made of Ti-4V-6Al alloy.

Table 1. Dimension of sonotrodes

Device	Length / mm	Tip diameter / mm
A	122	26
B	255	47

2.2. Calorimetric measurement

The calorimetric measurements were performed in an acrylic cylinder vessel, filled with 500ml distilled water. The vessel had a 100mm diameter, 146mm height and 10mm wall thickness. Temperature variation was measured with a thermocouple fixed at the centreline of sonotrode at a distance of 20, 40 and 60 mm from the vessel bottom. The vessel was wrapped with a heat insulating material to prevent heat loss. Immersion depth of sonotrode was 30mm.
Assuming that all the ultrasound energy is converted to heat, the heat increase of water bath can be considered equal to the acoustic energy transmitted into the liquid. The measurements were performed at 150C. Ultrasound irradiation time for device A and B were 1200s and 160s respectively.
Calorimetric measurements results showed that the lowest and the highest acoustic power for device A were 12W and 24W, and those for device B were 137W and 301W.

2.3. Reactor set up

In this study, a batch model reactor and circulatory model reactor were investigated respectively. Schematic drawings of both the reactors are showed in Fig. 1. For the batch experiment, a 2L beaker was used as the reactor filled with 1L of 0.1mol/L KI solution. To keep the solution temperature constant at a level of 15^{0}C a water-cooled coil was immerged in the beaker. Sonotrode immersion depth was 30mm. Ultrasound frequency was set in 20 kHz for both devices. Irradiation time was 60 minutes and 15 minutes, respectively. Solution samples were taken every 10 mins for device A and 3 mins for device B.

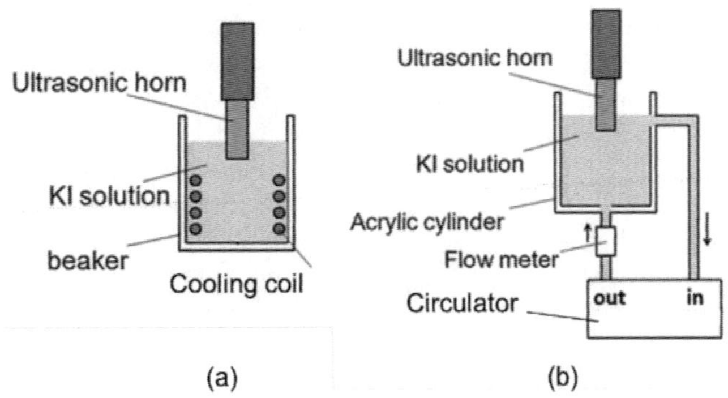

Fig. 1 Schematic drawings of (a) batch reactor and (b)circulatory reactor

In circulatory experiments, a 1L acrylic cylinder of 100mm in diameter and 200mm in height was used as a reactor equipped with inlet and outlet tubes as shown in Fig.2(b). The tubes were connected to a temperature-controllable fluid circulator filled with 6L of 0.1mol/L KI solution. The solution temperature was kept at the level of 15 ± 2^{0}C. Ultrasound irradiation time was set to 120 min and 15 min for device A and B, respectively. Solution samples were taken every 20 min for device A and 3 min for device B. These experiments were carried out at three flow rates, 2, 4 and 6 L/min measured by a flow meter.

2.4. Weissler reaction

In this study, Weissler reaction was applied to evaluate the formation rate of OH radicals. Weissler reaction has been widely used as a test reaction to investigate ultrasonic and hydrodynamic cavitation. The OH radicals, which generated in cavitation zone have a very short lifetime and react soon with iodide ion through a series of intermediate steps, to form tri-iodide complex I3-. The key reactions are assumed to be as follows [27]:

$$H_2O \rightarrow H\cdot + OH\cdot \quad (1)$$

$$OH\cdot + I^- \rightarrow OH^- + I \quad (2)$$

$$I + I^- \rightarrow I_2^- \quad (3)$$

$$I_2^- \rightarrow I_2 + 2I^- \quad (4)$$

$$I_2 + I^- \rightarrow I_3^- \quad (5)$$

Concentration of tri-iodide complex I_3^- was measured by an ultraviolet visual spectrophotometer (Shimadzu Co., Ltd, Japan), based on the maximum absorption peak of I_3^- at a wavelength for 353nm. Then, the rate of I_3^- ion formation was determined and used as a quantitative measurement of OH radical generation. As can be readily seen from the above reactions, two OH radicals must react with two I^- ions to produce one I_3^- ion. Therefore, the rate of OH radical formation can be obtained simply by multiplying the rate of I_3^- ion formation by 2.

3. Experimental results and discussion

3.1. Weissler reaction

In the batch type reactor, spectrophotometer measurements revealed that concentration of I_3^- ion increases linearly with treatment time when the other parameters are kept constant. This suggests that generation of OH radicals occurs at a constant rate at a fixed acoustic power. Based on these data, OH radical generation rate was determined at various values of acoustic powers. The results are shown in Fig. 2. It is clearly that OH radical generation rate increases significantly with increasing the acoustic power. However, this increase becomes slower when acoustic power exceeds 137W.

In the circulatory type reactor, concentration of I_3^- ion increased also linearly with treatment time. Two typical sets of data are plotted in Fig. 3 or the minimal (a) and maximal (b) acoustic powers examined in this study. In both cases, the time variations of I_3^- ion concentration was dependent on liquid flow rate, although this dependence appears to be different when device A and B were used.

Based on these data, the rate of OH radical generation for the circulatory reactor was calculated. The results are presented in Fig.4 for different flow rates. It is seen that the OH generation rate increases in the range of acoustic power from 0 to 137 W. After that, the generation rate is decreased with acoustic power or remains almost unchanged. Under the present experimental conditions, the maximal rate of OH radical generation was obtained at acoustic power of 137 W and flow rate of 4 L/min.

Comparison of the OH radical generations rates for batch and circular reactors reveal that the batch reactor provides a more efficient ultrasonic treatment than circulatory one. For example, the maximal rate in the circulatory reactor treatment is equal 14.8×10^{-9} mol/sec, while in the batch reactor this value reached 23.6×10^{-9} mol/sec at the same acoustic power.

In order to find out the reason for this difference, additional experiments were performed. The goal of these experiments was to measure the level of cavitation under the sonotrode tip. All these measurements were done using device B only.

3.2. Cavitation intensity test

It is well known from the relevant literature that cavitation level (or intensity) can be measured using aluminum foil test [28]-[30]. In this test, a piece of thin aluminum foil is fixed at a certain location in the cavitation zone. Cavitation bubbles, when collapse near the foil, poke tiny holes in the aluminum foil removing a part of aluminum from the foil. Obviously, the higher the cavitation intensity, the more holes are poked in the foil and the greater is the loss of foil weight. Thus, the cavitation intensity can be estimated from the rate of foil weight loss.

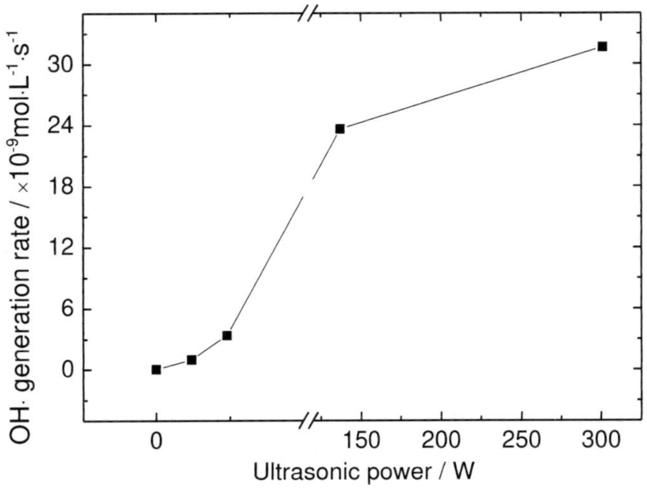

Fig. 2 Effefcts of output power on OH radical generation rate in batch reactor experiments

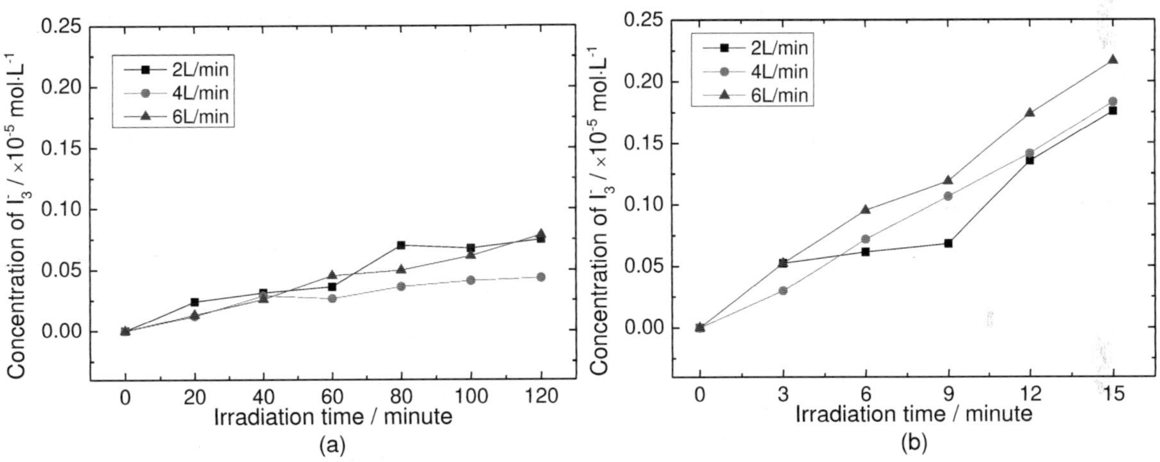

Fig. 3 Time variation of I_3^- concentration at various flow rates acoustic power (a)12W and (b)301W

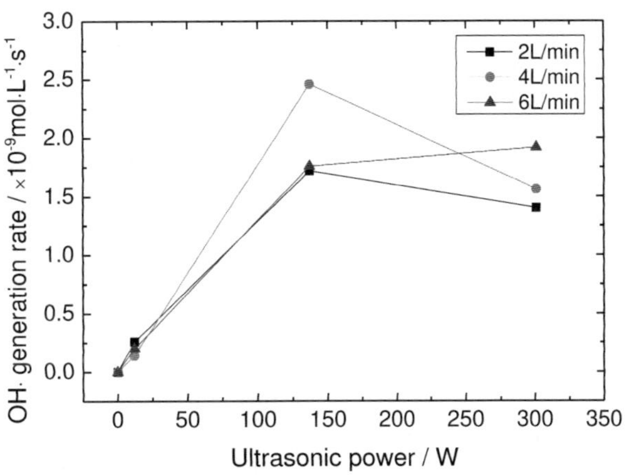

Fig. 4 Effects of output power on OH radical generation rate in circulatory reactor experiments

In these experiments, a piece of foil was fixed in a steel frame of 20×20 mm in size. The frame was positioned in water at a fixed distance Z from the sonotrode tip and parallel to its surface. Ultrasound waves were introduced in water to produce cavitation for the time interval of 20~600 sec, depending on distance Z. Then, the loss in foil weight was measured using a precision balance.
Figure 5 presents variation in the rate of foil weight loss U_e with the distance from the sonotrode tip for two acoustic powers, P=137W and 301W. Notice that the plot is made using a base 10 logarithmic scale for the y-axis because of a very strong dependence of U_e on Z. It is readily seen that at the higher acoustic power, a high-intense cavitation zone is formed under the sonotrode tip, however the intensity is drastically decreased with distance. At Z=20 mm, the intensity is reduced by 100 times relative to that at Z = 5 mm. On the other hand, at P = 137 W, reduction of U_e with Z is not so significant. At distances more that 20 mm, the cavitation intensity at P=137 W is approximately one order of magnitude higher than that at P=301 W. Local peaks, observed in both the curves, are conceivably due to formation of standing wave between sonotrode tip and aluminum foil.
Thus, too high acoustic power results in formation of a very intense but short cavitation zone in liquid. Alternatively, ultrasound waves of relatively low intensity produce a more uniform and longer cavitation field. This can be one of the reasons allowing us to explain the dependence of OH radical generation rate on acoustic power. As mentioned above, OH radicals can be produced in cavitation zone. In the batch-type reactor, liquid is involved in circulatory flow generated in the beaker by acoustic streaming. In this case, KI solution can pass through the cavitation zone many times. However, in the case of circulatory reactor, the residence time of liquid in the beaker is much shorter compared to the batch reactor case. That is why the above-mentioned decrease in the size of cavitation zone with acoustic power can lead to a situation where progressively smaller fraction of KI solution would be able to pass through cavitation zone.

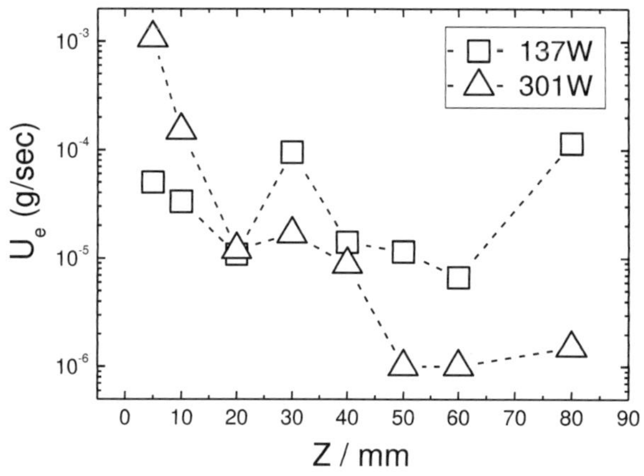

Fig. 5 Change of foil erosion rate with distance from sonotrode tip

Maikel M. van Iersel [31] has also claimed that shielding of the acoustic wave is more pronounced at higher cavity fractions. Accordingly, a large amount of the supplied energy is attenuated in the immediate vicinity of sonotrode tip and thus experimentally determined OH radical generation rate may tend to increase relatively little, or even decrease, with increasing the acoustic power.

4. Conclusion

In this study, effects of acoustic power on generations of OH radical in batch and circulatory reactor were investigated.

- In the case of batch reactor, OH generation rate increased with output power growth; reaction efficiency was much higher than in circulatory reactor at same output.

- In the case of circulatory reactor, the best reaction condition was obtained in 137W output and 4L/min flow rate; with the growth of output power, OH generation rate growed firstly, and then decreased after 137W output.
- Aluminum foil erosion test clarified that high acoustic power resulted in an intense but very short cavitation zone, which may suppress the generation of OH radicals.

Further research on flow pattern in circulatory reactor is need, especially on the coupling effects of acoustic streaming and bumped circulation flow.

5. References

[1] Ana R. Ribeiro, Olga C. Nunes, Manuel F.R. Pereira, Adrian M.T. Silva. An overview on the advanced oxidation processes applied for the treatment of water pollutants defined in the recently launched Directive 2013/39/EU. 2015. *Environmental International*. **75** 33-51

[2] Oller I, Malato S, Sanchez-Perez J.A. Combination of advanced oxidation processes and biological treatments for wastewater decontamination—a review. 2011. *Science of the Total Environment*. **20** 4141-66

[3] Hapman C.J, Torrens A. 1998. Direct radical oxidation process, New York: Wiley, 61-180.

[4] Anand G. Chakinala, Parag R. Gogate, Arthur E. Burgess, David H. Bremner. Industrial wastewater treatment using hydrodynamic cavitation and heterogeneous advanced fenton processing. 2009. *Chemical Engineering Journal*. **152** 498-502

[5] H. Destaillats, A.J. Colussi, J.M. Joseph, M.R. Hoffmann. Synergistic effects of sonolysis combined with ozonolysis for the oxidation of azobenzene and methyl orange. 2003. *Journal of physical Chemistry A*. **104** 8930–35

[6] W.H. Glaze, J.W. Kang. Advanced oxidation processes: description of a kinetic model for the oxidation of hazardous materials in aqueous media with ozone and hydrogen peroxide in a semibatch reactor. 1989. *Industrial & Engineering Chemistry Research*. **28** 1573–80

[7] W.H. Glaze, J.W. Kang, D.H. Chapin. The chemistry of water treatment processes involving ozone, hydrogen peroxide and ultraviolet radiation. 1987. *Ozone Science & Engineering*. **9** 335–42

[8] G.R. Peyton, W.H. Glaze. Destruction of pollutants in water with ozone in combination with ultraviolet radiation. 3. Photolysis of aqueous ozone. 1988. *Environmental Science Technology*. **22** 761–7

[9] O. Legrini, E. Oliveros, A.M. Braun. Photochemical processes for water treatment. 1983. *Chemical Review*. **93** 671–98

[10] D. Drijvers, H. Van Langenhove, M. Beckers. Decomposition of phenol and trichloroethylene by the ultrasound/H2O2/CuO process. 1999. *Water Resource*. **33** 1187–94

[11] J. Weiss. The free radical mechanism in the reactions of hydrogen peroxide. 1952. *Advance Catalysis*. **4** 343–65

[12] A. Mokrini, D. Oussi, S. Esplugas. Oxidation of aromatic compounds with UV radiation/ozone/hydrogen peroxide. 1997. *Water Science&Technology*. **35** 95–102

[13] C. Walling, A. Goosen. Mechanism of ferric ion catalyzed decomposition of hydrogen peroxide. Effect of organic substrates. 1993. *Journal of American Society*. **9** 2987–91

[14] B. Neppolian, H. Jung, H. Choi, J.H. Lee, J.W. Kang, "Sonolytic degradation of methyl tert-butyl ether: the role of coupled Fenton process and persulphate ion. 2002. *Water Resource*. **36** 4699–708

[15] Jun Liang, Sergey Komarov, Naohito Hayashi, Eiki Kasai, "Improvement in sonochemical degradation of 4-chlorophenol by combined use of Fenton-like reagents. 2007. *Ultrasonic Sonochemistry*. 14 201-7

[16] Anglada A, Urtiaga A, Ortiz I. Pilot scale performance of the electro-oxidation of landfill leachate at boron-doped diamond anodes. 2009. Environmental Science & Technology. **43** 2035-40

[17] Gerrity D, Gamage S, Holady J C. Pilot-scale evaluation of ozone and biological activated carbon for trace organic contaminant mitigation and disinfection. 2011. *Water Research.* **45** 2155–65
[18] Christopher E. Brennen. 2013. Cavitation and Bubble Dynamics, U.K.: Cambridge University Press, 5-7
[19] Manson, T.J., 1992. Practical Sonochemistry: User's guide in Chemistry and chemical Engineering, Ellis Horwood Series in Organic Chemistry, UK: Ellis Horwood, Chichester. 10-1.
[20] Parag R.Gogate. Application of cavitational reactors for water disinfection: Current status and path forward. 2007. *Journal of Environmental Management* **85** 801-15
[21] Cristina Leoneli, Timonthy J. Mason, "Microwave and ultrasonic processing: Now a realistic option for industry. 2010. *Chemical Engineering and Processing: Process Intensification* **49** 885-900
[22] Baharak Sajjadi, Abdul Aziz Abdul Raman, Shaliza Ibrahim. Acomparative fluid flow characterization in a low frequency/high power sonoreactor and mechanical stirred vessel. 2015. *Ultrasonic Sonochemistry.* **27** 359-73
[23] C. Eckart. tices and streams caused by sound waves. 1948. *Physical Review.* **73** 68-76
[24] W.L. Nyborg. Acoustic streaming due to attenuated plane waves. 1953. *Journal of the Acoustic Society of America.* **25** 68
[25] P.J. Westervelt. The theory of steady irrotational flow generated by a sound field. 1953. *Journal of the Acoustic Society of America.* **25** 60
[26] J. Lighthill. Acoustic streaming. 1978. *Journal of Sound Vibration.* **61** 391-418
[27] K.R. Morison and C.A. Hutchinson. Limitations of the Weissler reaction as model reaction for measuring the efficiency of hydrodynamic cavitation. 2009. *Ultrasonics Sonochemistry.* **16** 176-83
[28] S. Mitragotri, J. Farrell, H. Tang, T. Terahara, J. Kost, and R. Langer. Determination of threshold energy dose for ultrasoundinduced transdermal drug transport. 2000. *Journal of Controlled Release.* **52** 41-52
[29] Michael R. Bailey, David T. Blackstock, Robin O. Cleveland, Lawrence A. Crum. Comparison of electrohydraulic lithotripters with rigid and pressure-release ellipsoidal reflectors. II. Cavitation fields. 1999. *The Journal of the Acoustical Society of America.* vol. **106**, 1149.
[30] Hua Tang, Chiao Chun Joanne Wang, Daniel Blankschtein, Robert Langer. An investigation of the role of cavitation in low-frequency ultrasound-mediated transdermal drug transport. 2002. *Pharmaceutical Research.* **19** 1160-70
[31] Maikel M. van Iersel, Nieck E. Benes, Jos T.F. Keurentjes. Importance of acoustic shielding in Sonochemistry. 2008. *Ultrasonic Sonochemistry.* **15** 294-300

Acknowledgments
This work was supported in part by the Japanese Steel Foundation for Environmental Protection Technology (SEPT) and Chinese Scholarship Council (CSC).

IOP Conf. Series: Earth and Environmental Science **120** (2018) 012020 doi:10.1088/1755-1315/120/1/012020

Optimization of Lead Removal via Napier Grass in Synthetic Brackish Water using Response Surface Model

P Hongsawat[1*], P Suttiarporn[1], K Wutsanthia[1] and G Kongsiri[1]

1 Faculty of Science Energy and Environment, King Mongkut's University of Technology North Bangkok, Rayong campus, Thailand
E-mail: parnuch.h@sciee.kmutnb.ac.th

Abstract The efficiency of the lead (Pb) phytoremediation by Napier grass was studied on the plant's growth and plant's tolerance on the Pb toxicity in synthetic brackish water. It was found that the plant was high tolerance to high level of Pb concentration (10 mg/l) in synthetic brackish water. Which revealed on the possibilities of plant's growth under the presence of Pb contaminated condition. According to the Pb removal efficiency, the highest one (88.63±4.9%) was found at 10 ppm Pb concentration, 0.3 g/l NaCl concentration during the period 45 day. However, this study investigated the optimum condition for lead (Pb) removal from synthetic brackish water using phytoremediation treatment with Napier grass through a Box-Behnken Design. Three operational variables, i.e. Pb concentration (1, 5.5, 10 mg/l), NaCl (0.1, 0.3, 0.5 g/l) and period time (7, 26, 45 day), were determined. The results were provided evidence that the highest Pb removal efficiency (93.56%) from synthetic brackish water via Napier grass was Pb and NaCl concentration at 10 mg/l and 0.5 g/l during 45 day.

1. Introduction

High concentration of lead (Pb) is one of the most a detectable heavy metal in aquatic environment. Their contamination in marine and coastal environment (sediment, sea water, and brackish water) has been recognized as the major problem worldwide that affect to human health through both water and food consumption. Source of the various heavy metal releasing to the aquatic environment are both of anthropogenic source; leaching of municipal wastes, agricultural pesticides and effluent and wastewater from industrial manufacturing and mining, and naturally source such as leaching during soil erosion and flooding [1], [2].

According to the heavy metal remediate approach, phytoremediation has considered as a green alternative solution to reduce high concentrations of heavy metal due to the public acceptance, the restoration of contaminated environment with low costs and applicable in many ecosystems. However, the limitation of the phytoremediation is long time required for the clean up contaminated site and toxic plant left over after the remediation approach. Therefore, utilization and effective remediation of contaminated land synergistic bonding by using energy crop for bioenergy production has been considered as sustainable approach of phytoremediation [3], [4].

In recent year, cellulosic biomass including Napier grass has been identified as good candidates for bioenergy production due to their high growth rate and high biomass production. However, few studies have investigated their heavy metal removal efficiency [5].

Response surface methodology (RSM), a collection of mathematical and statistical approach for experimental design useful for analyzing and evaluating the effects of variable as well as searching optimum conditions of variable to predict targeted responses, has been proven to be successfully implemented in optimization of phytoremediation conditions. The effectiveness of southern cattail *(Typha domingensis)* for phytoremediation of heavy metals from municipal waste leachate was investigated using Central composite design in order to clarify the optimal conditions of the independent variables[6]. Therefore, this study objected to investigate the Pb phytoremediation efficiency and optimization of remediation conditions via Napier grass in synthetic brackish water using response surface methodology.

2. Materials and methods

2.1. Experimental design

Respond surface methodology (RSM) is a statistical method that uses quantitative data from appropriate experiments to determine regression model equations and operating conditions. RSM is a collection statistical techniques for modeling and analysis of problems that a response result influenced by several variables. Box-Behnken Design (BBD); the standard RSM design, was applied in this work to investigate the variables for removal of lead by Napier grass. BBD for three variables (initial Pb concentration, initial NaCl concentration, and time), each with two level (minimum and maximum), was used as experimental design model. In experimental design model, initial Pb concentration (1-10 mg/l), initial NaCl concentration (0.1-0.5 g/l), and time (7-45 day) were taken as input variables. Percentage removal of Pb was taken as the response of the system. Three independent variables were studied at three different levels, such as low (-1), medium (0) and high (+1) as shown in Table 1.

Table 1. Actual values of variables of the experimental design

Factor		Levels of variables		
		-1.00	0.00	1.00
Initial Pb concentration (mg/l)	X_1	1	5.5	10
Initial NaCl concentration (g/l)	X_2	0.1	0.3	0.5
Period Time (day)	X_3	7	26	45

2.2. Plant cultivation

Firstly, Napier grass was pre-cultivated in enrich nutrient soil for 30 days until the roots grew. Prior to the study the removal of lead in brackish water, stem's plant was conducted with approximately length of 20 cm. According to plant cultivation, it was cultivated in a hydroponic system using a half-strength Hoagland's nutrient solution supplemented with each sodium chloride (NaCl) and Lead (Pb) as the experimental design by BDD as expressed in Table 2.

2.3. Determination of Pb removal efficiency

According to the remaining Pb concentration, brackish water samples were collected and digested in nitric acid (HNO_3 65%) and hydrogen peroxide (H_2O_2 30%) solution using microwave digestion-model speedwave four, berghof. Then Pb concentration was determined using Microwave Plasma Atomic Emission Spectroscopy (MP-AES 4100, Algilent Technologies). Removal efficiency of Pb removal (%) was calculated as equation (1);

$$\text{Pb Removal Efficiency (\%)} = (C_i - C_t/C_i) \times 100 \quad (1)$$

Where C_i was the initial Pb concentration (mg/l) and C_t was Pb concentration at the period time indicated (mg/l).

2.4. Phytotoxicity of Pb

Plant's growth such as amount of leaves, amount of root and stem length was directly measurement subject to phytotoxicity of Pb via Napier grass in synthetic brackish water. The results were investigated using SPSS Version 11.5 software (SPSS Inc., USA). The data were subjected to ANOVA, and differences between means were determined using the least squares deviation (LSD) test.

3. Results and discussions

3.1. Phytotoxicity of Pb

To evaluate phytotoxicity of Pb via Napier grass in synthetic brackish water (NaCl concentration range between 0.1-0.5 g/l), the plant growth was direct measured as indicator such as the amount of root, shoot length and amount of leaf. In comparing between the beginning of cultivation and after treatment, a significantly increase in amount of root significant was found in Pb and NaCl concentration increased ($p < 0.05$) (Figure 1(a)). Beside the amount of leaf, the increased NaCl seem to increase their leaf that might because of the increasing osmotic pressure. That bring more nutrients to it's leaf and increase the plant growth. However, there were no significant differences in amount of leaf ($p < 0.05$) (Figure 1(c)). On the other hand, the shoot length was significant decreased after 45 day ($p < 0.05$) (Figure 1(b)). These results indicated that shoot has more sensitive to heavy metal than root [7]. Moreover, it was interesting that plant were high tolerated to high level of Pb concentration (10 mg/l) in synthetic brackish water.

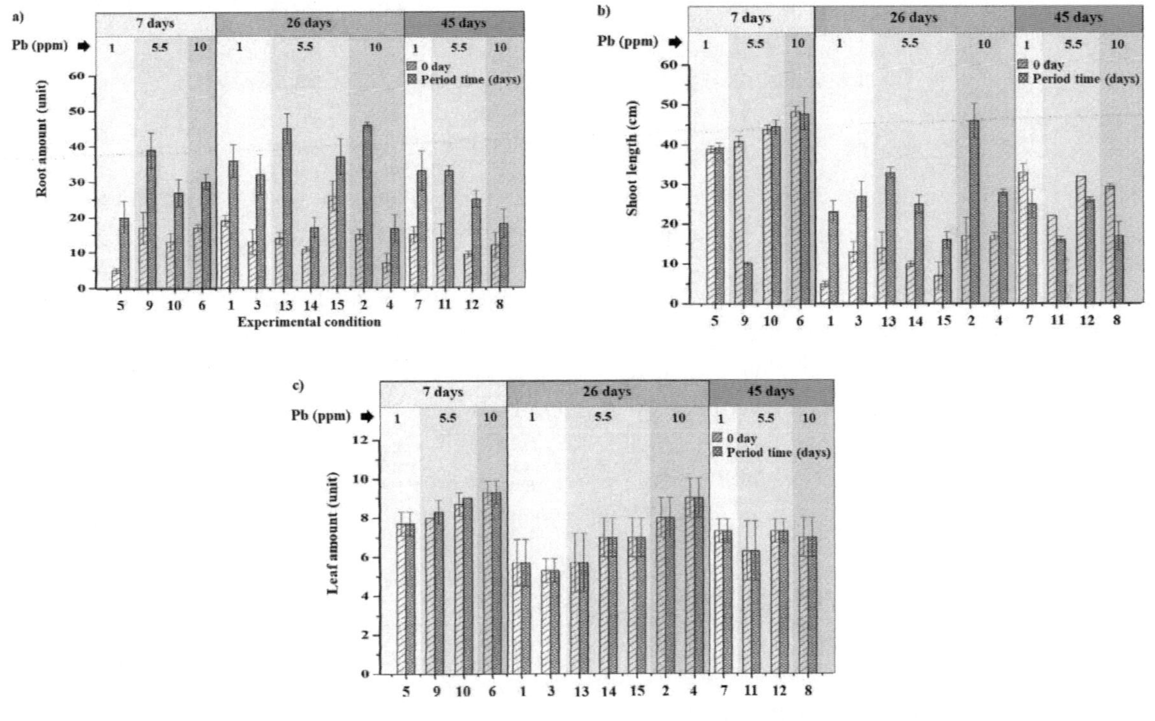

Figure 1. Pb phytotoxicity indicators; a) amount of root, b) shoot length and c) amount of leaf, via Napier grass in synthetic brackish water (NaCl concentration range between 0.1-0.5 g/l)

3.2. Phytoremediation of Pb

The Pb removal efficiency via Napier grass in synthetic brackish water has shown in Figure 2. The highest Pb removal efficiency (88.63±4.9%) found at 10 ppm Pb concentration, 0.3 g/l NaCl concentration during the period of 45 day. Moreover, it revealed that most of Pb content (80%) could be removed using phytoremediation within 7 day. According to the lower concentration of Pb and NaCl, plant removed Pb lower than the higher one. It is suggesting that the increased removal efficiencies were revealing in the increased Pb concentration. That might cause by the higher osmotic pressure can be enhance more mobilization of Pb from solution into the different part of plant in this present study. However, the accumulation of Pb in different part of plant should investigate in further study. Beside the heavy metal phytoremediation via Napier grass, it was the good representing plant for the removal of Cu from contaminated soil. In comparing to the other plant, S. monticola has found to be suited for Pb phytoremediation from contaminated watershed at low concentration (15 - 145 μg/l) [8].

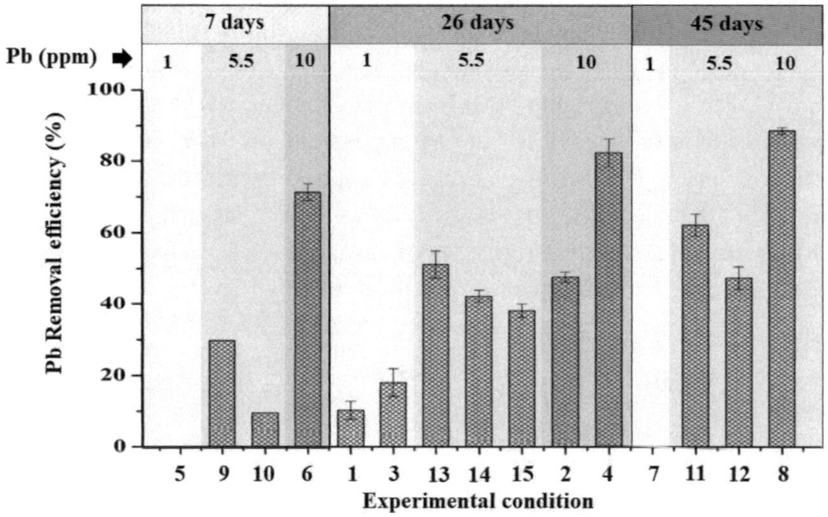

Figure 2. Removal efficiency of Pb via Napier grass in synthetic brackish water according to Box-Bechn design experimental condition

3.3. Optimization of the removal of Pb via Napier grass using response surface method

The experimental design matrix derived from BBD which involved 15 runs was shown in Table 2. Analysis of variance has been calculated to analyze the accessibility of the model. The model was significant at the 5% confidence level because probability values were less than 0.05. The lack of fit (LOF) F-test explains variation of the data around the modified model. LOF would be significant, if the model did not fit the data well. Generally, large probability values for LOF (>0.05) explained that the F-statistic was insignificant, implying a significant model relationship between variables and process responses.

To investigate the relationships between the response and the independent variables, BBD showed relationships as the polynomial equation as equation (2);

$$Y = 43.80 + 32.71X_1 + 0.94X_2 + 10.95X_3 - 0.70 X_1^2 - 3.47X_2^2 - 3.09X_3^2 + 6.77 X_1X_2 + 4.33X_1X_3 + 1.35X_2X_3 \quad (2)$$

Where Y is the percentage removal of Pb, X_1 is initial Pb concentration (mg/l), X_2 is initial NaCl concentration (g/l) and X_3 is period time (day). The regression coefficients of the full polynomial model using BBD are given in Table 3.

The R^2 value of model had 0.859. The optimum values of initial concentration of Pb, initial concentration of NaCl and period time from Box-Behnken design were found to be 10 mg/l, 0.5 g/l and 45 day, respectively. The maximum predicted Pb Removal Efficiency was found to be 93.56%.

Table 2. Experimental design and results for the Pb removal by Napier grass.

Condition	Initial Pb concentration (mg/l), X_1	Initial NaCl concentration (g/l), X_2	Time (day), X_3	% Pb Removal Efficiency
1	1.0	0.1	26	10.29
2	10.0	0.1	26	47.63
3	1.0	0.5	26	18.04
4	10.0	0.5	26	82.44
5	1.0	0.3	7	0.00
6	10.0	0.3	7	71.32
7	1.0	0.3	45	0.00
8	**10.0**	**0.3**	**45**	**88.63**
9	5.5	0.1	7	29.76
10	5.5	0.5	7	9.53
11	5.5	0.1	45	62.20
12	5.5	0.5	45	47.38
13	5.5	0.3	26	51.09
14	5.5	0.3	26	42.18
15	5.5	0.3	26	38.06

Table 3. Regression coefficient of full polynomial model.

Coefficient	Parameter estimate	p-Value
β_0	43.80	0.008*
β_1	32.71	0.004*
β_2	0.94	0.888
β_3	10.95	0.146
β_{11}	-0.70	0.943
β_{22}	-3.47	0.726
β_{33}	-3.09	0.755
β_{12}	6.77	0.486
β_{13}	4.33	0.651
β_{23}	1.35	0.886

**significant, if $P < 0.05$*

3.3. Influence of variables interaction on the Pb removal via Napier grass

The results of interactional between three independent variables and dependent variable are shown in three-dimention surface plots. The combined effect of initial of Pb and NaCl on removal efficiency of Pb is depicted in Figure 3(a). The percentage removal of Pb increased with increasing both of initial of Pb and NaCl. The effect of initial Pb and period time on Pb removal efficiency is shown in Figure 3(b). It was observed that percentage of Pb removal increased with increase in initial Pb and period time. The interaction between initial of NaCl and period time on Pb removal is presented in Figure 3(c). It was observed that an increase in the Pb removal occurred when the NaCl in the solutions changed from 0.3 to 0.5 g/l and the period time increase.

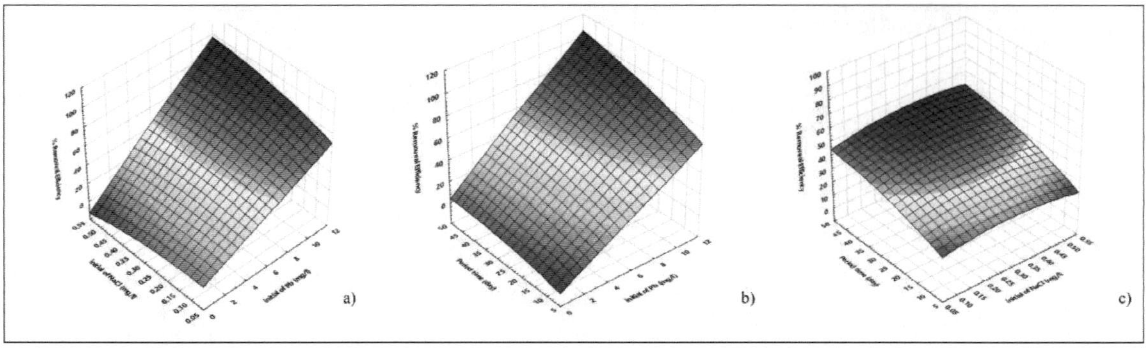

Figure 3. 3D-Respond surface plots showing the effect of variable parameter on the removal of Pb via Napier grass a) initial concentration of Pb and NaCl, b) initial Pb concentration and time and c) initial NaCl concentration and time

4. Conclusions

Napier grass; energy crop, was suggesting to be a good candidate for Pb phytoremedition under the brackish water that revealed in the high phytotoxicity (at 10 mg/l Pb concentration) and removal efficiency (88.63±4.9%). On the other hand, Napier grass was as high tolerated to high level of Pb concentration in synthetic brackish water. Which revealed on the possibilities of plant growth under the presence of Pb contaminated condition (10 mg/l). According to the experimental condition, the highest Pb removal efficiency via Napier grass in synthetic brackish water revealed at Pb and NaCl concentration at 10 mg/l and 0.5 g/l, respectively. BBD and RSM were used in the design of experiments, statistical analysis and optimization of the parameters. The optimum conditions for initial of Pb, initial of NaCl and period time 10 mg/l, 0.5 g/l and 45 day, respectively. At the optimized factors, the amount of removed pollutants Pb was 93.56%. Since this study was performed the optimum condition for Pb phytoremediation using Napier grass under brackish water, a further work will look on the continuous operation constructed system according to these prediction results.

5. References

[1] Carlos A H, Luisa M P and Alicia F-C 2015 *Ecol. Eng.* **82** 512-6

[2] Goham M M and Fahad A A-M 2017 *J. Aquat. Pollut. Toxicol.* **1** 1-5

[3] Maria A C G, Rachel A H-D, Adriane N S and Angela P V 2016 *Ecotoxicol. Environ. Saf.* **134** 133-47

[4] Vimal C P, Omesh B and Nandita S 2016 *Renew. Sustainable Energy Rev.* 54 58-73

[5] Xinghua L, Yixing S, Chenglong D and Qingsheng C 2009 *Biotech. Adv.* 27 633-40

[6] Amin M, Hamidi Abdul A, Mohammad Ali Z, Shuokr Q A and Mohamad Razip S 2013 *IJSRES.*
1 66-70

[7] Gangrong S and Qingsheng C 2009 *Biotech. Adv.* 27 555-61

[8] Gordon K, Brian M and Matin Q 2017 *Ecotoxicol. Environ. Saf.* **137** 225-32

Acknowledgments

This work was supported and funded by Faculty of Science Energy and Environment, King Mongkut's University of Technology North Bangkok, Thailand (SciEE002).

IOP Conf. Series: Earth and Environmental Science **120** (2018) 012021 doi:10.1088/1755-1315/120/1/012021

Effect of Staged Dissolved Oxygen Optimization on In-situ sludge Reduction and Enhanced Nutrient Removal in an A^{2}MMBR-M System

Shan-Shan Yang[1], Ji-Wei Pang[1,*], Xiao-Man Jin[1], Zhong-Yang Wu[1], Xiao-Yin Yang[1], Wan-Qian Guo[1], Zhi-Qing Zhao[1], Nan-Qi Ren[1,*]

1 Harbin Institute of Technology, Harbin 150090, PR China.
E-mail: crabpjw@126.com (Ji-Wei Pang); rnq@hit.edu.cn (Nan-Qi Ren)

Abstract. Redundant excess sludge production and considerable non-standard wastewater discharge from existing activated sludge processes are facing more and more challenges. The investigations on lower sludge production and higher sewage treatment efficiency are urgently needed. In this study, an anaerobic/anoxic/micro-aerobic/oxic-MBR combining a micro-aerobic starvation sludge holding tank (A^{2}MMBR-M) system is developed. Batch tests on the optimization of the staged dissolved oxygen (DO) in the micro-aerobic, the first oxic, and the second oxic tanks were carried out by a 3-factor and 3-level Box-Behnken design (BBD). The optimal actual values of X_1, X_2, and X_3 were DO_1 of 0.3-0.5 mg/L, DO_2 of 3.5-4.5 mg/L, and DO_3 of 3-4 mg/L. After the optimization tests, continuous-flow experiments of anaerobic/anoxic/oxic (AAO) and A^{2}MMBR-M systems were further conducted. Compared to AAO system, a 37.45% reduction in discharged excess sludge in A^{2}MMBR-M system was achieved. The COD, TN, and TP removal efficiencies in A^{2}MMBR-M system were respective 4.06%, 2.68%, and 4.04% higher than AAO system. The A^{2}MMBR-M system is proved a promising wastewater treatment technology possessing enhanced in-situ sludge reduction and improved effluent quality. The staged optimized DO concentrations are the key controlling parameters for the realization of simultaneous in-situ sludge reduction and nutrient removal.

1. Introduction

Activated sludge process, recognized as the widespread sewage treatment process for domestic, municipal and industrial wastewater treatment are the crucial method for sewage treatment [1]. However, redundant excess sludge yields from the existing activated sludge processes worldwide are facing more and more challenges [2], [3]. In terms of the study of Guo et al. [4], sludge reduction processes/technologies can be classified into two classifications: post sludge reduction through chemical, physical, biological or the combined treatment methods [5], [6]; and in-situ activated sludge reduction during sewage treatment process [7-10]. Thus, for the purpose of reducing the produced excess sludge yield at source, investigations on in-situ activated sludge reduction sewage treatment processes/technologies are urgently needed. On the other side, nowadays, the existing WWTPs in China, for instance, confront with considerable non-standard wastewater discharge [11]. Apparently, just blindly reduce the produced sludge yield is too blunt, also may deteriorate the quality of effluent as well as the operation of a sewage treatment plant, thus the study on lower sludge production and higher sewage treatment efficiency should be took into consideration in future researches.

During wastewater treatment processes, the operating parameter of the dissolved oxygen (DO) is proved extremely significant both for the performances of nutrient removal efficiencies and the production yield of the excess sludge [12-15]. In consideration of the effect of DO on chemical oxygen demand (COD), Janus and Ulanicki [16] predicted a slight increase in soluble microbial products (SMP) with increased DO concentration with the CES-ASM3 model, the production of SMP is the major components in effluent organic matters. Research from Holakoo et al. [17] found that the mixed liquor SMP concentrations increased with increases in DO concentrations. In the proposed staged aerated process of from the study of Yoo et al. [18], the optimal maximum DO concentration for nitrogen removal was determined to be around 2.0-2.5 mg/L. Wang et al. [19] found that the optimum DO concentration for nitrogen removal was 0.50 mg/L. Guo et al. [20] compared with high DO of 3 mg/L and lower DO of 0.4-0.8 mg/L in SBR, results showed that a simultaneous nitrification and denitrification was carried out in low DO. It was concluded that by selecting a proper DO lever is not only of benefit to the biological nitrogen removal technology, but also favorable to sludge population optimization. Katie et al [21] showed that simultaneous nitrification and denitrification via nitrite succeeded between 0.4-1.6 mg/L DO concentration in a pilot-scale-modified anaerobic-anoxic-oxic reactor. On the basis of the analyses above, a staged aerated and optimized DO concentrations applied are the key control parameters investigated for the realization of simultaneous nutrient removal and sludge reduction.

In this study, the anaerobic/anoxic/oxic (AAO) system which is the most widely used wastewater treatment technology due to the simultaneous biological nutrients (carbon, nitrogen, and phosphorus) removal without any chemicals, is applied. An improved AAO system, an anaerobic/anoxic/micro-aerobic/oxic-MBR combining a micro-aerobic starvation sludge holding tank (A^2MMBR-M) system, possessing good performances of enhanced sludge reduction and improved effluent quality, is investigated in this study. By using a 3-factor and 3-level Response surface methodology, batch tests on the study of the optimized DO concentrations in different reaction compartments were conducted. The objective of this paper is to optimize the DO concentrations in different reaction compartments of the A^2MMBR-M system to achieve remarkable sludge reduction effect and improved nutrient removal efficiency.

2. Material and methods

2.1. Sludge cultivation

Before conducting the batch tests and the continuous-flow experiments, the activated sludge, taken from Harbin sewage treatment plant was inoculated into two identical AAO systems (Fig. 1a) for 2 months sludge cultivation. After 2-month sludge cultivation, the characteristics and biophase of the activated sludge in the two AAO systems were maintained in stable state. One AAO system was modified into an A^2MMBR-M system (Fig. 1b), the other AAO system was performed as the control system. The composition of the synthetic wastewater used in these two AAO systems was as shown in the study of Yang et al. [22]. The specific operation and control parameters for the AAO systems were showed in Table 1.

(a)

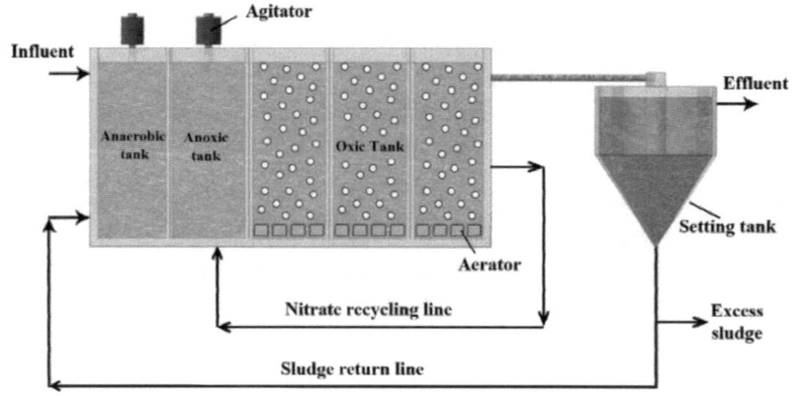

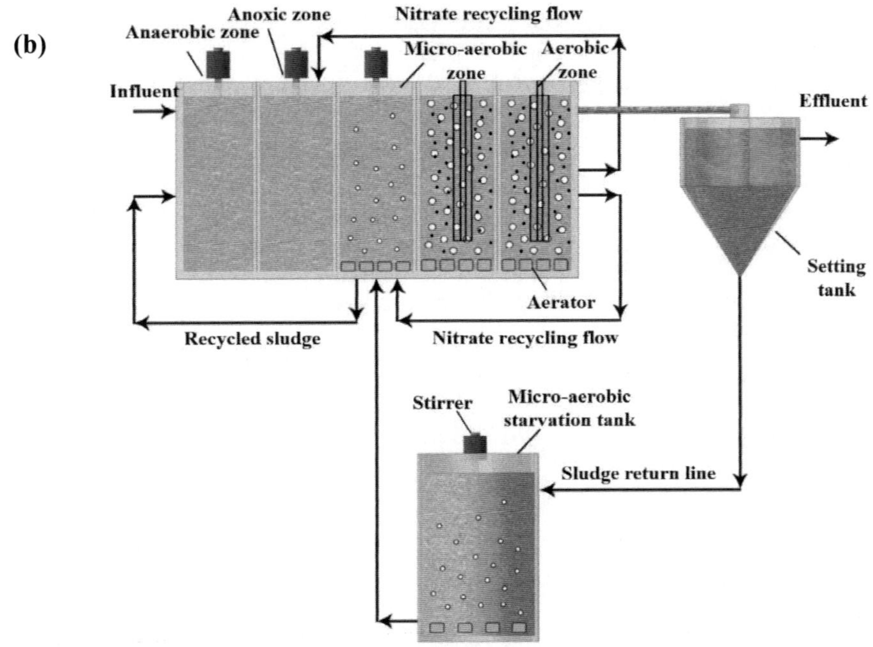

Fig. 1. The schematic diagrams of (a) an anaerobic/anoxic/oxic (AAO) system and (b) an anaerobic/anoxic/micro-aerobic/oxic-MBR combining a micro-aerobic starvation sludge holding tank (A^2MMBR-M) system

Table 1 Operation and control parameters for the AAO and the A^2MMBR-M systems

Control parameters	Unit	
Effective volume of anaerobic tank	L	8
Effective volume of anoxic tank	L	8
Effective volume of oxic tank	L	24 (3 × 8 L compartment)
Effective volume of setting tank	L	20
DO value in the first oxic tank	mg/L	2.0-3.0
DO value in the second oxic tank	mg/L	2.0-3.0
HRT in anaerobic tank	h	1.6
HRT in anoxic tank	h	1.6
HRT in oxic tank	h	4.8
Synthetic wastewater flow	L/h	5
Nitrate recycling line	%	100
Sludge return line	%	100
Mixed liquor suspended solids (MLSS),	mg/L	3500

2.2. Experiment design and operation

Before conducting the continuous-flow experiments, 17 batch tests were designed and carried out to optimize the DO concentrations in different reaction compartments by using a 3-factor and 3-level Response surface methodology. The schematic diagram of batch reactors was shown in Fig. 2. The composition of the synthetic wastewater used in batch tests were same as is applied in the continuous-flow experiments.

After obtaining the optimized staged DO concentrations, the lab-scale continuous-flow A^2MMBR-M system was performed (Fig. 1b), while the other AAO system was performed as the control (Fig. 1a).

For the A^2MMBR-M system, the effective volume of the micro-aerobic tank was 8 L, and the HRT in the micro-aerobic tank was performed at 1.6 h. The effective volumes of the first oxic and the second oxic tanks were respective 8L and 8 L, with 1.6 h HRT in each compartment. For the micro-aerobic starvation sludge of the A^2MMBR-M system, the HRT and the DO concentration were maintained at 9 h and 0.5 mg/L. During experiments period, effluent samples taken from the AAO and the A^2MMBR-M systems were collected to examine the production of the excess sludge yield and the efficiencies of the nutrient removals including COD, nitrogen, and phosphorus. Before examination, the collected samples were filtered through 0.45 μm filters for analyses. All the experiments were performed in triplicate. All the experiments were conducted at room temperature.

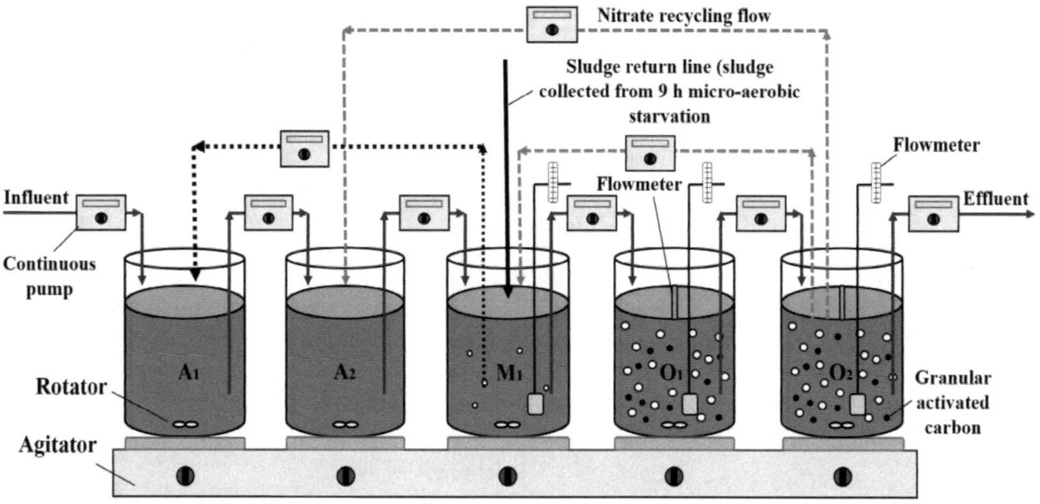

Fig. 2. The schematic diagram of batch reactors for the A^2MMBR-M system

2.3. Box-behnken design and statistical analysis

Response surface methodology (RSM), a mathematical and statistical technique is always used to analyze the mutual relationships between the response and the independent variables [23], and applied to identify the optimal operating conditions for a system. The important parameters of dissolved oxygen (DO) concentrations in the micro-aerobic tank, the first oxic tank, and the second oxic tank were respective the DO_1 (X_1, mg/L), DO_2 (X_2, mg/L), and DO_3 (X_3, mg/L). The most important response was the observed sludge production yield (Y_{obs}) (g·MLSS/g·COD·d, Y_1), COD removal efficiencies (%, Y_2), TN removal efficiencies (%, Y_3), and TP removal efficiencies (%, Y_4). A 3-factor and 3-level RSM using Box-Behnken design (BBD) were introduced to design tests to study the interrelationships and the optimal levels for these operating parameters. The ranges levels of the three independent variables were presented in Table 2. The relationship between the uncoded and coded values is shown in Eq. (1):

$$x_i = \frac{X_i - X_i^*}{\Delta X_i} \quad (1)$$

Eq. (2) shows a second-order polynomial model:

$$Y=\beta_0+\beta_1x_1+\beta_2x_2+\beta_3x_3+\beta_{11}x_1^2+\beta_{22}x_2^2+\beta_{33}x_3^2+\beta_{12}x_1x_2+\beta_{13}x_1x_3+\beta_{23}x_2x_3 \quad (2)$$

where Y is the response variable; X_i is the uncoded value of the independent variable; X_i^* is the uncoded value of the independent variable at the center point; and ΔX_i is the step change value; x_i is the coded value of the independent variable; β_0, β_1, β_2, β_3, β_{11}, β_{22}, β_{33}, β_{12}, β_{13}, and β_{23} represent the regression coefficients from the test data.

2.4. Analytical methods

In this study, the chemical oxygen demand (COD), soluble chemical oxygen demand (SCOD), mixed liquor suspended solids (MLSS), total phosphorus (TP), and total nitrogen (TN) were measured in accordance with standard methods [24]. ORP and dissolved oxygen (DO) values were monitored by a portable ORP/DO meter with ORP and DO probes (Germany WTW Company ORP/Oxi 340i main engine, Germany). The analyses of TP, TN, and NH_4^+-N were carried out every other day. The measurements of COD and MLSS were tested daily. The daily produced excess sludge yields were calculated.

Table 2 Experimental ranges and levels of the independent variables.

Independent variables	Range and levels		
	-1	0	1
DO_1 (mg/L), x_1	0.1	0.3	0.5
DO_2 (mg/L), x_2	2	4	6
DO_3 (mg/L), x_3	2	4	6

3. Results and discussion

3.1. Optimization of operating variables and their reciprocal analysis

The design matrix and results obtained based on the experimental BBD design are showed in Table 3.

Table 3 Response surface BBD and experiments.

Run	Actual values of parameters			Response			
	X_1	X_2	X_3	Y_1	Y_2	Y_3	Y_4
1	0.1	4	6	0.22	89.12	84.33	91.76
2	0.3	6	6	0.24	88.38	86.76	92.58
3	0.5	6	4	0.21	90.87	91.29	94.24
4	0.1	6	4	0.2	90.45	83.46	91.35
5	0.3	4	4	0.16	95.21	89.76	93.81
6	0.3	2	2	0.12	92.94	81.52	91.44
7	0.5	2	4	0.15	94.56	91.16	93.36
8	0.3	4	4	0.16	95.21	89.76	93.81
9	0.5	4	2	0.15	93.58	90.82	94.66
10	0.3	4	4	0.16	95.21	89.76	93.81
11	0.1	4	2	0.13	93.67	82.94	89.78
12	0.3	2	6	0.18	92.78	87.52	92.14
13	0.3	4	4	0.16	95.21	89.76	93.81
14	0.5	4	6	0.23	89.47	92.67	94.82
15	0.3	6	2	0.19	94.29	88.77	92.15
16	0.3	4	4	0.16	95.21	89.76	93.81
17	0.1	2	4	0.14	94.52	83.65	90.02

Second-order polynomial models for coded responses of Y_1 were established as described in Eq. (3):

$$Y_{coded}=+0.16+6.250E\text{-}003x_1+0.031x_2+0.035x_3+0.000x_1x_2-2.500E\text{-}003x_1x_3-2.500E\text{-}003x_2x_3+7.500E\text{-}003x_1^2+7.500E\text{-}003x_2^2+0.015x_3^2 \quad (3)$$

Table 4 shows the analysis of variance for the experimental model equations to examine the significance and the adequacy of the second-order polynomial equation. $P<0.0001$ indicates a high significance of the corresponding variable. A high squared regression coefficient, R^2 of 0.9763 indicates a high degree of correlation between the predicted and actual responses, indicating that the model could fit the response well. In Fig. 3, the contour 2D curves and the response surface 3D plot were depicted to represent the interaction between the independent variables and to determine the optimal levels of each independent variable for observing the optimal response levels.

Table 4 Analysis of variance (ANOVA) results for the response surface quadratic mode.

Source	Statistics

	Sum of squares	df	Mean Square	F-value	P-value	
Model	*0.020*	*9*	*2.171E-003*	*31.99*	*< 0.0001*	*significant*
x_1	*3.125E-004*	*1*	*3.125E-004*	*4.61*		
x_2	*7.812E-003*	*1*	*7.812E-003*	*115.13*		
x_3	*9.800E-003*	*1*	*9.800E-003*	*144.42*		
$x_1 x_2$	*0.000*	*1*	*0.000*	*0.000*		
$x_1 x_3$	*2.500E-005*	*1*	*2.500E-005*	*0.37*		
$x_2 x_3$	*2.500E-005*	*1*	*2.500E-005*	*0.37*		
x_1^2	*2.368E-004*	*1*	*2.368E-004*	*3.49*		
x_2^2	*2.368E-004*	*1*	*2.368E-004*	*3.49*		
x_3^2	*9.474E-004*	*1*	*9.474E-004*	*13.96*		
Residual	*4.750E-004*	*7*	*6.786E-005*			
Lack of Fit	*4.750E-004*	*3*	*1.583E-004*			
Pure Error	*0.000*	*4*	*0.000*			
Cor Total	*0.020*	*16*				

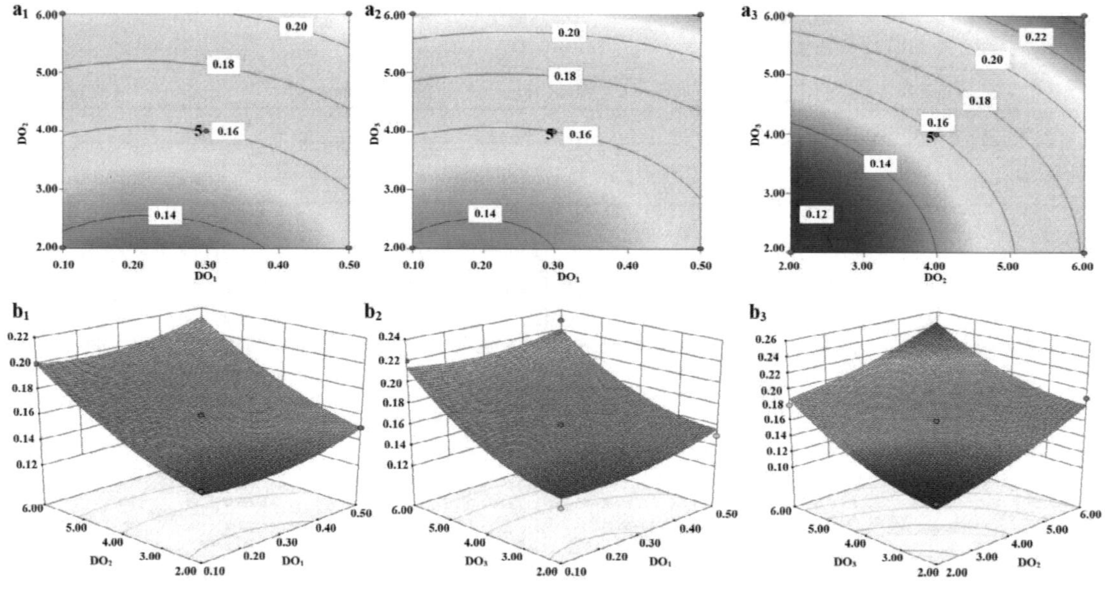

Fig. 3. The effect of the three DO concentrations (DO_1, DO_2, and DO_3) and the mutual interaction of the DO values on Y_{obs}: (a) 2D contour curves and (b) 3D response surface plots.

Second-order polynomial models for coded responses of Y_2 (COD removal efficiencies, %) were established as described in Eq. (4):

$$Y_{coded}=+95.21+0.09x_1-1.35x_2-1.84x_3-1.44x_2x_3-1.62x_1^2-0.99x_2^2-2.13x_3^2 \quad (4)$$

Second-order polynomial models for coded responses of Y_3 (TN removal efficiencies, %) were established as described in Eq. (5):

$$Y_{coded}=+89.76+3.95x_1+0.8x_2+0.9x_3-2.00x_2x_3-0.41x_1^2-1.96x_2^2-1.66x_3^2 \quad (5)$$

Second-order polynomial models for coded responses of Y_4 (TP removal efficiencies, %) were established as described in Eq. (6):

$$Y_{coded}=+93.81+1.77x_1+0.42x_2+0.41x_3-0.11x_1x_2-0.46x_1x_3-0.45x_1^2-1.12x_2^2-0.61x_3^2 \quad (6)$$

For the other three significant responses, p values are all significant proving that a high significance of the corresponding variable with the three responses. Furthermore, R^2=0.9596 for Y_2, R^2=0.9703 for Y_3, and R^2=0.9828 for Y_4 demonstrate a high degree of correlation between the predicted and actual responses. On the basis of the results, the obtained second-order polynomial models were adequate and significant. Based on Eqs. (3) to (6), the optimal actual values of X_1, X_2, and X_3 were calculated and determined to be DO_1 of 0.3-0.5 mg/L, DO_2 of 3.5-4.5 mg/L, and DO_3 of 3-4 mg/L for simultaneously obtaining lower sludge production and good performances of nutrient removal. The three optimal DO_1, DO_2, DO_3 in the three compartments in the A^2MMBR-M system chosen as the optimal staged DO levels.

3.2. Performances on in-situ sludge reduction and nutrient removal

In comparison with the performances of AAO system, continuous-flow studies on the impact of the optimal staged DO concentrations (the three optimal DO_1, DO_2, DO_3 in the three compartments) on sludge reduction and nutrient removal in the continuous-flow A^2MMBR-M system were further discussed. As depicted in Fig. 4a, the average Y_{obs} values for the AAO and the A^2MMBR-M systems were respective 0.26 and 0.17 g·MLSS/g·COD·d, indicating a 37.45% reduction in discharged excess sludge in A^2MMBR-M system. As for the removal efficiencies in COD, TN and TP in AAO system (COD, TN, and TP removal efficiencies were respective 90.35%, 87.73%, and 90.29%), the COD, TN, and TP removal efficiencies in A^2MMBR-M system were 94.02%, 90.08%, and 93.94% in Figs. 4b-d. Although the AAO system is the most widely used wastewater treatment technology due to the simultaneous biological nutrients removal without any chemicals. Compare to AAO system, both higher sludge reduction and better nutrient removal efficiencies could be observed in the A^2MMBR-M system, implied that the staged DO optimization applied in A^2MMBR-M system benefit for the organic substrate competition between phosphorus accumulating organisms and denitrifying bacteria [12], [13]. In light of the previous studies [14], [15], higher nitrogen and phosphorus removal by nitrification, denitrification, and denitrifying phosphorus removal were positively correlated to the staged optimized DO concentrations in the A^2MMBR-M system. Furthermore, a micro-aerobic starvation sludge holding tank introduced in the sludge return line would induce the energy uncoupling metabolism [23], and hence lower sludge production would be obtained in A^2MMBR-M system. Then the returned low micro-molecule compounds from 9 h micro-aerobic starvation treatment would further increase the organic matters removal efficiencies [22]. Therefore, possessing good performances of nutrient removal and lower excess sludge production, the A^2MMBR-M system is proved a promising wastewater treatment process. The staged optimized DO concentrations applied in A^2MMBR-M system are the significant controlling parameters for realizing the simultaneous in-situ sludge reduction and enhanced nutrient removal.

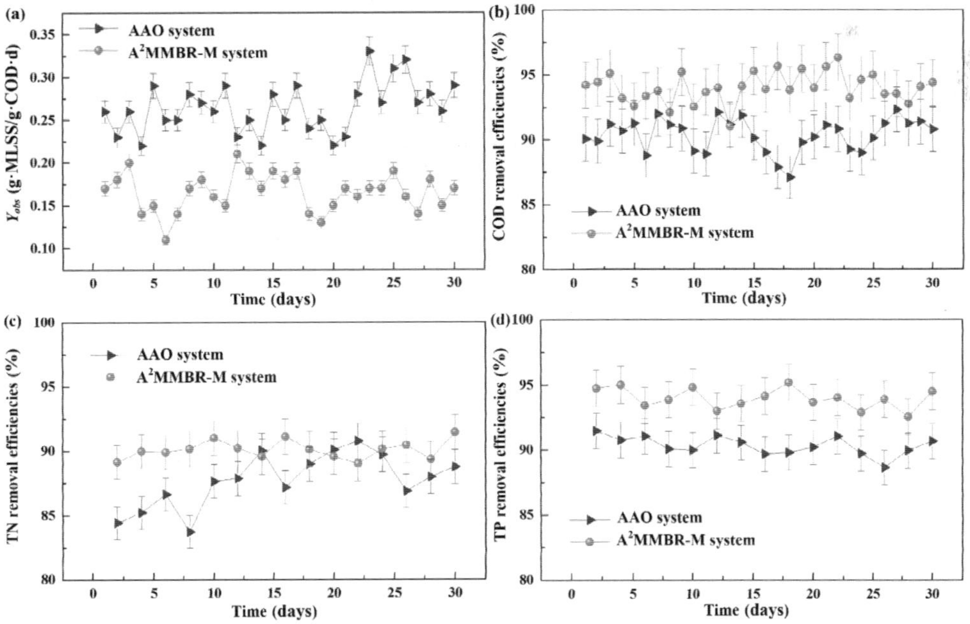

Fig. 4. Changes in (a) Y_{obs}, (b) COD removal efficiencies, (c) TN removal efficiencies, and (d) TP removal efficiencies in AAO and A^2MMBR-M systems.

4. Conclusion

In this study, a 3-factor and 3-level BBD was introduced to design and study the optimal levels for the staged DO_1, DO_2, DO_3 in the three micro-aerobic, the first oxic, and the second oxic tanks of the A^2MMBR-M system on sludge reduction and nutrient removal. Based on the equations obtained in Eqs. (3) to (6), the optimal actual values of X_1, X_2, and X_3 were calculated and determined to be DO_1 of 0.3-0.5 mg/L, DO_2 of 3.5-4.5 mg/L, and DO_3 of 3-4 mg/L. After obtaining the optimized staged DO levels by BBD, comparative studies on the performances of sludge reduction and nutrient removal in AAO and A^2MMBR-M systems were further conducted. In comparison of the AAO system, a 37.45% reduction in discharged excess sludge in A^2MMBR-M system was obtained. The removal efficiencies of COD, TN, and TP for the A^2MMBR-M system were respective 94.02%, 90.08%, and 93.94%, which increased by 4.06%, 2.68%, and 4.04% comparing with the performances of AAO system. On the basis of the results obtained in the batch and the continuous-flow experiments, a staged aerated and optimized DO concentrations applied are the key controlling parameters for the realization of simultaneous nutrient removal and sludge reduction.

5. References

[1] Hreiz R, Latifi M A, Roche N 2015 *Chem. Eng. J.* **281** 900-920.
[2] Yang S S, Guo W Q, Cao G L, Zheng H S, Ren N Q 2012 *Bioresour. Technol.* **124** 347-354.
[3] Tian Y, Zhang J, Wu D, Li Z P, Cui Y N 2013 *Water Res.* **47** 279-288.
[4] Guo W Q, Yang S S, Xiang W S, Wang X J, Ren N Q 2013 *Biotechnol. Adv.* **31** 1386-1396.
[5] Yang S S, Guo W Q, Zhou X J, Meng Z H, Liu B, Ren N Q 2011 *Bioresour. Technol.* **102** 43-51.
[6] Kim D H, Jeong E, Oh S E, Shin H S 2010 *Water Res.* **44** 3093-3100.
[7] Wang L, Qiang Z M, Li Y G, Ben W W 2017 *J. Environ. Sci.* **56** 263-271.
[8] Romero-Pareja P M, Aragon C A, Quiroga J M, Coello M D 2017 *Ultra. Sonochem.* **36** 336-342.
[9] Fang F, Hu H L, Qin M M, Xue Z X, Hu Z R 2015 *Bioresour. Technol.* **185** 1-6.
[10] Wang B, Peng Y Z, Guo Y Y, Zhao M Y, Wang S Y 2016 *Bioresour. Technol.* **214** 284-291.
[11] Zhang Q H, Yang W N, Ngo H H, Guo W S, Jin P K, Dzakpasu M, Yang S J, Wang Q, Wang X C, Ao D 2016 *Environ. Int.* **92-93** 11-22.
[12] Keene N A, Reusser S R, Scarborough M J, Grooms A L, Noguera D R 2017 *Water Res.* **121** 72-85
[13] Meng F, Yang A Q, Zhang G M, Wang H Y 2017 *Bioresour. Technol.* **241** 993-997.
[14] Niu T H, Zhou Z, Shen X L, Qiao W M, Jiang L M, Pan W, Zhou J J 2016 *Water Res.* **90** 369-377.
[15] Fan H T, Qi L, Liu G Q, Zhang Y K, Fan Q, Wang H C 2017 *J. Environ. Sci.* **55** 224-235
[16] Janus T, Ulanicki B, 2010 *Desalination* **261** 117-125.
[17] Holakoo, L., Nakhla, G., Bassi, A.S., Yanful, E.K., 2007. *Chemosphere* **66** 849-857.
[18] Yoo H S, Ahn K H, Lee H J, Lee K H, Kwak Y J, Song K G 1999 *Water Res.* **33** 145-154.
[19] Wang J L, Peng Y Z, Wang S Y, Gao Y Q 2008 *Chinese J. Chem. Eng.* **16** 778-784.
[20] Guo J H, Peng Y Z, Wang S Y, Zheng Y N, Huang H J, Wang Z W 2009 *Bioresour. Technol.* **100** 2796-2802.
[21] Katie A T, Natalie B, Ralf C R 2003 *Biotechnol. Bioeng.* **83** 706-720.
[22] Yang S S, Guo W Q, Chen Y D, Peng S M, Du J S, Zheng H S, Feng X C, Ren N Q 2016 *Water. Res.* **88** 524-537.
[23] Ocholi O, Menkiti M, Auta M, Ezemagu I 2017 *Egyptian J. Petrol.* https://doi.org/10.1016/j.ejpe.2017.04.001
[24] APHA (American Public Health Association) 2005 **21 ed.** *APHA*, Washington DC, USA.

Acknowledgments
The authors gratefully acknowledge the financial support by the National Nature Science Foundation of China (Grant No. 51708154), the Special Financial Grant from the China Postdoctoral Science Foundation (Grant No. 2016T90302), the class General Financial Grant from the China Postdoctoral Science Foundation (Grant No. 2016M601434), and the State Key Laboratory of Urban Water Resource and Environment (Grant No. HC201621-01).

IOP Conf. Series: Earth and Environmental Science **120** (2018) 012022 doi:10.1088/1755-1315/120/1/012022

The Removal of Cu (II) from Aqueous Solution using Sodium Borohydride as a Reducing Agent

N T Sithole, F Ntuli, T Mashifana

1 University of Johannesburg, Department of Chemical Engineering, P.O. Box 17011, Doornfontein 2088, South Africa
E-mail: nastassias@uj.ac.za

Abstract. The removal and recovery of metals from wastewater has been a subject of significant importance due the negative impact these toxic metals have on human health and the environment as a result of water and soil pollution. Increased use of the metals and chemicals in the process industries has resulted in generation of large quantity of effluents that contains high level of toxic metals and other pollutants. The objective of this work was to recover of Cu in its elemental form as metallic powder from aqueous solution using $NaBH_4$ as a reducing agent. Reductive precipitation was achieved in a batch reactor at 65°C using Cu powder as a seeding material. This study also investigated the effect of concentration of sodium borohydride ($NaBH_4$) as a reducing agent. The amount of $NaBH_4$ was varied based on mole ratios which are 1:1, 1:0.25 and 1:0.1 to recover Cu from synthetic wastewater. The results obtained showed that sodium borohydride is an effective reducing agent to recover Cu from wastewater. The optimum concentration of $NaBH_4$ that gives the best results the 1:1 molar ratio with over 99% Cu removal.

1. Introduction

The removal and recovery of metals from wastewater has been a subject of significant importance due to the adverse impact these toxic metals have on human health and the environment as a result of water and soil pollution [1]. Increased use of the metals and chemicals in the process industries has resulted in generation of large quantities of wastewater that contains high levels of toxic metals and other pollutants [1]. The presence Cu in the environment poses environmental disposal problems due to its non-biodegradability in the environment [2]. Unlike organic pollutants, the majority of which are susceptible to biological degradation, Cu ions do not degrade into harmless end products. Cu exists in most wastewater sources including, printed circuit board manufacturing, plating, wire drawing, copper polishing and paint manufacturing, wood preservatives and printing operations [3]. Copper is a toxic metal which may be accumulated in the human or ecological food chain through consumption or uptake and may be hazardous to human health or the environment. Drinking water with high Cu levels ($\geq$1,300 ppb) may cause vomiting, diarrhea and stomach cramp [4]. The problem of recovering base metals like Cu as a pollutant from waste water is an important process and is becoming more important with the increase of industrial activities [5].

Several treatment methods such as precipitation, adsorption, ion-exchange and membrane technologies have been developed for metal remediation. Of these methods precipitation is the mostly widely used because it is the most economical method and easier to implement and operate on a large scale [6]. However, traditional precipitation methods using lime, sulfides or hydroxides recover metals in the form of a sludge which is not reusable and has to be disposed in landfills creating a potential environmental hazard and resulting in loss of valuable minerals. Due to the fast depletion of mineral reserves globally the current focus in effluent treatment is now on the recovery and re-use of these heavy metals rather than removal and disposal. So far none of this methods is capable of recovering metals in a form that is suitable for re-use hence there is a need to develop alternative technologies to meet technology based treatment. In this study, reductive precipitation was investigated as a possible method for remove and recover Cu in its elemental form.

2. Materials and methods

2.1. Equipment

A batch reactor with an overhead stirrer was used to ensure adequate suspension of the seeding material in solution. A thermometer was used to measure temperature. The atomic absorption spectroscopy (AAS;Tescan Vega 3 XMU l) was used to measure the concentrations of Cu in solution after reduction. X-Ray fluorescence (XRF;Rigaku ZSX Primus II) was used to determine the elemental composition of the powder product. Malvern Particle Analyser Mastersizer (2000) was used to measure the particle size distribution (PSD) of the powder using the laser diffraction technique. The powder morphology was captured using a Scanning Electron Microscopy (SEM; Thermo scientific ICE 3000 series) to determine the effect of sodium borohydride.

2.2. Materials

$CuSO_4.5H_2O$ salts supplied by Sigma Aldrich was used to prepare synthetic 0.5g/l Cu^{2+}solution. Cu powder supplied by Sigma Aldrich were used as seeding materials. Reverse osmosis water was used to prepare all the samples. NaOH supplied by Rochelle chemicals was used to adjust the pH. HNO_3 and H_2SO_4 supplied by Sigma Aldrich was used to prepare the residual solution for AAS analysis.acid respectively.

2.3. Methods

2.95 g of $CuSO_4.5H_2O$ was measured and dissolved in RO water to make a 0.5 g/L solution of Cu^{2+}. 500 ml of the solution was mixed with 0.161 g of $NaBH_4$ and 30 g of Cu powder making a 1:1 stoichiometric solution. The resulting solution was then heated to 65°C and then the pH was adjusted to 12 by adding 70 ml of 1 M NaOH. At the end of 3 min agitation and heating was stopped and the solution was allowed to cool. After settling the clear supernatant was decanted carefully leaving the seeding material and reduced metal in the beaker. The above procedure was repeated twice with fresh Cu solution and the same seeding material. After the third experiment the seeding material was discharged from the beaker. The seeding material was filtered using a Buchner vacuum filtration unit and dried for particle size distribution, elemental and morphology analysis. The reduced solution was then filtered and 100 ml of the sample was analysed for the residual metal concentration using Atomic Absorption Spectroscopy. To establish the effect of concentration of reducing agent the amount of $NaBH_4$ was varied based on mole ratios which are 1:1, 1:0.25 and 1:0.1.

3. Results and discussions

3.1. Effect of $NaBH_4$ concentration

Figure 1 shows the variation in % removal of Cu^{2+} in different molar ratios with the number of batch reduction.

Figure 1 shows that the % removal decreases from the 1st batch to the 3rd batch for all the molar ratios. This can be attributed to the plating of Cu which did not catalyze the reduction reaction meaning Cu was not autocatalytic. The highest percentage removal obtained was 99.7 % at 1:1 molar ratio as compared to the 1:0.25 and 1:0.1 molar ratios. This is due to the decrease in the NaBH4 concentration.

3.2. Reduction rate of Cu^{2+}

Figure 2 shows the variation of Fe^{3+} reduction rate from solution with successive number of densifications.

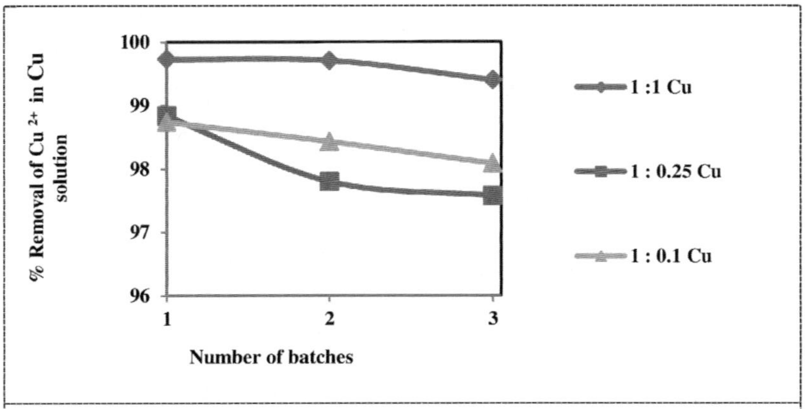

Figure 1. % Removal of Cu^{2+}using Cu powder as a seed material.

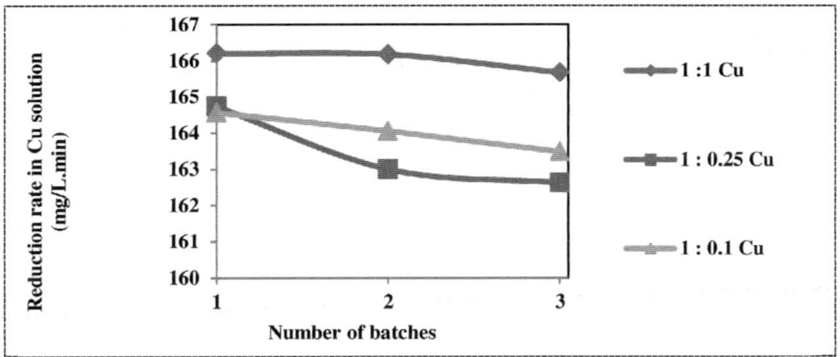

Figure 2. Reduction rate of Cu^{2+}

The reduction rate of copper in the Cu solution generally decreased from the 1st to the 3rd batch in Fig. 2. This is due to the fact that Cu plating decreases the effective surface activity of the seed decelerating the rate of reduction.

3.3. Effect of reduction on Cu powder purity

The composition of the powder produced was analysed using the X –ray Flourescence.Table 1shows the XRF analysis of Cu seed before and after reduction crystallization.

The % Cu content before reduction was 99.2% and after reduction there was rapid decrease in Cu content the lowest being at 1:1 molar ratio followed by the 1:0.25 and then the 1:0.1 molar ratios. It can be seen that as the concentration NaBH4 is reduced the purity of the powder increases. There was a dramatic decrease at 1:1 as compared to 1:0.25 and 1:0.1.This is attributed by the incorporation of decomposition products like sodium and borane being incorporated into the powder.

3.4. Effect of reductive precipitation on PSD

Figure 3 and Figure 4 shows the PSD analysis of BOFS before and after electroless plating.

Figure 3 shows that there was a shift in modal size from that of the seed indicating a generation of smaller sized particles which is evidence of breakage or nucleation. µm. The number distribution was uni-modal as depicted in Figure 4. Furthermore the trend was the same before and after reduction. The only difference is that smaller particles dominated at 1:1 molar ratio followed by 1:0.25 then 1:0.1.This trend resembles that as the smaller particles dominate, the higher the reducibility in the residual concentration hence 1:1 molar ratio was preferentially favored over 1:0.25 and 1:0.1.

Table 1. Copper powder purity in copper solution

	Before reduction	After reduction		
Elements	Cu seed	1:1	1:0.25	1:0.1
Na_2O	0.000	1.680	0.695	0.462
MgO	0.051	0.053	0.000	0.000
Al_2O_3	0.090	2.940	0.238	0.175
SiO_2	0.112	0.186	0.075	0.055
P_2O_5	0.042	0.044	0.059	0.053
SO_3	0.015	1.110	0.164	0.114
Cl	0.017	0.018	0.027	0.025
K_2O	0.000	0.005	0.000	0.000
Fe_2O_3	0.366	0.048	0.054	0.0368
NiO	0.106	0.000	0.053	0.0646
CuO	99.20	79.10	98.50	99.00
ZnO	0.000	0.057	0.000	0.000
Ag_2O	0.000	0.035	0.043	0.035
Cr_2O_3	0.013	0.000	0.000	0.000
CaO	0.012	0.026	0.000	0.019
MnO	0.000	0.012	0.000	0.000
PdO	0.000	0.000	0.029	0.000

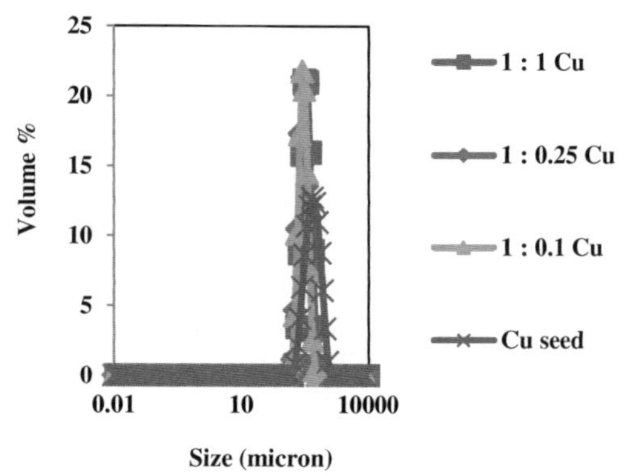

Figure 3. Evolution of PSD by volume distribution using Cu powder

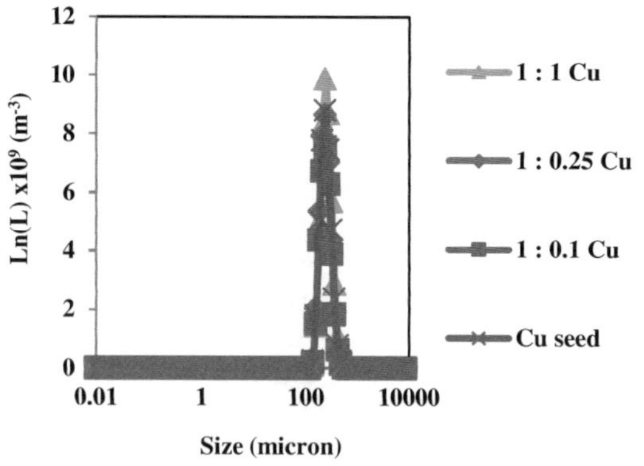

Figure 4. Evolution of PSD by number distribution using copper powder

3.5. Effect of electroless plating on particle morphology

Figure 5 shows the scanning electron micrographs of the Cu powder before and after reductive precipitation.

Figure 5. SEM of the Cu powder (a) before plating, (b)Cu powder after reductive precipitation.

The particles exhibited a nearly spherical structure and wide particle size range before and after reduction. However, it was seen that the particle size increased significantly after reduction. This is typical of particle aggregation.

4. Conclusion

Sodium borohydride was found to be an effective reducing agent as results in an increase in the rate of reduction. The optimum concentration of $NaBH_4$ that gave the best results was the 1:1 molar ratio which achieved over 99% removal of Cu. The treated effluents (wastewater) after reduction process is environmentally friendly as it meets a general limit for effluent discharge which is stipulated in the Johannesburg acceptance standards for effluents as 20mg/L for copper. Reductive precipitation using sodium borohydride recovers Cu in elemental form as metallic powder which can be re-used which is more economical than other methods such as precipitation agents which produces by-products such as sludge which are toxic and cannot be re-used resulting in loss of valuable minerals.

5. References

[1] Abdullah M.A, and Devi Prasad A.G 2009 Kinetic and equilibrium studies for the biosorption of Cr (VI) from aqueous solutions by potato peel waste *International Journal of Chemical Engineering Research* 1(2) p 51-62.
[2] Gupta V.K and Sharma S 2001 Process development for the removal of lead and chromium from aqueous solutions using red mud—an aluminium industry waste *Water research* 35(5) p1125-1134.
[3] Baysal A, Ozbek N and Akman S, 2013 Determination of trace metals in waste water and their removal processes *In Waste Water-Treatment Technologies and Recent Analytical Developments.* InTech.
[4] Tasker P. A, Tong C.C, and Westra A.N 2007 Co-extraction of cations and anions in base metal recovery *Coordination chemistry reviews* 251(13) p 1868-1877.
[5] Gaballah I, and Kilbertus G 1998 Recovery of heavy metal ions through decontamination of synthetic solutions and industrial effluents using modified barks", *Journal of Geochemical Exploration* 62(1) p 241-286.
[6] Sithole T, Freeman N, and Falayi T 2015 The removal of Ni & Cu from a mixed metal system using sodium borohydride as a reducing agent *South African Journal of Chemical Engineering* 20(1) p 16-29.

Acknowledgments
The authors would like to thank the NRF for funding for financial assistance and the University of Johannesburg for providing resources to conduct the study.

IOP Conf. Series: Earth and Environmental Science **120** (2018) 012023 doi:10.1088/1755-1315/120/1/012023

Removal of Manganese from Solution using Polyamide Membrane

M Mathaba, N Sithole, T Mashifana

1 University of Johannesburg, Department of Chemical Engineering, P.O. Box 17011, Doornfontein 2028, South Africa
E-mail: nastassias@uj.ac.za

Abstract. The work demonstrates the performance of polyamide membrane in the removal of manganese ions from single salt aqueous solution simulating real acid mine drainage. The membrane was tested using a dead-end filtration cell with manganese sulphate was used to prepare a feed solution. The membrane flux and metal rejection was evaluated. Effect of operating parameters such as pH, initial feed concentration and pressure on membrane performance was investigated. The pressure was varied between 10 and 15 bar and it was observed that increasing the pressure increases the membrane flux. Acidic pH conditions contributed to the removal of the contaminate as Mn^{2+} ions are freely at low pH. The percentage rejection was found to be 63.5 to 77.6 % as concentration is increased from 290 ppm to 321 ppm for a feed solution. The membrane showed satisfactory results in removing metal ions from solution.

1. Introduction

The mining industry makes use of different type of utilities to extract valuable desired products from ore; minerals such as gold, copper, nickel etc. However, the common problem encountered by most industries is that of acid drainage which contains high concentration of toxic substances such as cyanides and heavy metals which have serious human health and ecological implications [1]. Generally, acid mine drainage is produced, usually but not exclusively, in iron sulphide-aggregated rocks and can be accelerated when naturally-occurring bacteria such as acidothiobacillus assist in the breakdown of sulphide minerals. Acid Mine Drainage (AMD) is wastewater produced when sulphide-bearing material from underground mining is exposed to oxygen and water. It is characterized by low pH and high concentration of heavy metals and other toxic elements that can severely contaminate surface and ground water, as well as soils. Mining operations promote AMD formation during rock exploration which exposes the sulphide bearing rocks and during communition which increases the surface area [2]. The process begins with pyrite (Fe2S) oxidation which releases Ferrous iron (Fe^{2+}), Sulphate ions (SO_4^{2-}) and Hydrogen ions (H^+). Fe^{2+} undergoes further oxidation to form Fe^{3+}, which will react with water to form ferric hydroxide $Fe(OH)_3$, an insoluble orange precipitate. Although there are numerous reasons for the toxicity of AMD to receiving water bodies the major impact is because of the proton acidity of the AMD, which will lead to a decrease in pH of the recipient water should it have insufficient neutralization capacity [3].

The need to address acid mine drainage inspired research to explore treatment technologies to treat the effluent before discharging into the environment. Active and passive treatments are traditional technologies which have been practiced for decades to deal with the problem. Active treatment depends on addition of alkaline chemicals which neutralizes the acidity of AMD and passive treatment relies on biological, geochemical and gravitational processes. Adding alkaline materials such as lime

cause the pH to increase and cause the metals present to precipitates. This will result in the production of iron rich sludge that may also contain several metals depending on the chemistry of the mine water treated [4]. Mostly, the sludge produced is collected and disposed of through burial or injection into abandoned mine [5]. However; little effort has been focused on the beneficial use of precipitated metals in these systems, especially in passive systems as oxides, carbonates, or sulphides. Recovery of these minerals will not be as profitable as the mine itself, due to slow reaction kinetics of the process [6]. Passive treatment systems involve using sulphate-reducing bacteria or limestone or both to neutralize acidity and precipitate dissolved metals. The systems are sometimes called wetlands or bioreactors [7]. It relies on sulphate-reducing bacteria (SRB) that can be found in natural environments where anoxic conditions prevail [1]. Passive treatment systems have been shown to be more economical than hydrated lime or similar neutralizing reagent methods [6]. However, the activity of Sulphate reducing bacteria (SRB) is controlled by the reactive mixture composition and this limit the long-term efficiency of the process [1].
The demand for clean water and environmental consciousness has prompted researchers to explore cost effective technologies which could treat acid mine drainage up to drinking standards. Membrane technology has emerged as a promising technology to substitute conventional methods [8] Advantages such as easy operation, inexpensiveness, low energy consumption, high separation efficiency and no need for integrated steps for further treatment, makes membrane technology an effective and attractive technology [9]. Reverse osmosis membranes are preferred during AMD treatment for the removal of heavy metals, salts and even ions through electrostatic repulsion and filtration due to smaller pore sizes [10]. These membranes are known as polymers which enclose repeating amide, -CO-NH-, linkages. They are well-known for their high strength, abrasion resistance, and resilience.

2. Experimental

2.1. Materials and equipment
A non-porous polyamide membrane manufactured by GE healthcare Companies was commercially purchased from Sigma Aldrich. A packet acquired contained 100 pieces with pore sizes of 0.45μm and 0.047m diameter. Atomic Absorption Spectrometer (Thermo scientific ICE 3000 series) was used to analysis metal ions. pH and conductivity were measured using Metler Toledo dual meter (Sevenduo pH /conductivity meter with a Metler Toledo inLab Pro ISM pH electrode and inLab 738 ISM conductivity probe).

2.2. Permeation batch tests
Membrane performance (water flux and metal ion rejection) was tested in a dead-end filtration with a capacity of 300 mL and effective filtration area of 14.6 cm^2. Pure water flux (J, L/m^2 h) was determined by $J = \frac{A}{Vt}$ and metal ion rejection by $R = \frac{C_{feed} - C_{permeate}}{C_{feed}}$. Feed solution was prepared by dissolving $MnSO_4.2H_2O$ in water to make 321mg/L solution and the pH was altered using H_2SO_4 to 3.2. The solution was fed through the dead-end cell using nitrogen gas to push the solution through and filtrated were collected and analysed using AAS for metal content.

3. Results and discussions

3.1. Membrane flux
Figure 1 shows the variation in % removal of Cu^{2+} in different molar ratios with the number of batch reduction.

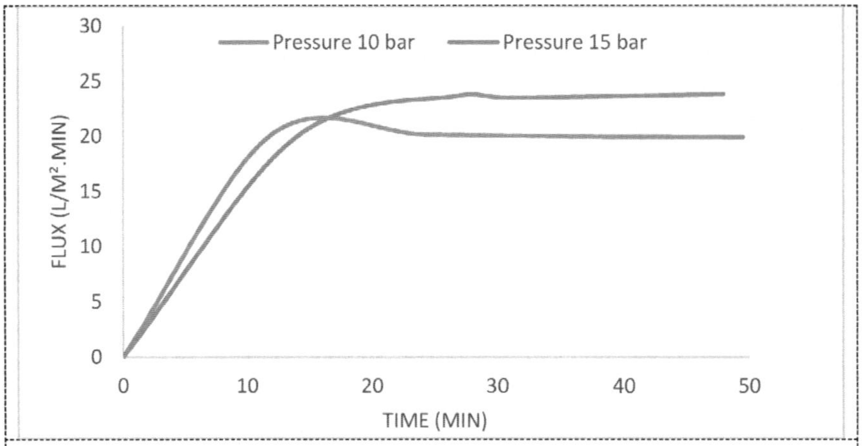

Figure 1. Effect of pressure on membrane flux.

Figure 1 shows the effect of pressure on membrane flux. This was done to calculate the water permeability of the membrane using pure water. The experiments were run for 30 min using 3 replicated samples collected after every 10 min. Pressure of 10 bar and flow rate of 0.034 L/hr resulted in a calculated water flux of 20.17 L/m2.hr. In addition, at 15 bar pressure and flow rate of 0.048 l/hr, the flux was found to be 24.78 L/m2.hr. This is because operational conditions, such as feed pressure, temperature, and feed flow rate, influence the convective transport of molecules toward the membrane surface. The hydrophilicity of polyamide membranes had great influence as the chemical structure had affinity for water molecules and easily let them pass through the membrane.

3.2. Effect of pH and feed concentration on flux

.

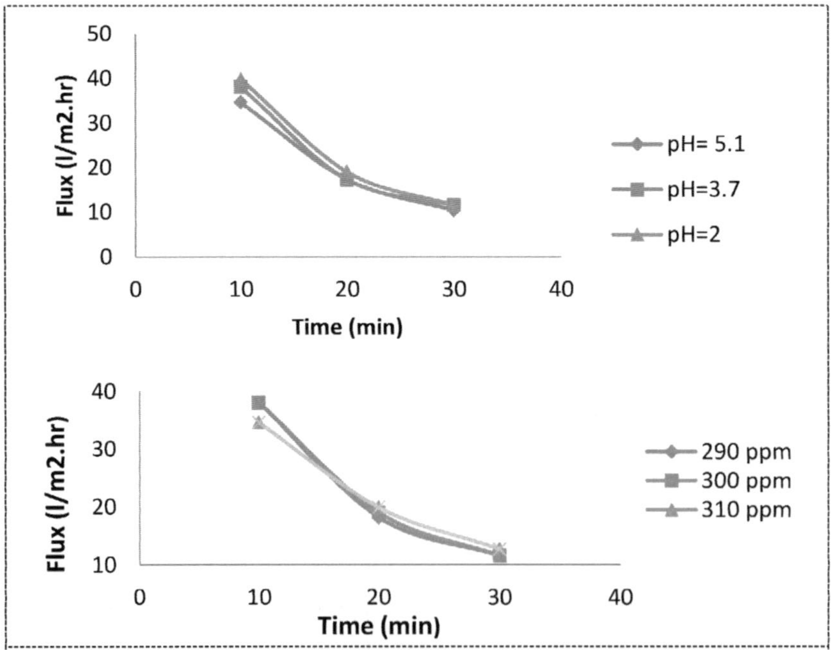

Figure 2. Effect of pH and initial metal concentration

The influence of solution pH and metal concentration on membrane flux was investigated by manipulating the pH to 2, 3.7 and 5.1 using sulphuric acid. The feed concentration was varied from

IOP Conf. Series: Earth and Environmental Science **120** (2018) 012023 doi:10.1088/1755-1315/120/1/012023

290 to 321 ppm. The trend which agrees with what was reported [11], the functional groups of polyamide membrane becomes positive under acidic conditions and this creates an electrostatic repulsion between the membrane and the metal contaminate. Over time, more metal ions are repelled from the membrane surface creating back pressure which pushes solution backwards and reducing membrane flux. Accumulation of metal ions on membrane surface due to filtration causes what is termed concentration polarization. Over time, the membrane effective pore sizes get blocked and this reduces membrane flux.

3.3. Effect of pH and feed concentration on metal rejection.

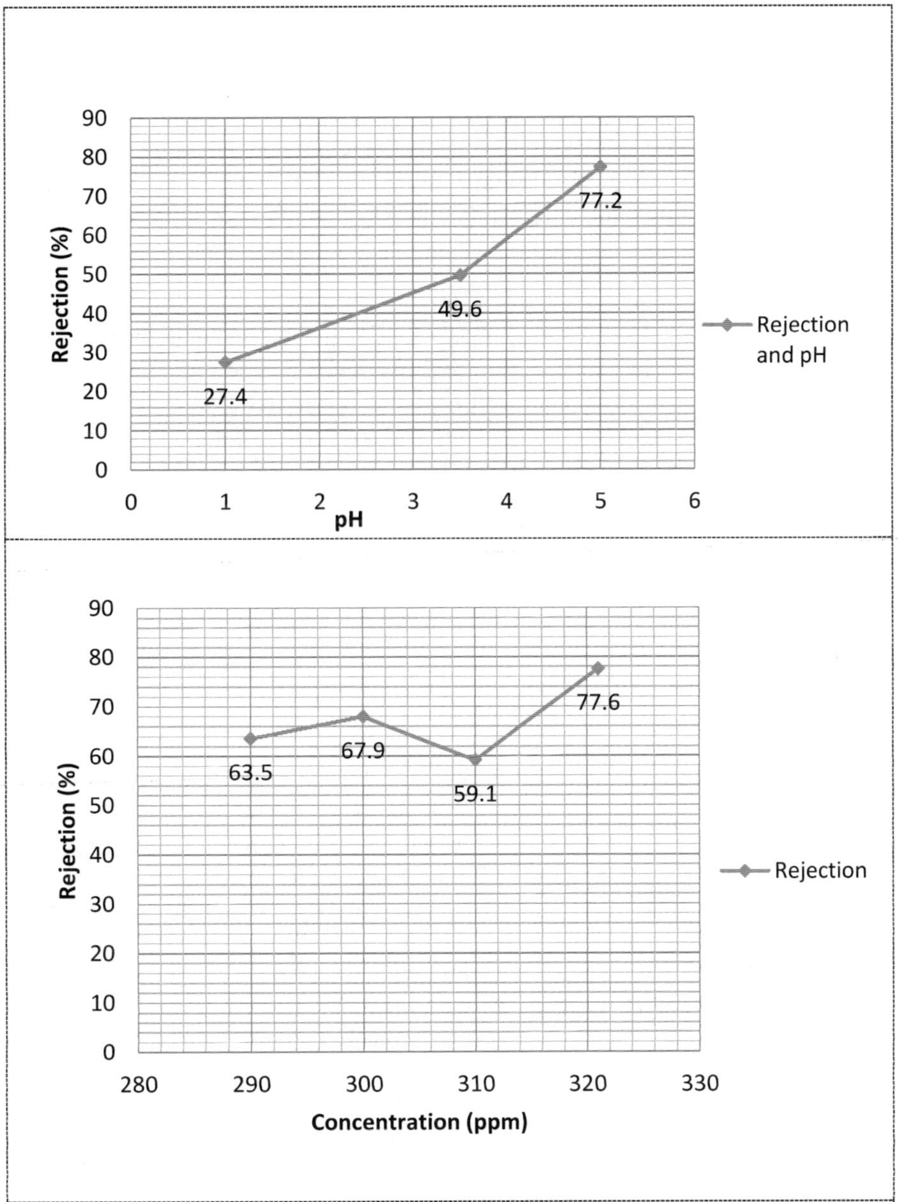

Figure 3. Influence of pH and feed concentration on rejection

Metal ion removal from solution could be due to absorption by the membrane through filtration and/or the attachment of metal ions to functional groups on the membrane surface. Reverse osmosis membrane constructed using polyamide material are known for high metal ion rejection owing to their small effective pore sizes. This trend agrees with Riley et al. [12] in a study on Polyamide reverse osmosis membrane fouling and its prevention. They deduced that when reaction reaches equilibrium at pH level of 7, manganese occurs as MnO_2 (c), an insoluble crystal. As the pH reduces to below 6, it becomes oxidised and remains (at equilibrium) is in its most stable form as a $MnSO_4$ (aq). Reducing the pH further to below 2 (~1.7) reduces $MnSO_4$ to $Mn^{2+,}$ ions free to move about in solution. Hence more high rejection was observed towards neutral pH. Polyamide membranes have very small. The percentage rejection was found to be 63.5 to 77.6 % as concentration is increased from 290 ppm to 321 ppm for a feed solution pH of 2. At this pH, it is suggested that manganese crystals (MnO_2) have oxidised to form $MnSO_4$ which is also is oxidizing to form (Mn $^{2+}$). These large molecules ($MnSO_4$) may not have found free pathways through the membrane therefore it is suggested that they are remaining sliding on the surface of the membrane. An irregular trend is observed when the concentration of the feed has reached 310ppm when the pH is lowest 1.5. At this level, the rejection is reduced to 59.1 % for a concentration of 310ppm. This trend can suggest that the solution is reaching equilibrium where all the $MnSO_4$ molecules oxidise to Mn^{2+} at pH below 2 and some ions can pass though the cross-linked membrane structure.

4. Conclusion

The study showed maximum flux of 24.78 L/m^2.hr after 30 min at an operation pressure of 15 bar. Polyamide materials which are normally used to fabricate reverse osmosis membranes, are known to have very small effective pore sizes, hence high pressure is needed to push molecules through the membrane. The feed solution pH alters the membrane surface charge in which under acidic conditions, the functional groups become positively charged and this creates an electrostatic repulsion between the membrane surface and cationic metal contaminates which reduces membrane flux but effectively increases membrane rejection. Additionally, the solution pH influences the stability of the manganese metal in which under acidic condition it moves freely as Mn^{2+} and is able to pass through the membrane but increasing the pH towards neutral, it exist as an insoluble oxide which is large enough to get trapped by the membrane. Further studies should investigate the anionic metal rejections by polyamide to see the effect of membrane charge under acidic conditions. Also, checking the anti-fouling behaviour of the membrane could be helpful to understand the long-term sustainability of the membrane.

5. References

[1] Neculita C-M, and Zagury G.J 2007 Passive treatment of acid mine drainage in bioreactors using sulfate-reducing bacteria *Journal of Environmental Quality 36(1),* 1-16.

[2] Akcil A and Koldas S 2006 Acid Mine Drainage (AMD): causes, treatment and case studies *Journal of Cleaner Production, 14*, 1139-1145.

[3] Ford K 2003 Passive treatment systems for acid mine drainage.

[4] Gusek J and Clarke-Whistler K 2005 Where Does the Recovery of Metal Resources from Passive Treatment Systems Fit in Sustainable Development Initiatives Associated with Large Mining Projects? Annual Meeting of the American Society for Mining and Reclamation Breckenridge, USA.

[5] Hallberg K 2010 New perspectives in acid mine drainage microbiology *Hydrometallurgy 104(3),* 448-453

[6] Hedin R.S 2003 Recovery of marketable iron oxide from mine drainage in the USA. Land Contamination and Reclamation *11(2),* 93-98

[7] Hilal N, Al-Zoubi H, Mohammad A.W and Darwish N.A 2005 Nanofiltration of highly concentrated salt solutions up to sea water salinity *Desalination*. 184: 315–326.

[8] Johnson D. B and K B Hallberg 2005 Acid mine drainage remediation options: a review. Science of the total environment *338(1)*, 3-14
[9] Shen M, Kete S and Lueptow R.M 2016 Rejection mechanisms for contaminants in polyamide reverse osmosis membranes *Journal of Membrane Science*, *509*, pp.36-47.
[10] Zhong C, Xu Z, Fang X and Cheng L 2006 Treatment of Acid Mine Drainage (AMD) by Ultra-Low-Pressure Reverse Osmosis and Nanofiltration Environmental *Engineering Science*. 24 (9): 1297-1306.
[11] Shockravi A, Vatanpour V, Najjar Z, Bahadori S and Javad A 2017 A new high performance polyamide as an effective additive for modification of antifouling properties and morphology of asymmetric PES blend ultrafiltration membranes *Microporous and Mesoporous Materials 246*, pp.24-36.
[12] Riley R.L, Ridgway H and Ishida K 2000 Polyamide reverse osmosis membrane fouling and its prevention: oxidation-resistant membrane development, membrane surface smoothing and enhanced membrane hydrophilicity *Separation Systems Technology, Inc., San Diego, CA, Orange County Water District, Fountain, CA*

IOP Conf. Series: Earth and Environmental Science **120** (2018) 012024 doi:10.1088/1755-1315/120/1/012024

Heavy Metals and Radioactivity Reduction from Acid Mine Drainage Lime Neutralized Sludge

T Mashifana, N Sithole

University of Johannesburg, Department of Chemical Engineering, P.O. Box 17011, Doornfontein 2088, South Africa
E-mail: tmashifana@uj.ac.za

Abstract. The worldwide known treatment processes of acid mine drainage result into the formation of hydrous ferric oxides that is amorphous, poorly crystalline and into the generation of hazardous voluminous sludge posing threat to the environment. Applicable treatment technologies to treat hazardous solid material and produce useful products are limited and in most cases nonexistence. A chemical treatment process utilizing different reagents was developed to treat hazardous acid mine drainage (AMD) sludge with the objectives to conduct radioactivity assessment of the sludge generated from lime treatment process and determine the reagent that provides the best results. Leaching with 0.5 M citric acid, 0.4 M oxalic acid, 0.5 M sodium carbonate and 0.5 M sodium bicarbonate was investigated. The leaching time applied was 24 hours at 25 ^{o}C. The characterization of the raw AMD revealed that the AMD sludge from lime treatment process is radioactive. The sludge was laden with radioactive elements namely, ^{238}U, ^{214}Pb, ^{226}Ra, ^{232}Th, ^{40}K and ^{214}Bi. 0.5 M citric acid provided the best results and the hazardous contaminants were significantly reduced. The constituents in the sludge after treatment revealed that there is a great potential for the sludge to be used for other applications such as building and construction.

1. Introduction

Acidic mine effluent such as AMD is commonly treated through lime neutralization. This process remains the most widely applied technology due to the high efficiency in removing dissolved metals and the fact that lime costs are low as compared to other alternatives chemicals. The technology is effective in raising the pH of the water and precipitating the metals to below regulatory limits. However, one of the major challenges with simple lime neutralization treatment process is the production of a voluminous, hard to settle, radioactive and hazardous sludge, laden with metals. The two major drawbacks regarding the AMD sludge are the volume of the sludge generated from the treatment processes, the long-term chemical stability and the negative potential impact the sludge poses to the environment, ground water and human beings.

Various industries produces significant quantities of acid mine drainage sludge, lignite and fly ash, however the beneficial reuse technologies for these materials are limited [1]. The growing global urbanization of society coupled with increasingly stringent sludge reuse/disposal regulations and increasing public pressure, is forcing both public and private sludge generators to re-evaluate their sludge management strategies [2]. Conventionally, the waste sludge is disposed by means of incineration, landfilling or ocean disposal as well as reused as soil conditioner in agriculture. Legislation requires that sludge from neutralization plants be discharged into lined ponds to prevent metal leachate from polluting ground water [2]. The volume of sludge to be disposed also influences the cost and processes that produce sludge. An estimated amount of 20 t/d of sludge is produced from

IOP Conf. Series: Earth and Environmental Science **120** (2018) 012024 doi:10.1088/1755-1315/120/1/012024

1 Ml/d of discard leachate when neutralized with lime or limestone [3]. The principle of lime neutralization lies in the insolubility of heavy metals in alkaline conditions. Metals such as iron (Fe), zinc (Zn), and copper (Cu) are precipitated when pH is adjusted to a set point of about 9.5 [4]. When dry lime is added to the waste stream, the hydrated lime reacts or dissociates to increase pH. Hydrolysis reactions occur causing the metals present to precipitate as hydroxides [4]. The following two equations illustrate these reactions:

$$CaO + H_2O \rightarrow Ca(OH)_2 \tag{1}$$

$$Ca(OH)_2 \rightarrow Ca^{2+} + 2OH^- \tag{2}$$

A common by-product of lime neutralisation process is gypsum (calcium sulphate bi-hydrate). Gypsum precipitation occurs as acidic drainage is often rich in sulphate and calcium added from the lime will bring the solubility product well above saturation. Gypsum is a major sludge component and contributes significantly to the volume of sludge generated [3].

$$Ca(OH)_2 + H_2SO_4 \rightarrow CaSO_4 \cdot 2H_2O \tag{3}$$

Another common by-product of lime neutralization is calcium carbonate. The inorganic carbon for this reaction can either come from the AMD itself or as a result of carbon dioxide from air, which is dissolved during aeration. This carbon dioxide converts to carbon bicarbonate and then partially to carbonate due to the high pH. The carbonate fraction will precipitate with high calcium content of the slurry to form calcite (calcium carbonate). This calcite can play an important role in the stability of the final sludge product as it provides neutralizing potential to the sludge, as it is stored [5].

Radioactive materials which occur naturally and where human activities increase the exposure of people to ionizing radiation are known by the acronym 'NORM'. NORM results from activities such as burning coal, making and using fertilizers, oil and gas production [6]. NORM levels are typically expressed in one of two ways: Becquerel per kilogram (or gram) indicates level of radioactivity generally or due to a particular isotope, while parts per million (ppm) indicates the concentration of a specific radioisotope in the material [6]. The materials may be original (such as uranium and thorium) or decay products thereof, forming part of characteristic decay chain series, or potassium-40. The two most important chains providing nuclides of significance in NORM are the thorium series and the uranium series [6]:

In the acid mine drainage treatment process, using lime to reduce acidity in the effluent and precipitate some heavy metals from the water, a radioactive sludge is produced and may emit radioactive gas. Several hundred of mine dumps and tailing dams, each containing millions of tonnes of waste are located along a long line of Witwatersrand. They are exposed to the elements with winds blowing the finer particles away and heavy summer rains wash larger quantities of acidic and radioactive particles into surrounding watercourses. The material utilized for building contains small amounts of radioactive substances. These are radionuclides in uranium (^{238}U), thorium (^{232}Th) decay series and radioactive potassium isotope (^{40}K), which mostly originate naturally from rock and soil. In addition to the natural occurring radionuclides, some industrial by-products also contain the radioactive substance such as cesium (^{137}Cs). Incorporating the by-product into building and construction material, the final produced product may contain these radionuclides [7]. To assess whether an action level is exceeded, the activity index must be calculated from activity concentration measurements of the material. For the activity indexes, radionuclides such as radium (^{226}Ra), thorium (^{232}Th), potassium (^{40}K) and cesium (137Cs) are considered. Other radionuclides may need to be taken into consideration in special cases [8]. It is critical to assess the radiological hazard associated with the exposure to the radiation from ^{40}K, ^{226}Ra, and ^{232}Th, to account for the collective effect of the activity concentrations of these radionuclides in a material.

This study was undertaken to evaluate the radioactivity of AMD sludge generated from a lime neutralization treatment process in a gold mine and determine the best reagent to remove or reduce the radioactivity in the material and investigate the possibility of utilizing the raw or treated product for other applications.

IOP Conf. Series: Earth and Environmental Science **120** (2018) 012024 doi:10.1088/1755-1315/120/1/012024

2. Experimental

This section present the experimental procedure and methodology followed.

2.1. Method

AMD sludge sample was obtained from a gold mine in Randfontein. Salts were used to prepare different concentrations of leaching reagents, namely 0.4 M citric acid, 0.5 M oxalic acid, 0.5 M sodium bicarbonate and 0.5M sodium carbonate. AMD sludge samples were first prepared prior conducting the experimental work. The raw sample was dried in an oven for 24 hours at 50 ^{o}C. The dried sample was then milled in a rod mill for 2 hours. Sieves were utilized to remove all the solid particles that could not be milled, i.e. rocks in the AMD sludge. A representative sample was obtained using Eriez Magnetic Rotary riffler for characterization. The specific gravity of the material was determined using a gas pycnometer. A pH meter was used to measure pH. The prepared reagents of citric acid, oxalic acid, sodium carbonate and sodium bicarbonate were added to AMD sludge at 15% solid loading.

A thermostatic shaker was utilized to conduct the leaching experiments. 4 flat bottom beakers containing the material were placed in a thermostatic shaker at 35 ^{o}C, and agitated at 170 rpm for 24 hours. After 24 hours, the thermostatic shaker was switched off, the samples were filtered, the solution stored for further characterization and the solids were dried in an oven at 50 ^{o}C. After drying the solid products were analyzed using XRF and XRD to check the effect of the different reagents on the leaching of AMD sludge. Relative density was determined using a gas pycnometer by weighing 5 g of AMD sludge into a pycnometer cup. The cup was then inserted into the measuring equipment. After the relative density was measured, the reading displayed in g/cm3 was recorded. To determine the pH of AMD sludge, AMD sludge/water mixture was prepared by adding 50 g of sludge into 100 ml of deionized water under continuous stirring and measuring the pH after 30 minutes.

AMD sludge composites were prepared to determine the unconfined compressive strength (UCS) of the raw and citric acid treated sludge. The composites were casted and cured at the temperatures of 40°C. The strength was allowed to develop over 14 days. After the curing period, the UCS was determined. The treated AMD sludge was stabilised with fly ash and lime at the ratio of 50:30:20 to enhance the pozzolanic reaction and cured at 40 ^{o}C for 14 days.

2.2. Analysis

X-ray Floroscence (XRF, Rigaku ZSX Primus II) was used to determine the semi quantitative chemical composition of AMD samples. Mineral species were determined by X-ray diffraction (XRD, Rigaku Ultima IV)). Radionuclides activity concentrations in the sample were determined using Gamma Ray Spectroscopy (GRS). Using GRS, the samples were packed, hermetically sealed and stored for about six weeks prior to counting so as to ensure radioactive equilibrium between ^{226}Ra and its short-lived progeny.

3. Results and discussions

The measured specific activities in the raw samples collected for AMD sludge are presented in table 1. The activity of radionuclides in the samples is given in Bqkg^{-1} dry weight. The world average activity concentrations of 226Ra, ^{232}Th and ^{40}K are 35, 30 and 400 Bqkg^{-1}, respectively [8]. The activity in AMD sludge ranges from 32 Bqkg^{-1} to 153 Bqkg^{-1}. The highest concentration was observed in ^{238}U, followed by ^{226}Ra, and the lowest concentration was that of ^{232}Th. All the major natural occurring radionuclides were detected in AMD sludge, ^{238}U, ^{226}Ra, ^{232}Th and 40K. The largest contribution of the radioactivity in the sample is due to ^{238}U. ^{226}Ra is drastically above the world's average activity of 35 Bqkg^{-1}. ^{40}K is within the standards of 400 Bqkg^{-1} and ^{232}Th slightly exceeding the stipulated concentration of 30 Bqkg^{-1}.

Table 1. Radioactivity of AMD sludge.

Detected nuclides	Radioactivity concentration (Bq/kg)
^{214}Pb	91
^{214}Bi	98
^{226}Ra	150
^{40}K	43
^{238}U	153
^{232}Th	32

The AMD sludge showed to have a pH of 10.24, relative density of 2.66 g/cm^3, liquid limit of 49.71 and plastic limit of 14.53. The coefficient of gradation (C_c) and uniformity coefficient of the material was found to be 4.327 and 1.716, respectively. Numerous studies conducted previously reveals that pH values for the sludge from various lime neutralization treatment processes ranges from 8.2-10.8 [9]. Mostly the aged sludge showed a lower pH as compared to their counterparts. The measured pH for AMD sludge is 10.24, an alkaline material due to the basic neutralization reagent used (lime), this value agrees with results obtained by other researchers for the sludge from neutralization processes [9].

Most of sludge have a specific gravity of 1.0, i.e. they are almost equal to the weight of the water [6]. The specific gravity of AMD sludge was found to be 2.6568. For sand, if Cu is greater than 6 and Cc is between 1 and 3, it is considered well graded. However, for a gravel to be well-graded, Cu should be greater than 4 and Cc must be between 1 and 3 [7]. The Cu and Cc for AMD sludge is 4.327 and 1.716, respectively. Therefore, it is not well graded, but gravel. The liquid limit for AMD sludge is 49.71 and plastic limit is 14.53, respectively.

3.1. Chemical treatment of AMD sludge with different reagents

Comparing the final product with the raw material the sulfur relative proportion in citric acid medium was reduced by 33%, Figure 1(a). Calcium and iron relative proportion was increased by 8% and 5%, respectively. Sulphur reduction shows that it was leached out during the treatment process. Other elements leached are Mg, Al, Si and K, Figure 2(e). Citric acid drastically decreases the sulfur relative proportion contained in AMD sludge and calcium and iron proportion slightly went up.

Using oxalic acid as a leaching reagent Figure 2(b), sulfur relative proportion was reduced from 5.45 wt% to 3.17 wt%%. Calcium and iron relative proportion was increased from 28.57 wt% to 31.05 wt% and 24.83 wt% to 26.685 wt%, respectively. Oxalic acid dissolved the sulfur initially available in the material and its relative proportion decreased by 42%, while calcium and iron relation slightly increased by 8% and 7.5%, respectively.

The results obtained when sodium carbonate was employed as a leaching agent, Figure 2(c), indicates that sulfur relative proportion was drastically reduced by 82.1%. Calcium relative proportion increased by 7.73% and iron increased by 5.52%. Leaching with sodium bicarbonate, Figure 2(d), resulted in the sulfur reduction of 82.72% in the final product, hence the increase in calcium relative proportion by 6.26% and 5.5% for iron. For both sodium carbonate and sodium bicarbonate, optimal leaching of sulfur is evident, more than 80% reduction in sulfur relative proportion is observed.

Other elements detected in AMD sludge by XRF are sodium, magnesium, aluminum, silicon, potassium and manganese as shown in Figure 1(e). Magnesium, aluminum, silicon and potassium were leached out in citric acid medium and manganese relative proportion slightly increased by only 2%. Using oxalic acid, only magnesium was leached out and the relative proportion for other elements slightly increased. The elements contained in AMD sludge behaved the same when sodium carbonate and sodium bicarbonate were used. In mediums, sodium, magnesium, aluminum, silicon and manganese relative proportion increased and the only reduction in relation was observes in potassium.

There was a drastic increase in relative proportion of sodium when sodium carbonate and sodium bicarbonate were utilized.

Sulfur was leached out from AMD sludge with all the leached agents employed, and the highest reduction was obtained with sodium carbonate and sodium bicarbonate. Calcium and iron relative proportion slightly increased with all the leaching reagents.

3.2. Elemental analysis of AMD sludge treated with different

The chemical components of cement include SiO_2, Al_2O_3, Fe2O_3, CaO, SO_3 and MgO. Therefore, the chemical constitution of the AMD sludge is typical of construction materials. The SO_3 is very high due to the precipitation of sulphate normally available in high concentration in the acid mine effluent and the SiO_2 very low in the AMD as expected. Depending on the application of the material produced, a treatment process that introduces more or increase SiO_2, Al_2O_3 and suppress SO_3 contents is desirable. Using the different reagents sodium carbonate and sodium bicarbonate managed to drastically reduce the sulphate composition by 80% and 81%, respectively. The best reagent that increased the aluminium composition to the highest is sodium carbonate. All the reagents used increased SiO_2 content; however, both sodium carbonate and sodium bicarbonate showed best results. Al_2O_3 composition was enhanced best in sodium carbonate medium.

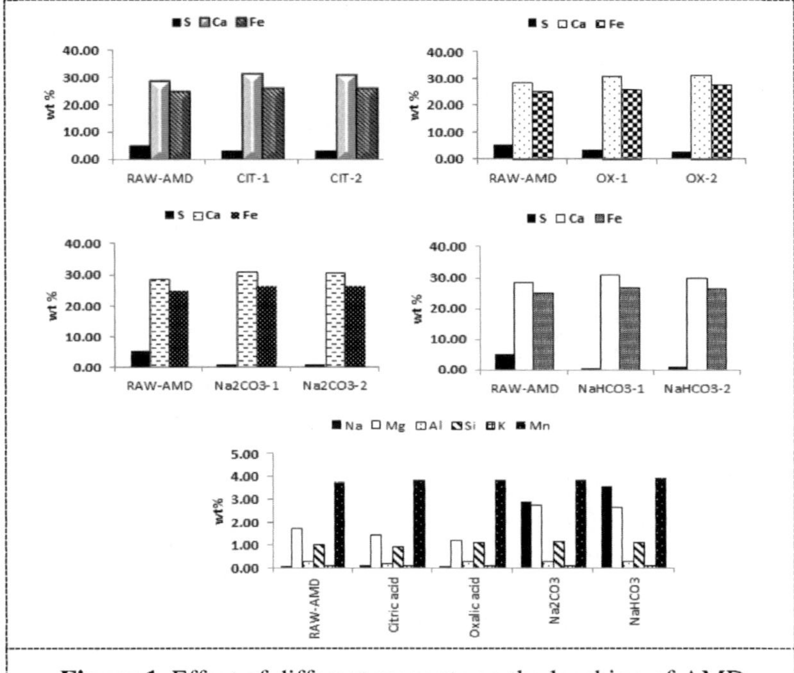

Figure 1. Effect of different reagents on the leaching of AMD sludge at 35°C mixing for 4 hours (a) citric acid, (b) oxalic acid, (c) sodium carbonate, (d) sodium bicarbonate.

3.3. Mineralogical study of AMD sludge treated with different reagents

Figure 2 indicates the XRD results for the AMD sludge treated with citric acid, oxalic acid, sodium carbonate and sodium bicarbonate.

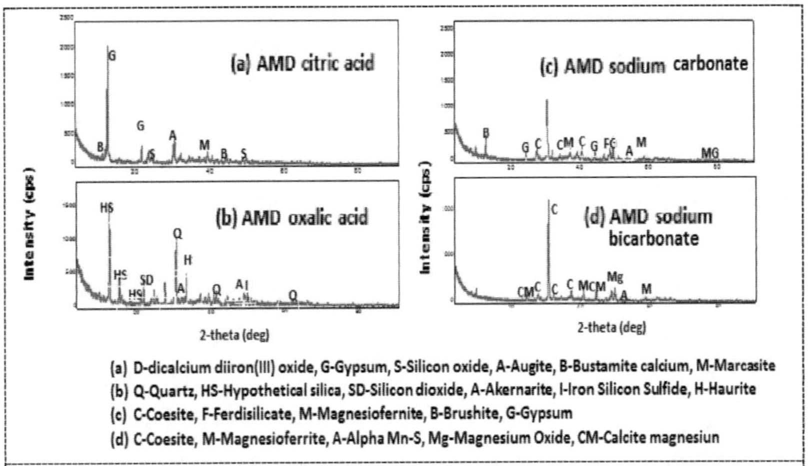

Figure 2. XRD results for AMD sludge after treatment at 35 °C mixing for 4 hours (a) citric acid, (b) oxalic acid, (c) sodium carbonate, (d) sodium bicarbonate.

The XRD results in figure 2 (a) indicates that the predominant mineralogical phase in the final AMD sludge treated with citric acid was Augite aluminia (Ca $(MgAlFe)Si_2O_6$) with a content of 34.26%, followed by 29.78% marcasite (FeS_2), 14.58% silicon oxide (SiO_2), 14.34% (di-calcium di-iron oxide ($Ca_2Fe_2O_5$) with a content, 3.86% gypsum ($Ca(SO_4)$ $(H_2O)_2$) and 3.16% bustamite calcian ($Ca1.18Mn0.19SiO_3$). All the detected phases after the leaching process contain compounds that were initial present in the raw AMD sludge. Citric acid reacted with AMD sludge to form a predominant compound bonded with calcium-magnesium-aluminum-iron and silicate. The final product also contained a calcium sulphate compound with a different compound from the anhydride initially detected in the raw material.

The predominant phase when oxalic acid was used as a leaching agent is quartz (SiO_2) as determined by XRD, Figure 2 (b). The final product contains a total of 87.91% content of SiO_2 in different forms (51.03% quartz, 19.54% hypothetical silica and 17.31% silicon dioxide). The material initially contained 8.29% SiO_2 and 16.6% $MgSi_2$ and oxalic acid increased the total quartz content to 87.91%. This increase is also evident in the XRF results obtained, the Si relative proportion increased from 1.07 to 1.145 wt%. Calcium sulphate reacted with oxalic acid and produced a compound containing Ca-Mg-Si-O_7 with a content of 6.76%, the compound contains elements initially available in the raw AMD sludge. Manganese oxide and iron oxide initially contained in the AMD sludge both reacted with the leached sulfur to form sulfide compounds, MnS_2 and Fe_2SiS_4 respectively.

The XRD results using for sodium carbonate, figure 2 (c) shows that the predominant phase is 78.81% SiO_2. The final product contained 71% more coesite as compared to the initial content. Sodium carbonate as a leaching agent promoted the formation of SiO_2. Iron oxide reacted with silicon and formed iron silicate, $FeSi_2$. Magnesium relative proportion was drastically increased from 1.67 wt% to 2.725 wt%, raw AMD sludge contained 16.6% Mg_2Si and during leaching, magnesium reacted with iron to form 3.99% MgFe. Sulphur contained in the raw AMD sludge was associated with calcium (Anhydrite, $CaSO_4$) with a content of 15.87%. After the leaching process, a calcium sulphate compound with hydrate was formed with 0.14% content. In the presence of sodium carbonate, calcium sulphate reacted with phosphorus to produce 15.46% brushite ($CaHPO_4$ $(H_2O)_2$), the leaching agent exposed phosphorus contained in the raw material as detected by XRF. The final product produced when sodium carbonate was employed as a leaching medium is predominated by SiO_2 with a content of 78.81%.

Calcite magnesium ($MgCaCO_3$) is the predominant compound in the final product with a content of 51.198%, followed by 41.05% coesite (SiO_2), in a sodium bicarbonate medium figure 2 (d). This shows an increase of 32.76% SiO_2 in content from raw material to the final product. It is evident that sodium bicarbonate increased the formation of coesite. Magnesium reacted to form three different compounds containing magnesium in the final product, namely, magnesioferrite (MgFe), magnesium oxide (MgO) and calcite magnesium ($MgCaCO_3$). MgO was initially available in raw AMD sludge as detected by XRF and its relative proportion increased drastically by 57%. The manganese oxide reacted with the sulphur leached out to produce 0.45% MnS.

3.4. Radioactivity analysis of AMD sludge and unconfined compressive strength

The treated AMD sludge was measured for radioactivity and the results are presented in figure 3.

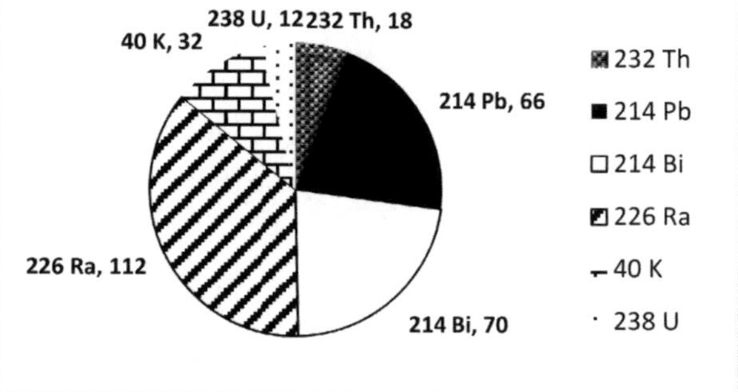

Figure 3. Radioactivity of AMD sludge treated with citric acid.

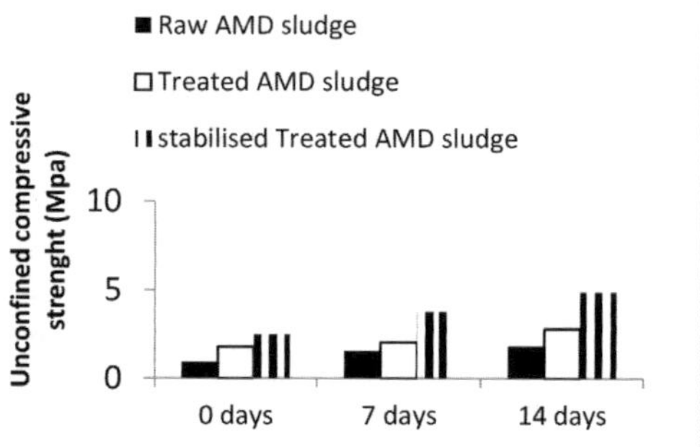

Figure 4. Unconfined compressive strength of AMD sludge.

The radioactivity results of the citric acid treated AMD sludge show a significant decrease in the radionuclides. Uranium the dominant contributor to overall radioactivity was reduced by 28%. All the other contaminants were also reduced; this resulted to a lower radioactivity as compared to the raw AMD sludge. It is crucial and it is highly recommended to assess the radiological hazards associated with exposure to the radiation from ^{40}K, ^{226}Ra, and ^{232}Th. This is conducted to account for the collective effect of the activity concentrations of the radionuclides in a material.

Figure 4 depicts the UCS for the raw, treated AMD sludge and fly ash lime stabilized AMD over 14 days curing. The results show that the strength for both the raw and citric acid treated AMD sludge improved over 14 days. The effect of curing time on UCS was investigated by measuring the compressive strength of specimens. The highest strength obtained is 2.79 MPa for the treated AMD sludge and 1.83 MPa for the raw AMD sludge. There was a significant strength gain after 14 days curing. The chemical composition of both the raw and treated AMD sludge showed that there was insufficient pozzolans in the materials to trigger the pozzolanic reaction for strength development. To further improve the UCS of the treated AMD, the sludge was stabilised with fly ash and lime and the UCS of 4.9 MPa was obtained. The strength obtained for the treated sludge is close to the minimal applicable strength for masonry brick, therefore there is a great potential for this material to be utilized in building and construction and it can be applied for load bearing. A minimum of 3.5 MPa is required on burnt masonry clay [10-12]. The results showed that the treated sludge can attain load bearing strengths without using a binder such as cement or lime.

The composite developed from the treated AMD sludge is presented in figure 5. The identified constituents in the treated AMD composite are Gypsum (G), $CaSO_{4}.{}_2H_2O$ (32.35%), calcium aluminium sulphate, $Ca_6Al_2(SO4)_3(OH)_{12}\cdot 26H_2O$ (33.18), wollastonite (W), CaSiO3 (17.93%), hedenbergite (H), $CaFeSi_2O_6$, (1.48%) and keatite (K), SiO_2 (15.01%). The constituents in the mineralogy are associated with of silicate, iron and calcium. The new hydration products formed are, calcium aluminium sulphate, wollastonite and hedenbergite. Due to the addition of fly ash, high content of silicate oxide is observed, a compound which has silicate as the predominant compound. The results also reveal that lime addition activated the alumina and silica phases and the final material contained keatite with higher proportion than the raw material.

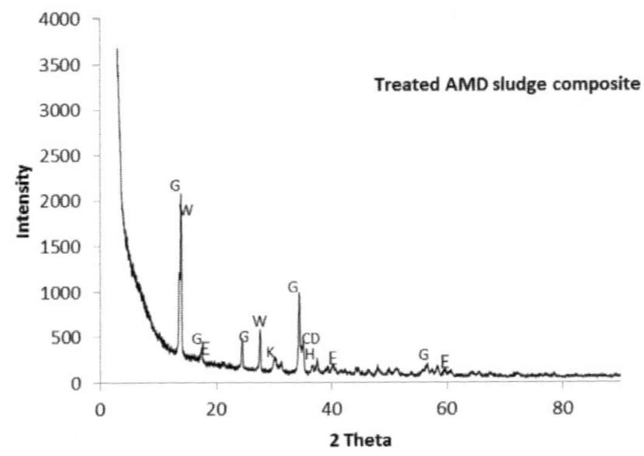

Figure 5. XRD analysis of AMD composite

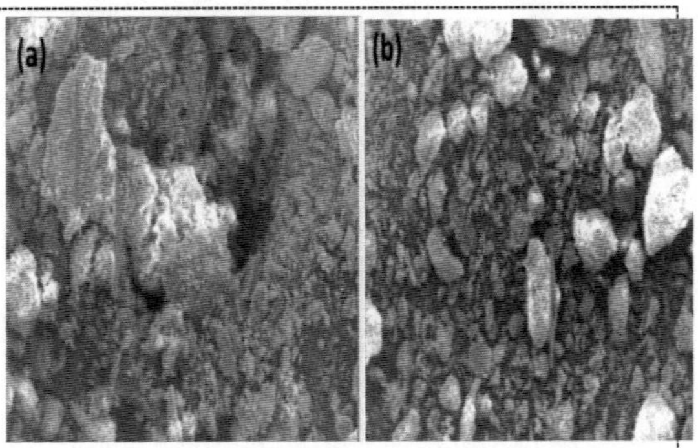

Figure 6. SEM micrograph of AMD sludge: (a) raw AMD sludge (b) treated AMD sludge

As presented in figure 6, the raw AMD sludge showed more flakes like particles that are irregular shaped. Fairly stronger and intact particles were observed in the treated AMD sludge, this might have led to the voids between the larger particles to be filled by the smaller grains particles in the treated AMD sludge and contribute to higher strengths development.

4. Conclusion

The activity concentrations showed that the highest contributor to radioactivity is ^{238}U. The measured activity concentration levels for radium, thorium and potassium revealed that radium was above the average world activity, thorium and potassium slightly exceeds the required standards. No radionuclides were detected in the raw AMD sludge by XRF and XRD; however there was evidence of the radionuclides when GRS was used to characterize the raw material. There was a drastic increase in sodium relative proportion when sodium carbonate and sodium carbonate was utilized; this is due to the sodium available in $NaCO_3$ and $NaHCO_3$ salt, as the XRD results did not reveal any Na contained in the raw AMD sludge. Using citric acid as a leaching reagent extracted most metals from the AMD sludge as the relative proportion of most metals was reduced as compared to oxalic acid, sodium carbonate and sodium bicarbonate. Citric acid proved to be the best organic extracting agent for hazardous metals contained in the AMD sludge. It is the most preferred because it is environmental friendly as it a naturally occurring organic complex, showed consistent removal efficiency and it is cost effective as compared to oxalic acid, sodium bicarbonate and sodium carbonate. In respect to unconfined compressive strength, the final treated AMD sludge proved to have a potential to be used for load bearing. Other application for the sludge may be in the treatment of acid mine drainage/wastewater that is acidic and contaminated with high concentration of heavy metals [13].

5. References

[1] Wang YR, Tsang DCW, Olds WE and Weber PA 2013 Utilizing acid mine drainage sludge and coal fly ash for phosphate removal from dairy wastewater *Environmental Technology*. 34(24), p 3177-3182,

[2] Liu Y and Tay JH 2001 Strategy for minimization of excess sludge production from the activated sludge process *Biotech. Advances*. 19 (2), p 97–107.

[3] Maree JP, Strydom WF, Adlem CJL, de Beer M, van Tonder GJ and van Dijk B 2005 Neutralisation of Acid Mine water and sludge disposal *Division of Water: Environment and Forestry Technology CSIR.*

[4] Zinck J 2005 Review of disposal, reprocessing and review options for acidic drainage treatment sludge *Mine Environment Neutral Discharge Program.*

[5] Aube B and Zinck JM,2003 Lime treatment of acid mine drainage in Canada *desktop Publishing*, Brazil, p 23-40.
[6] Ahmari S and Zhang L 2012 Production of eco-friendly bricks from copper mine tailings through geopolymerization *Constr Build Mater*. 29, p 323–331.
[7] Saeed A and Zhang L 2012 Production of eco-friendly bricks from Cu mine tailings through geopolymerization *Constr Build Mater.*29. p 323–331.
[8] World nuclear association 2009 Natural Occurring Radioactive Materials".
[9] Malatse M and Ndlovu S 2015 The viability of using the Witwatersrand gold mine tailings for brickmaking The Southern African Institute of Mining and Metallurgy, 115, p 321–180.
[10] Chen Y, Zhang Y, Chen T, Zhao Y and Bao S 2011 Preparation of eco-friendly construction bricks from hematite tailings *Constr Build Mater*. 25. p 2107–2111.
[11] Bennet JM, Sudhakar M, and Natarajan C 2013 Development of coal ash -GGBS based geopolymer bricks *European International Journal of Science and Technology*, 2, (5). p 133–139.
[12] South African National Standards (SANS 227:2007), Burnt clay masonry units.
[13] Cui MC, Jang M, Cho SH, Khim J, Cannon FS. 2012 A continuous pilot-scale system using coal-mine drainage sludge to treat acid mine drainage contaminated with high concentrations of Pb, Zn, and other heavy metals. *Journal of Hazardous Material*;215:122–128.

Acknowledgments
The authors would like to thank the University of Johannesburg for providing resources to conduct the study and for financial assistance.

IOP Conf. Series: Earth and Environmental Science **120** (2018) 012025 doi:10.1088/1755-1315/120/1/012025

Kinetic Study on the Removal of Iron from Gold Mine Tailings by Citric Acid

T Mashifana[1], N Mavimbela[1] and N Sithole[1]

[1]University of Johannesburg, Department of Chemical Engineering, P.O. Box 17011, Doornfontein 2088, South Africa
E-mail: tmashifana@uj.ac.za

Abstract. The Gold mining generates large volumes of tailings, with consequent disposal and environmental problems. Iron tends to react with sulphur to form pyrite and pyrrhotite which then react with rain water forming acid rain. The study focuses on the removal of iron (Fe) from Gold Mine tailings; Fe was leached using citric acid as a leaching reagent. Three parameters which have an effect on the removal of Fe from the gold mine tailings, namely; temperature (25 °C and 50 °C), reagent concentration (0.25 M, 0.5 M, 0.75 M and 1 M) and solid loading ratio (20 %, 30 % and 40 %) were investigated. It was found that the recovery of Fe from gold mine tailings increased with increasing temperature and reagent concentration, but decreased with increasing solid loading ratio. The optimum conditions for the recovery of Fe from gold mine tailings was found to be at a temperature of 50 °C, reagent concentration of 1 M and solid loading of 20 %. Three linear kinetic models were investigated and Prout-Tompkins kinetic model was the best fit yielding linear graphs with the highest R^2 values.

1. Introduction

Africa is a mineral-rich country with metals such as gold, copper, and platinum group metals being exploited to a significant extent in the country's mining history. Mining generates large volumes of tailings, with consequent disposal and environmental problems. Research shows that the largest quantity of gold that has been mined in South Africa (98%) has come from the Witwatersrand goldfields. The cluster of gold mines located in the Witwatersrand Basin generates a significant amount of mine tailings, which have adverse effects on the environment and ecological systems. In addition, disposal costs are very high [1].

Un-reclaimed gold mining sites are a major problem in the entire world with tailings being a source of contamination for communities living close to these mine dumps. Tailings consist of ground rock and process effluents that are generated in a mine processing plant. Tailings storage is associated with ever increasing challenges. Innovations allow low grade ores to be exploited resulting in high waste volumes which require safe storage. Environmental laws are advancing, putting pressure on mining industries in handling of tailings. Due to this, mine operations are forced to treat tailings after recovering precious metals from ores prior to disposal. Rosner, [2] analysed heavy metal concentrations in stream sediments and rivers affected by gold mines in the Witwatersrand region and the Free State Province. Acid mine drainage and the leaching of toxic metals such as Co, Cu, Fe, Mn, Ni and Zn resulted in an increase in metal concentrations [2].

An investigation on the bioleaching of heavy metals from mine tailings by Aspergillus fumigatus showed that in the one-step bioleaching process where the fungi was cultivated in the presence of the tailings, concentration of oxalic acid was the highest and in the two-step process where the metabolic products of fungal growth, which have been separated from its biomass, were used, citric acid was

dominant [3]. In the one-step process, the highest removals of As, Fe, Mn, and Zn were observed at the lowest tailings concentration. In general, it was found that heavy metals removal efficiency decreased with increased tailings concentration in both bioleaching processes.
In 2014 Vapur et.al., conducted an investigation using oxalic, citric and glycolic acids as principal organic acids used in leaching slimes sized (-75 μm) feldspar ores in order to reduce iron amount [4]. The parameters investigated were; temperature, pulp density, leaching time and acid molarity. The overall test results showed that the highest iron removal was obtained as 67.90 % using oxalic acid. The toxic metals in gold mine tailings have a negative impact on human health and the environment and this necessitates the recovery of these metals [5].
The objective of this study was to find an optimal, cost effective treatment process for gold mine tailings (focusing on the removal of iron) and develop a kinetic model for the leaching of Fe from the gold mine tailings.

2. Experimental

2.1. Material
The mine tailing was collected from a gold mine and dried in the oven for 24 hours at 50 °C. Different concentrations (0.25 M, 0.5 M. 0.75 M and 1 M) of citric acid were prepared as the leaching reagents. The thermostatic shaker was used as a leaching equipment to leach Fe from the gold tailings at different temperatures (25 °C and 50 °C), at a constant speed of 200 rpm. Solid loading of 20 %, 30 % and 40 % we investigated.

2.2. Method
The sample was then milled to 80 % passing 75 μm. The leaching period of 24 hours was allowed, samples were taken in the interval of 2 hours. After 24 hours; the slurry was filtered, the solids were air dried for 24 hours and the effluent was taken for analysis. The prepared reagents with respective concentrations were mixed with the gold mine tailing in the solid liquid ratio of 20%, 30% and 40%, in a thermostatic shaker for the leaching of the material. The initial conditions (pH and temperature) of the slurry were recorded. The leaching period of 24 hours was allowed, with sampling in the interval of 2 hours. After 24 hours; the slurry was filtered, the solids air dried for 24 hours and the effluent taken for analysis.

2.3. Analysis
X-ray Fluorescence (XRF) was used for elemental composition of the material. X-ray diffraction (XRD) was used to study the mineralogy and Scanning Electron Micrograph (SEM) was utilized to study the crystal structure and morphology images of the materials. To analyze the solution samples, an atomic absorption spectrophotometer (AAS) was used.

3. Results and discussions
Table 1 shows that the tailings consisted highly of silica (Si 71.8444 mass %). This is of no surprise as the sample is from Gold Mine which is located in the West Rand and is known to be dominated by ores containing silica. The tailings also consisted of Fe (7.6945 %), Al (8.7896 %), Mg (4.5297), S (3.4468 %) and traces of other metals. Table 1 shows that the tailings consisted highly of silica (Si 71.8444 mass %). This is of no surprise as the sample is from Gold Mine which is located in the West Rand and is known to be dominated by ores containing silica. The tailings also consisted of Fe (7.6945 %), Al (8.7896 %), Mg (4.5297), S (3.4468 %) and traces of other metals.

Table 1. XRF results for raw gold mine tailings.

Element	Weight %	Element	Weight %
Mg	4.53	Mn	0.10

Al	8.79	Fe	7.69
Si	71.84	Ni	0.10
P	0.11	Cu	0.01
S	3.45	Zn	0.05
Cl	0.17	Rb	0.01
K	1.59	Zr	0.03
Ca	0.76	Pb	0.50
Ti	0.41	U3O8	0.01

3.1. XRD and SEM of gold mine tailing before treatment

Figure 1 shows that the gold tailings consisted highly of quartz (SiO_2), chalcocite (Cu_2S) and pyrite (FeS_2) complimenting the XRF results. Since Fe and Cu are both transition metals and tend to behave the same, it was concluded that only one of the two elements will be analysed. The study focused only on the leaching of Fe instead of Si and since Cu tends to behave in the same manner as Fe, it was assumed that it will give the similar results as those of Fe.

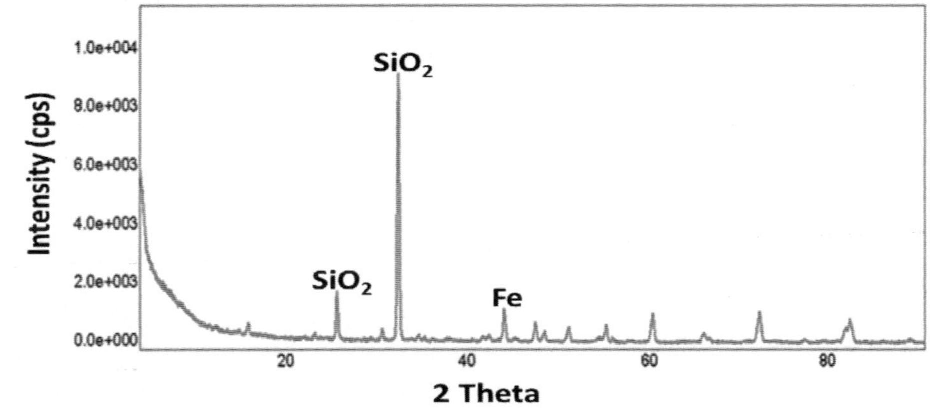

Figure 1. XRD of gold mine tailings.

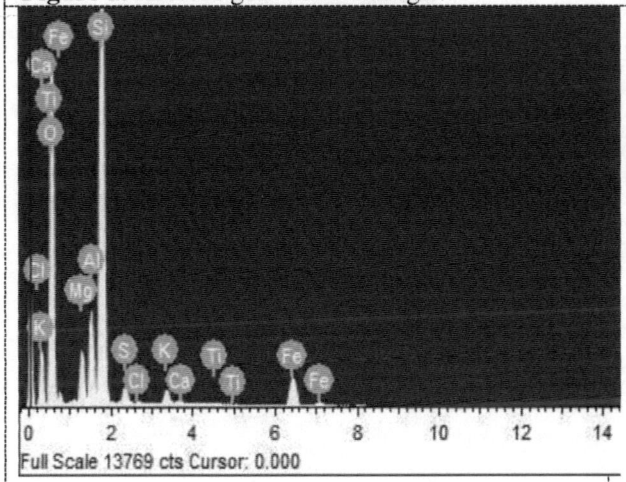

Figure 2(a). EDS of gold mine tailings.

Figure 2 (b). SEM of gold mine tailings.

IOP Conf. Series: Earth and Environmental Science **120** (2018) 012025 doi:10.1088/1755-1315/120/1/012025

Figure 2(a) shows the elements present in gold mine tailings. The spectrum shows that the tailings were predominated by Si and Fe. The result correspond to the XRD results, the SEM, figure 2(b) , shows that the tailings consisted highly of silicon (greyish rock-like structure) which appeared as quartz in the gold tailings and Fe (black spaces in between the quartz).

3.2. Effect of citric acid concentration on leaching rate

The effect of citric acid concentration on the leaching rate of Fe from the gold tailings was studied at the reagent concentrations of 0.25 M, 0.5 M, 0.75 M and 1 M. The tests were conducted at a temperature of 25 °C and at a constant speed of 200 rpm. The percentage of the Fe leached is dependent on the concentration of the citric acid used to leach the gold tailings; the higher the citric acid concentration, the higher the recovery of heavy metals from the gold tailings. In Figure 3, the leaching temperature was kept constant at 25 °C, and leaching at a speed of 200 rpm and 20 % solids. The results show that the recovery of Fe increases with increasing reagent concentration and time; the concentration of the iron leached is highest at 1 M compared 0.25 M, 0.5 M and 0.75 M.

3.3. Effect of temperature on leaching rate

Leaching is a chemically controlled process in which temperature mostly affects the reaction rate [6]. The rate of leaching is directly proportional to temperature; it increases with increasing temperature as molecules tend to gain more energy [6]. Although increasing the leaching temperature increases the solubility of the gold tailings and the leaching rate, the solvent's (citric acid) boiling point must not be exceeded as this might result in loss of solvent by evaporation and reduced metal recovery [7]. The effect of temperature on the recovery of gold tailings was studied at temperatures 25 °C and 50 °C at a constant speed of 200 rpm and reagent concentrations of 1 M. Figure 4 shows the leaching of Fe from gold tailings leached at 20 % solids. The figure shows that the recovery of Fe from the tailings increases with an increase in temperature. The recovery of Fe is significantly higher (47.08%) at 50 °C compared to the Fe leached at 25 °C (20.99%).

3.4. Effect of solid liquid ratio

The effect of solid-liquid ratio on the amount of Fe leached is shown in figure 5. The results show that the amount of the Fe leached decreases with an increase in the solid-liquid ratio. For the same amount of the reagent used at the same temperature and reagent concentration, it is expected that the concentration of the Fe leached decreases at a higher solid loading than at lower solid loadings. A higher volume of the reagent is needed to leach the 40 % solids compared to the 20 % solids. In the 20 % solids, molecules can move freely, resulting in high recovery.

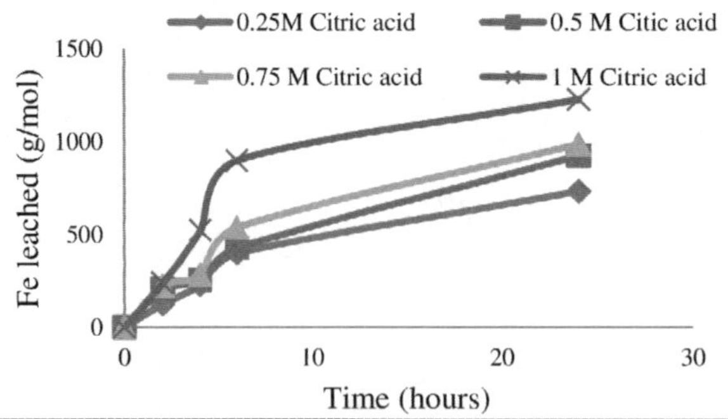

Figure 3. Comparison of Fe leached at different concentrations.

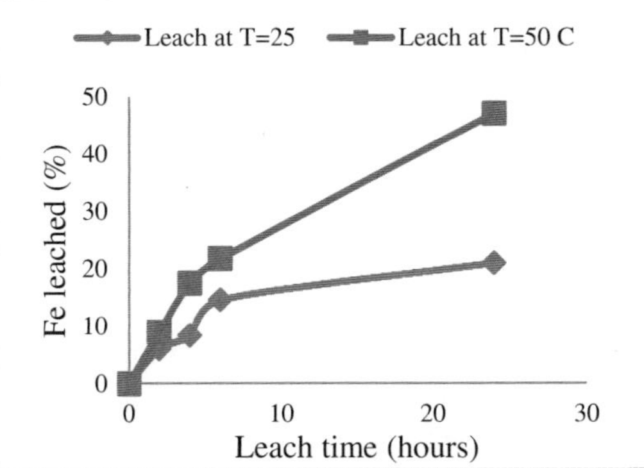

Figure 4. Comparison of Fe leached at different temperatures..

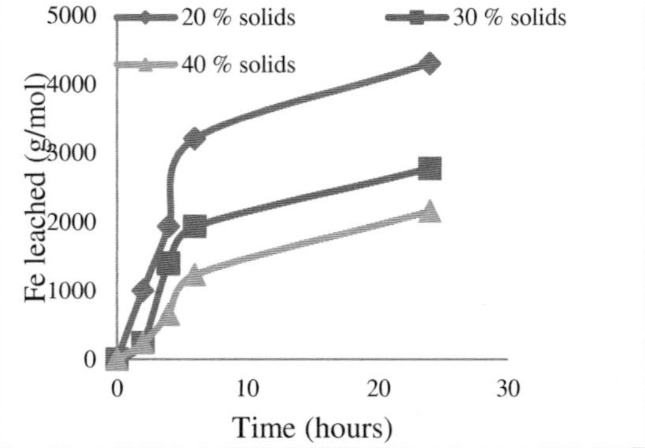

Figure 5. Comparison of Fe leached at different solid loading.

3.5. Leaching kinetics

Different kinetic models were drawn. Of all the kinetic models studied, only three models with the highest R^2 values are discussed. Based on the effect of temperature, reagent concentration and solid-liquid ratio, kinetic models were drawn. Figure 6 shows the results obtained with the Prout-Tompkins kinetic model. The Prout-Tompkins kinetic model gives linear graphs with an R^2 value of 0.9957 at a temperature of 50 ºC and an R^2 value of 0.9968 at a temperature of 25 ºC.

Figure 7 shows results obtained with the Interface kinetic model. Figure 7 shows that the Interface kinetic model had an R^2 value of 0.9856 at a temperature of 50 ºC and 0.9928 at a temperature of 25 ºC. Figure 8 shows the results obtained with the Second Order kinetic model. The results show that the Second Order kinetic model had an R^2 value of 0.9249 at a temperature of 50 ºC and 0.882 at a temperature of 25 ºC.

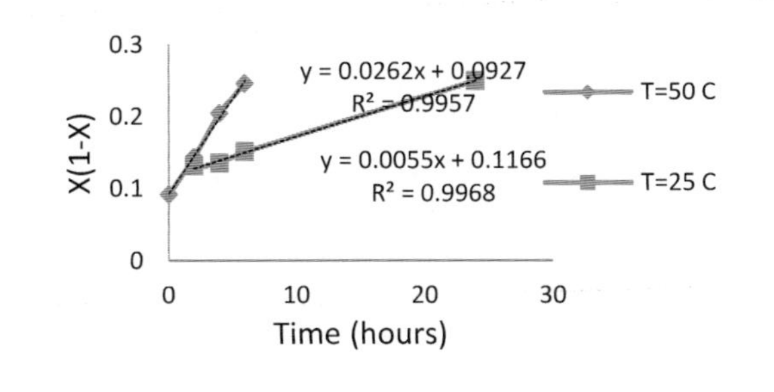

Figure 6. Prout-Tompkins kinetic model.

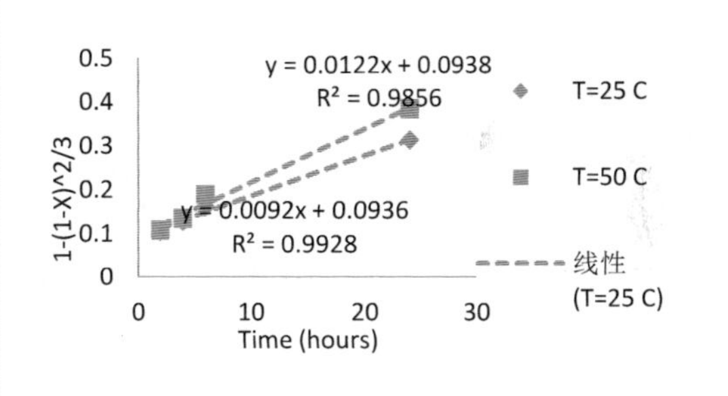

Figure 7. Interface kinetic model.

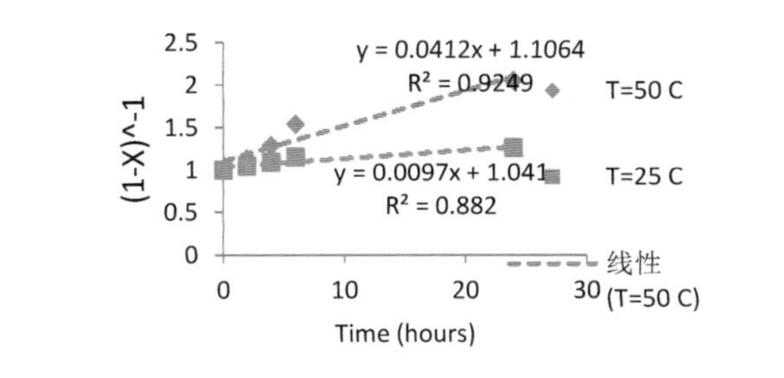

Figure 8. Second order kinetic model.

Table 2 gives a summary of the three models that gave the best fit.

Table 2. Summary of R^2 values for three kinetic models

Kinetic model	R^2 , T= 25 °C	R^2, T=50 °C
Prout-Tompkins	0.9968	0.9957
Interface	0.9928	0.9856
Second order	0.8820	0.9246

The Prout- Tompkins gave the best line of fit with an R2 of 0.9957 at a temperature of 50 ºC and an R^2 of 0.9968 at a temperature of 25 ºC. All three models show that the rate of reaction depends on the Fe leached and the amount of citric acid used. The higher the concentration of the reagent used, the higher the conversion.

4. Conclusion

The effect of citric acid concentration on leaching of the Fe was investigated at concentrations: 0.25 M, 0.5 M, 0.75 M and 1 M at a temperature varying from 25 ºC to 50 ºC and a solid loading of 20 %, 30 % and 40 % at a constant stirring speed of 200 rpm. It was found that the concentration of the metals leached from the gold tailings increased with increasing reagent concentration and temperature, but decreased with increasing solid-liquid ratio. The highest concentration of the iron leached was found to be at a reagent concentration of 1 M at 20 % solids and a temperature of 50 ºC. This is because when increasing the leaching temperature, solubility of the tailings also increases, resulting in high recovery of the metal being leached. Increasing the solid-liquid ratio had a negative effect on the leaching rate of the metals being leached. The leaching rate decreased with an increase in solid loading at a constant slurry volume. The kinetic model that was found to be fit for the overall investigation at all temperatures was found to be the Prout-Tompkins model. Since citric acid is an organic that is biodegradable and nature friendly and inexpensive, it is highly recommended for the removal of iron from gold tailings.

5. References

[1] Malatse S and Ndlovu M 2015 The viability of using the Witwatersrand gold mine tailings for brickmaking *Journal of the Southern African Institute of Mining and Metallurgy*, 115G.
[2] Rosner T 1999 The environmental impact of seepage from gold mine tailings near Johannesburg South Africa. Pretoria : University of Pretoria, pp 3-7.
[3] Seh-Bardan BJ, Othman, Wahid S, Husin A and Sadegh-Zadeh F 2012 Column bioleaching of arsenic and heavy metals from gold mine tailings by Aspergillus fumigatus". *CLEAN–Soil Air Water*, 40(6), pp.607-614.
[4] Vapur H, Demirci H., Top. S and Altiner M 2014 Removal of feldspar ores by leaching with organic acids *Turkey: Department of Mining Engineering*, pp.761.
[5] Phetla TP, Ntuli F and Muzenda E 2012 Reduction crystallization of Ni, Cu, Fe and Co from a mixed metal effluent *Journal of Industrial and Engineering Chemistry*, 18, pp. 1171–1177.
[6] Eksteen J and Oraby E 2015 The leaching and absorption of gold using low concentrated amino acids and hydrogen peroxide: Effect of catalyst ions, sulphide minerals and amino acid type *Minerals Engineering,* pp. 309-315.
[7] Zhong SP 2015 Leaching kinetics of gold bearing pyrite in H_2SO_4-$Fe_2(SO_4)_3$ system *Transition of Nonferrous Metals Society of China*, pp. 3461-3466.

Acknowledgments
The authors would like to thank the University of Johannesburg for providing resources to conduct the study and for financial assistance.

Modification of Indirect Solar Dryer for Simplicia Production

C W Purnomo[1,2] and S Indarti[1,3]

1 Agrotechnology Innovation Center (PIAT), Universitas Gadjah Mada, Berbah Sleman, Yogyakarta, Indonesia
2 Chemical Engineering Department, Univeristas Gadjah Mada, Jl Grafika no 2, Yogyakarta, Indonesia
3 Faculty of Agriculture, Universitas Gadjah Mada, Sleman, Yogyakarta, Indonesia
E-mail: chandra.purnomo@ugm.ac.id

Abstract. Simplicia is natural ingredient for herbal medicine that has been subjected to drying only. This study aims to develop an appropriate drying equipment to produce dried leaves of sambung nyawa (Gynura procumbens (Lour.) Merr.) or also called "longevity spinach". Typical indirect solar drier was modified to be able to process more fresh leaves in order to speed up the production. The modification was done using double solar collector and wind powered ventilation. The double solar collector was applied in order to collect more solar energy for larger dimension of drying chamber, while the wind-ventilator was installed to provide forced convection of hot air flow inside the dryer. The drying kinetic including the drying constants were investigated using three common thin layer drying equations to model the drying behavior of the leaves. The moisture ratio (MR) depletion with respect to the drying time (t) of the leaves can be well represented by equation of $MR = 1.1732\exp(-0.0993t) - 0.1732\exp(-17.3871t)$.

1. Introduction

Solar dryer is a common technology to be used to preserve agriculture product. This dryer aims to replace the direct sun drying method in open environment which suffers from various drawbacks such as uneven moisture removal, loss of the crop and less hygienic [1]. In terms solar energy collection, in general there are two type of solar dryer which are direct and indirect solar dryer. The direct solar dryer is using transparent cover which allows solar energy to enter inside the drying chamber to dry up the materials while indirect dryer employs a solar energy collector outside the drying chamber which produce hot air that will flow passing the material inside the chamber. Direct dryer could have higher temperature inside the chamber especially when using natural air flow compared with indirect drying [2]. However, it may suffer from overheating of delicate materials like medicinal herbs. Indirect solar dryer is more favorable to be used for drying simplicial herbs due to ease of temperature and moisture control to avoid active ingredient such as essential oils deterioration [3].

Meanwhile in term of air flow system inside the dryer, there are two type which are natural and forced convection. The natural convection utilize the different air density due to the temperature different between hot air inside the drying chamber and cold air outside the dryer to create air circulation. Forced convection is suitable for large scale production for fast drying mechanism. However this type require electricity to run the blower or exhaust fan.

Despite the advantage of the solar dryer compared with direct sun dryer, the equipment is still gained not much attention to the ordinary farmers due to high investment cost and time consuming operation [4]. Thus the design of the dryer has to be continuously improved to minimize the drawbacks. A typical indirect solar dryer has only one solar collector to be placed North or South direction for longer collection time of solar energy from morning until late afternoon. Typically, this solar dryer type only has a single drying chamber. In this study, indirect conventional solar heater was modified by additional of an extra solar collector in order to be able to enlarge the drying chamber. The solar collector is placed in the same direction of sun movement (East to West). The modification also provide a longer service time of the dryer throughout the day compared with a single solar collector dryer.

The other modification is by adding wind-ventilator on top of the chamber in order to create a negative air pressure to induce the air flow inside the dryer. Then, the modified dryer performance is evaluated by drying fresh longevity spinach leaves.

2. Experimental method

The leaves of longevity spinach (*gynura procumbens*) was freshly harvested and collected from Mangunan Region, Yogyakarta, Indonesia (7°92'51"S and 110°39'78"E). Before drying, the harvested leaves was spread on perforated stainless steel trays to form a single thin layer covering all the tray surface. Each tray was loaded with 0.22 kg of fresh leaves. Drying experiments were carried out using an indirect solar dryer with double collector which is constructed from stainless steel frames and walls as shown in Figure 1. The drying process were operated closer to the plantation site under forced convection by wind powered induced fan (ventilation). The experimental data were manually recorded every 2 h from 8 a.m. to 4 p.m. over three days and duplicated to obtain four data sets. The dryers were shut down during the night, and experimentation was resumed the next day at 8 a.m.

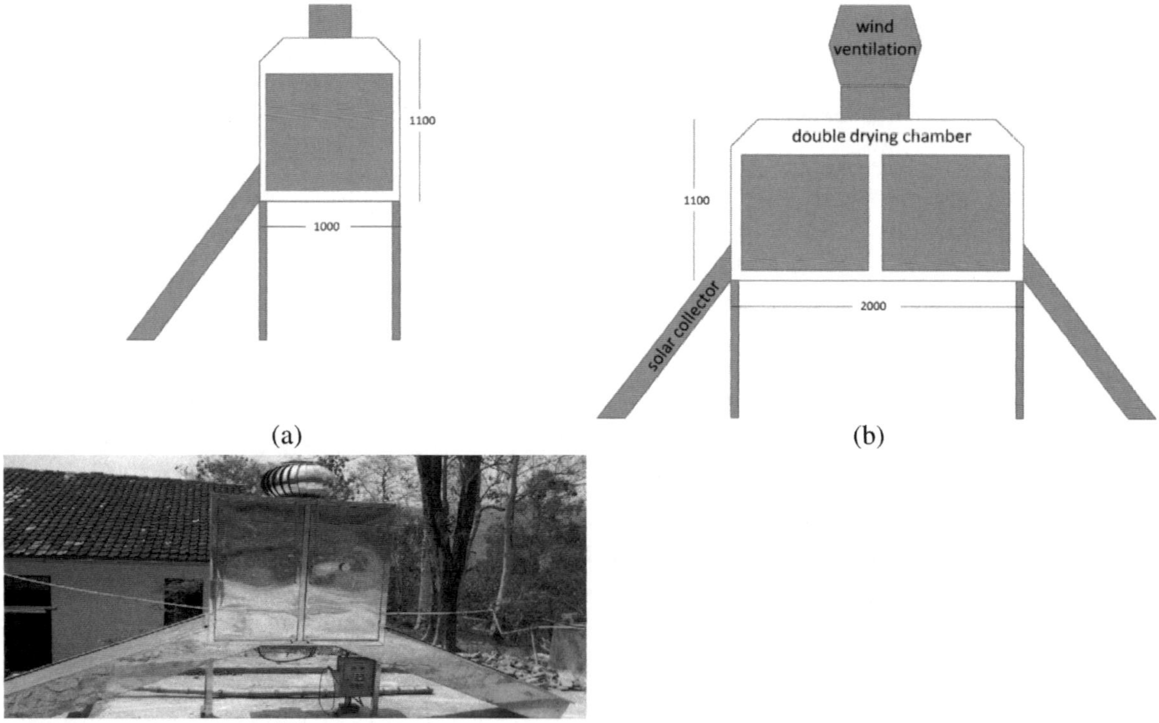

(a) (b)

(c)

Figure 1. Typical indirect solar dryer: (a) conventional single collector dryer, (b) and (c) double collector dryer (in this study)

The drying calculation followed a previously reported work [2]. Moisture ratio (*MR*) was calculated using the following equation:

$$MR = \frac{M}{M_o} \quad (1)$$

where M, and M_o (kg water/kg solid) are the moisture content at a particular time, and initial moisture content respectively. The equilibrium moisture content can be omitted due to the continuous fluctuation of the relative humidity of the drying air during solar drying and the relatively small value compared to Mo and M. Three empirical models were used to fit the drying curves which are Newton, Henderson and Pabis, and Verma et al. [5] as shown in equation (2), (3) and (4) respectively. The data fitting was performed using MATLAB with minimum sum of square error (SSE) method.

$$MR = \exp(-kt) \quad (2)$$

$$MR = a\exp(-kt) \quad (3)$$

$$MR = a\exp(-kt) + (1-a)\exp(-gt) \quad (4)$$

Then, Fick's second law was solved analytically in order to evaluate the D_{eff} (effective diffusivity) by plotting experimental drying data in terms of *ln(MR)* versus drying time (t) to give a straight line with a slope (k_o) which D_{eff} can be calculated by equation (4) with L is the thickness of the dried material.

$$D_{eff} = 4\frac{k_o L^2}{\pi^2} \quad (5)$$

.

3. Results and discussion

The internal temperature of drying chamber throughout a week is provided in Figure 2. The optimum temperature is about 42 °C at 12 p.m. while a large fluctuation is observed at 2 hours before and after the maximum temperature. Variety of cloud coverage and wind speed could be the cause of the temperature fluctuation inside the drying chamber. The temperature range is ideal for drying medicinal herb since a higher drying temperature can destroy active components inside the simplicia.

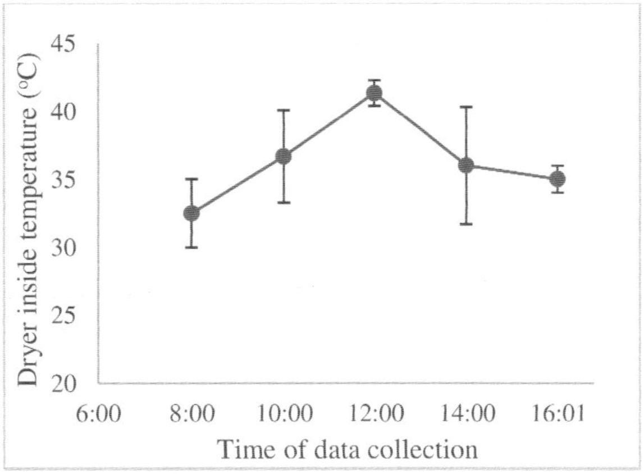

Figure 2. Temperature range in the drying chamber during data collection

The data fitting of drying curve is shown in Figure 3. The moisture was decreased rapidly during the beginning of the drying time due to the existence of surface water on the solids [6]. It can be visually observed that equation (3) fit the data better that the others. This also confirmed by the smallest SSE value by using this equation to fit the data. The obtained parameters and SSE values are listed for each equation in Table 1.

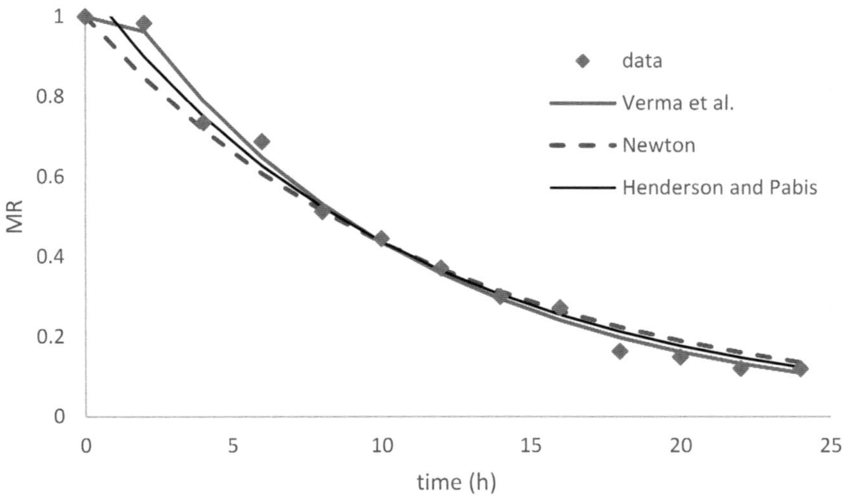

Figure 3. Data fitting plot with the model equations parameters obtained by Matlab

From Table 1, it is shown that the constants are quite close with the reference especially for the first two equation. The *k* values are slightly lower than the dryer of whole plant mints [2] but higher than stevia leave dryer [7]. This indicates that the rate of drying in this study is comparable to the literature which is promising since the dryer has no electric powered equipment for air circulation. This lower rate also could be due to the difference of the dried materials since the leaves of longevity spinach is thicker and larger than mint. Smaller leaves will have larger surface area than larger one which will accelerate the mass transfer of moisture during drying process. It is reflected by the value of effective diffusivity. The calculated D_{eff} of this study is 2.7×10^{-12} m^2s^{-1} which is smaller than the literature of 2.1×10^{-11} m^2s^{-1} [2]. The best fitted equation is Verma equation according to the lowest SSE value.

Table 1. The constants of the model equations of MR and SSE values

Model	**Constants**	**SSE**	**Constants**	
	(this study)		**[2]**	**[7]**
Newton	k = 0.0832	0.0327	k = 0.1646	k = 0.0210
Henderson and Pabis	a = 1.0772; k = 0.0902	0.0214	a = 0.9718; k = 0.1575	a = 1.0117; k = 0.0212
Verna et al.	a = 1.1732; k = 0.0993; g = 17.3871	0.0082	a = 0.4357; k = 0.0693; g = 0.3215	*n.a.*

4. Conclusion

The new design of indirect dryer with two solar collector has shown an acceptable performance to be used for herb leaves dryer. The double collector provide enough drying air for larger drying chamber configuration that will also double the production rate of the simplicia. Verma equation is the most suitable model for representing the drying rate of the longevity leaves using the modified dryer.

5. References

[1] V. Tomar, G. N. Tiwari, and B. Norton, "Solar dryers for tropical food preservation : Thermophysics of crops , systems and components," *Sol. Energy*, vol. 154, pp. 2–13, 2017.
[2] Y. I. Sallam, M. H. Aly, and A. F. Nassar, "Solar drying of whole mint plant under natural and forced convection," *J. Adv. Res.*, pp. 171–178, 2015.
[3] A. K. Bhardwaj, R. Chauhan, R. Kumar, M. Sethi, and A. Rana, "Case Studies in Thermal Engineering Experimental investigation of an indirect solar dryer integrated with phase change material for drying valeriana jatamansi (medicinal herb)," *Case Stud. Therm. Eng.*, vol. 10, no. August, pp. 302–314, 2017.
[4] A. Lingayat, V. P. Chandramohan, and V. R. K. Raju, "Design , Development and Performance of Indirect Type Solar Dryer for Banana Drying," *Energy Procedia*, vol. 109, no. November 2016, pp. 409–416, 2017.
[5] D. Arslan, M. M. Özcan, and H. O. Menges, "Evaluation of drying methods with respect to drying parameters , some nutritional and colour characteristics of peppermint (Mentha x piperita L .)," *Energy Convers. Manag.*, vol. 51, pp. 2769–2775, 2010.
[6] R. Ben, M. Combarnous, "Study of orange peels dryings kinetics and development of a solar dryer by forced convection, *Sol. Energy*. vol. 85, pp. 570–578, 2011.
[7] M. Castillo, I. Pilatowsky, B. Castillo, E.C. López, A. López, "Solar drying of Stevia (Rebaudiana Bertoni) leaves using direct and indirect technologies," *Sol. Energy*. vol. 159 pp. 898–907, 2018.

Acknowledgments

This study was supported by Teknologi Tepat Guna (TTG) Grant UGM 2017.